REMOTE SENSING OF GLOBAL CROPLANDS FOR FOOD SECURITY

Taylor & Francis Series in Remote Sensing Applications

Series Editor

Qihao Weng

Indiana State University
Terre Haute, Indiana, U.S.A.

REMOTE SENSING OF GLOBAL CROPLANDS FOR FOOD SECURITY

Edited by

PRASAD S. THENKABAIL
JOHN G. LYON
HUGH TURRAL
CHANDRASHEKHAR M. BIRADAR

CRC Press
Taylor & Francis Group
Boca Raton London New York

CRC Press is an imprint of the
Taylor & Francis Group, an **Informa** business

CRC Press
Taylor & Francis Group
6000 Broken Sound Parkway NW, Suite 300
Boca Raton, FL 33487-2742

First issued in paperback 2017

ISBN 13: 978-1-138-11655-9 (pbk)
ISBN 13: 978-1-4200-9009-3 (hbk)

This book contains information obtained from authentic and highly regarded sources. Reasonable efforts have been made to publish reliable data and information, but the author and publisher cannot assume responsibility for the validity of all materials or the consequences of their use. The authors and publishers have attempted to trace the copyright holders of all material reproduced in this publication and apologize to copyright holders if permission to publish in this form has not been obtained. If any copyright material has not been acknowledged please write and let us know so we may rectify in any future reprint.

Visit the Taylor & Francis Web site at
http://www.taylorandfrancis.com

and the CRC Press Web site at
http://www.crcpress.com

Contents

SECTION I Background and History of Global Irrigated Area Maps (GIAM)

*Hugh Turral, Prasad S. Thenkabail, John G. Lyon, and
Chandrashekhar M. Biradar*

*Y.J. Li, Prasad S. Thenkabail, Chandrashekhar M. Biradar,
Praveen Noojipady, Venkateswarlu Dheeravath, Manohar
Velpuri, Obi Reddy P. Gangalakunta, and Xueliang L. Cai*

SECTION II Global Irrigated Area Mapping (GIAM) Using Remote Sensing

*Prasad S. Thenkabail, Chandrashekhar M. Biradar, Praveen
Noojipady, Venkateswarlu Dheeravath, MuraliKrishna Gumma,
Y.J. Li, Manohar Velpuri, and Obi Reddy P. Gangalakunta*

SECTION III GIAM Mapping Section for Selected Global Regions

Songcai You, Shunbao Liao, Suchuang Di, and Ye Yuan

SECTION V   Rain-Fed Cropland Areas of the World

SECTION VI   Methods of Mapping Croplands Using Remote Sensing

SECTION VII   Accuracies and Errors

SECTION VIII   Way Forward in Mapping Global Irrigated and Rain-Fed Croplands

Series Foreword

Remote sensing refers to the technology of acquiring information about the Earth's surface (land and ocean) and atmosphere, using sensors onboard airborne (aircraft, balloons) or spaceborne (satellites, space shuttles) platforms. After World War II, the technology of remote sensing gradually evolved into a scientific subject. Its early development was mainly driven by military uses. Later, remotely sensed data became widely applied for civic usages. The range of remote-sensing applications includes archaeology, agriculture, cartography, civil engineering, meteorology and climatology, coastal studies, emergency response, forestry, geology, geographic information systems, hazards, land use and land cover, natural disasters, oceanography, water resources, and so on. Most recently, with the advent of high spatial-resolution imagery and more capable techniques, commercial applications of remote sensing are rapidly gaining interest in the remote-sensing community and beyond.

The "Taylor & Francis Series in Remote Sensing Applications" is dedicated to recent developments in the theories, methods, and applications of remote sensing. Written by a team of leading authorities, each book is designed to provide up-to-date developments in a chosen subfield of remote-sensing applications. Each book may vary in format, but often contains similar components, such as review of theories and methods, analysis of case studies, and examination of the methods for applying remote-sensing techniques to a particular practical area. This book series may serve as guide or reference books for professionals, researchers, scientists, and alike in academics, governments, and industries. College instructors and students may also find them excellent sources for textbooks or as a excellent supplement to their chosen textbooks.

This book, *Remote Sensing of Global Croplands for Food Security*, presents a major effort in global mapping of irrigated and rain-fed cropland areas using remote-sensing techniques. The chapters collected in this volume illustrate various estimation and mapping methods, and lessons learned in different regions of the world. These mapping methods reflect the application of evolving sensor technology to map large geographical regions of different physical, climatic, economic, and cultural settings. Estimating and mapping of croplands at the global and regional scales are of prime significance in food security, land and water resources management, and global sustainability. This is especially true when considering the increasing human population and changing climate. The evolving sensor technology, image processing methods, and computing power continue to drive users' interest on better quality and globally consistent datasets and maps of global croplands from remotely sensed data.

Dr. Thenkabail has rich experience and excellent achievements in the area of hyperspectral remote sensing of agriculture and vegetation. By coordinating the efforts of a group of leading figures as contributors and co-editors, this edited volume is expected to have a great impact on global mapping of land and water resources.

I hope that the publication of this book will promote better use of remote-sensing data, science, and technology, and will facilitate the monitoring and assessment of the global environment and sustaining our common home—the Earth.

Qihao Weng, PhD
NASA, Huntsville, AL

Preface

Global Mapping of Irrigated and Rain-Fed Cropland Areas Using Remote Sensing

Thomas R. Loveland
U.S. Geological Survey Earth Resources Observation
and Science Center

This book, *Remote Sensing of Global Croplands for Food Security*, contributes to the understanding of fundamental technical capabilities needed to address an increasingly important global resource issue. Why is this topic important? Irrigation is the largest use of the global water supply and is tightly linked to the production of food, fuel, and fiber for an expanding population (Seckler et al., 1998). Crop water use, especially when associated with irrigation, is also in direct competition—and often conflict—with other needs, including ecological, municipal, and industrial requirements.

The book also represents an exciting leap forward in the impressive maturing of the remote sensing data users community. The centerpiece of this book, Global Irrigated Area Mapping (GIAM), provides the first sensor-based map of irrigated areas, and the most objective and complete estimate of global irrigated areas ever made. This is an important evolution in understanding global water use and demands, and will contribute to improving water management decision making and policies. The foundation for meeting long-term food, fiber, and fuel needs of the growing global populace is an accurate understanding of current resources utilization. GIAM, and the ideas contributed in the book, is a major step forward.

Since Earth Resources Technology Satellite 1 (ERTS-1, later renamed Landsat 1) ushered in the modern era of global land observations and monitoring, there has been accelerating progress in applying remotely sensed data to the mapping and measurement of the Earth's characteristics. Over the past 35 years, we have seen amazing improvements in sensor technologies, incredible advances in computing, and impressive innovations in analytical procedures. Perhaps most important, though, is the boom in intellectual talent that now has the resources, vision, and capabilities needed to apply the wealth of Earth observations data to societal problems. Those advances have enabled more complex assessments, covering larger geographic areas in a relatively short period of time, using increasingly rigorous analysis methods. Certainly, the collection of research published in this book is testimony to the collective maturity of technical and scientific expertise and technology.

The first attempts to generate global sensor-based maps of land use using Advanced Very High Resolution Radiometer (AVHRR) data began almost 15 years ago. Those efforts were successful because they not only created needed global land data, but also because those efforts demonstrated the potential for global land mapping and monitoring. The approaches of the early efforts were simple in comparison to the global mapping challenges required to accurately map irrigated areas. The first 1-km resolution global land cover map, the Global Land Cover Characteristics Database completed in 1998, was a pathfinder that highlighted the challenges associated with access to high-quality remotely sensed data, mapping methods, and validation approaches (Loveland et al., 2000). The global cropland mapping initiative reported in this book still relies on AVHRR imagery, as was used in the first global land cover efforts; however, advances in the state-of-the-practice were necessary to meet the challenges for generating globally consistent, locally relevant, and acceptably accurate irrigation maps. The augmentation of AVHRR data with SPOT Vegetation imagery used to produce GIAM illustrates the emerging trend and necessity to use multiple sources of remotely sensed data to produce timely and accurate maps of land characteristics. In addition, it also highlights the challenges and value associated with rigorous quantification of the accuracy of the resulting maps. Because of the rigorous effort to document mapping uncertainties, the results will be more useful to those faced with grand challenges associated with managing freshwater supplies and feeding a growing global population.

It is important to stress that GIAM is more than a single map of global irrigated areas. This book shows the depth of research that has been stimulated through the push to improve the characterization of irrigated lands. Several key research lessons from the chapters of this book should stimulate large-area mapping of other landscape variables. Those lessons include the importance of addressing the geographic variability of irrigation practices and characteristics, the importance of a multisource Earth observation strategy, and the need for state-of-the-art mapping and validation techniques. While each of these topics has been addressed for other purposes in many journals and books, the collection of chapters in this volume provides a unique means to understand and compare strategies for mapping cropland and irrigation status.

The need to recognize that geographic variability of irrigation and other land uses due to different environmental characteristics, cultural settings, and irrigation-related technologies is particularly important and will affect the strategies and types of remotely sensed data used to document regional to global irrigated areas. The case studies presented here, addressing irrigated lands mapping in China, India, Pakistan, England, the Middle East, the United States, and other regions, are a useful starting point in understanding how place influences approaches. The book also highlights the trend in large-area remote sensing to use multisource and multiresolution remotely sensed data to more accurately map land use and land cover. The capabilities of a number of old and new Earth observation systems are presented within the context of irrigated lands mapping, including such standards as AVHRR and Landsat, but also SPOT Vegetation, Moderate Resolution Imaging Spectroradiometer (MODIS), Multiangle Imaging Spectroradiometer (MISR), and Advanced Spaceborne Thermal Emission and Reflection Radiometer (ASTER). Equally important is linkage to field-level conditions. This book provides numerous examples of the use of field data for calibration and validation of the irrigated lands maps and measurements.

GIAM represents an important global cropland and irrigated areas baseline, which future efforts will build upon. The efforts reported in this book are crucial to the next steps in improving our understanding of global land use characteristics. Future generations will undoubtedly improve upon this first step, with improved spatial resolution, cropland and irrigated characteristics, and accuracy. The exciting and realistic challenge is for GIAM to evolve to finer scales that take advantage of the global availability of the continuing Landsat record and other similar international data sets.

After reading the chapters of this book, it will be clear where the state-of-the-practice in irrigated land mapping is today. However, there are also important insights into where the next innovations are that will further improve the irrigation information needed by decision makers. For example, the research on landscape evapotranspiration estimation and understanding crop water requirements points toward the need to couple environmental processes with irrigated lands mapping, and the investigation of urban irrigation is an often-missed topic that may be an important term in future water budget equations.

Collectively, this volume offers considerable insight into the challenges and opportunities for continuous global monitoring of irrigated lands. As populations increase, water consumption rises, and populations expand, and the necessity for understanding all aspects of water utilization and food, fuel, and fiber production will increase.

The late Dr. Jack Estes and Wayne Mooneyhan often talked about the "mythical map" (Estes and Mooneyhan, 1994). They argued that a major factor hindering global research and applications was the lack of adequate maps for many areas of the world. Depending upon scale, thematic content, and timeliness, this is equally true for both the developed and developing world. They also stressed the need for "science quality" maps with known and traceable lineage. Much has improved since Estes and Mooneyhan challenged the remote-sensing community to step forward and advance global land mapping. In order to improve or even maintain current global environmental conditions, efforts are needed to replace mythical maps with real maps of the areal extent of key components of the Earth system. Fortunately, the research reported in this book represents significant steps in retiring the mythical map of irrigated lands.

REFERENCES

Estes, J.E. and Mooneyhan, D.W., 1994. Of maps and myths. *Photogrammetric Engineering and Remote Sensing* 60(5): 517–24.

Loveland, T.R., Reed, B.C., Brown, J.F., Ohlen, D.O., Zhu, J., Yang, L., and Merchant, J.W., 2000. Development of a global land cover characteristics database and IGBP DISCover from 1-km AVHRR data. *International Journal of Remote Sensing* 21(6/7): 1303–30.

Seckler, D., Amerasinghe, U., Molden, D., de Silva, R., and Barker, R. 1998. *World water demand and supply, 1990 to 2025: Scenarios and issues.* Research Report 19, International Water Management Institute, Colombo, Sri Lanka.

Acknowledgments

The idea of a book on remote sensing of global cropland areas was always on my mind, ever since I began working on and, later, leading the projects on global irrigated area mapping (GIAM) and the global map of rain-fed cropland areas (GMRCA) at the International Water Management Institute (IWMI). During this period, we noticed a pressing need—the absence of a book on the global cropland areas. Precise estimates of areas are crucial for assessments, such as the actual areas available for growing food, productivity of these lands, people fed by them, and water used by them. Areas are also needed as inputs to economic models, national and global planning, and developing trade scenarios. Yet, very little attention is given to proper and accurate assessments of areas: Where are they spatially? and What are their characteristics (e.g., what crops are grown in these croplands, when, and how often)? Currently, irrigated or the rain-fed cropland areas are at best rough estimates and not precise science. This book does not promise to make cropland area estimates precise, but provides a definitive road map to move toward that using advanced remote-sensing data and methods. The book presents the first ever global estimates of irrigated and rain-fed cropland areas using remote sensing, presents advanced methods of mapping, provides perspectives of top experts from different countries on irrigated and rain-fed cropland areas, touches on innovative subject matter (e.g., urban irrigation, minor irrigation), looks into the history of agricultural development, dwells on determining water use by croplands, and highlights issues related to uncertainties, errors, and accuracies. The goal of the book is to produce the first definitive and authoritative knowledge base on the state of the art of irrigated and rain-fed cropland areas, their mapping methods using remote sensing, and to set up a road map for higher resolution work in the near future. It is expected that the book will remain a baseline reference work for future irrigated and/or rain-fed cropland area mapping and area assessments. Such a body of work, as presented in this book, required active participation and contributions by the best experts on the subject matter from around the world. So, I have many people to thank and acknowledge.

The first steps for a book on remote sensing of global croplands were taken during an International Workshop on GIAM held in November 2006 in Colombo, Sri Lanka. Many of the authors of different chapters in this book attended the workshop. The idea of the book, discussed on the sidelines of this workshop, was pursued vigorously soon after. A few authors who did not attend the workshop also enthusiastically agreed to participate. Soon we had 20 chapters, starting from the history of irrigated areas, to cropland (irrigated and rain-fed) mapping methods, to innovative studies from the individual nations, regions, and the world.

I want to acknowledge the contributions and support of so many scholars who have contributed to this book in many different ways.

First and foremost, Professor Frank Rijisberman, former Director General of the International Water Management Institute (IWMI). He strongly supported the

GIAM and GMRCA projects at IWMI and was supportive of the book project. He was a leader with vision, energy, grace, dignity, and fairness—truly very rare qualities. In many ways, he was a very special leader—ideal to lead an institute with a global mandate on water focused on developing countries. Science development requires visionary leaders like him. I remember his support with gratitude. Frank currently works for Google.org.

A book project such as this needs patience and persistence. There were times I had given it up due to many other commitments and distractions. At times likes these, you need a motivator. There is no better motivator than Professor John G. Lyon (co-editor of this book)—my long-term perennial guru. His encouragement at various stages of the book project helped me stay focused on goals. He found a publisher and helped me edit chapters. He fast-forwarded the whole project in many ways. I suspect that without him joining the book project it would still be limping along. Thanks Dr. Rijisberman and Dr. Lyon for your great support.

A hearty thank you to all contributing authors. This book belongs to them. It is their intellectual insight that has made this book possible. My special thanks to Dr. Songcai You (Chinese Academy of Sciences, China), Dr. Obi Reddy Ganguntla (Indian Council of Agricultural Research, India), Dr. Jesslyn Brown (U.S. Geological Survey, USA), Dr. Seelan Santhosh Kumar and Dr. Bethany Kurz (University of North Dakota, USA), Dr. Cristina Milesi (California State University at Monterey Bay and NASA Ames, USA), Dr. Jerry Knox (Cranfield University, UK), Dr. Eddy de Pauw (International Center for Agricultural Research in Dry Areas, Syria), Dr. Wataru Takeuchi (University of Tokyo, Japan), Dr. Roland Geerken (Yale University, USA), Dr. Gabriel Senay (U.S. Geological Survey, USA), Dr. Mobin-ud-Din Ahmad (Council for Scientific and Industrial Research Organization, Australia), Dr. Francis Canisius (Canada Center for Remote Sensing, Canada), Dr. Vijendra Boken (University of Nebraska at Kearney, USA), Dr. Russell Cogalton (University of New Hampshire, USA), Dr. Chandrashekhar M. Biradar (University of Oklahoma, USA), Yuanjie Li (University of Maryland, USA), and Nilantha Gamage (University of Melbourne, Australia). As you can see, this is an excellent team of international researchers who are very knowledgeable in the subject matter. We are very lucky to have them all come together and contribute to this book. I have not mentioned the various co-authors of chapters although their contributions are highly appreciated.

My biggest thanks go to my former GIAM and GMRCA team members at IWMI (like me, most of them have moved to new institutions now): Dr. Chandrashekhar M. Biradar (University of Oklahoma, USA), Mr. Praveen Noojipady (University of Maryland, USA), Manohar Velpuri (University of South Dakota, USA), Venkateswarlu Dheravath (United Nations Joint Logistics Center, Sudan), Yuanjie Li (University of Maryland, USA), MuraliKrishna Gumma (IWMI, India), Mitchell Schull (Boston University, USA), Ranjith Alankara (IWMI, Sri Lanka), and Sarath Gunasinghe (IWMI, Sri Lanka). They formed the GIAM and GMRCA core team and were instrumental in producing the global irrigated and rain-fed cropland maps and statistics. They created many of the maps and tables reproduced in this book, and have actively contributed in writing Chapters 3 and 15. I thank them all for their wonderful spirit, teamwork, and hard work. It is vital to have young energy to create new knowledge. The many computing hours they put in is much appreciated.

I should also thank my co-editors. Dr. Chandru Biradar has made a special contribution to the GIAM, GMRCA, and the book project. Chandru was always someone with lots of enthusiasm and energy, working on these projects from the beginning. He also contributed as one of the editors of this book. Chandru richly deserves a special thanks for all his hard work and commitment. I should also thank my other co-editor Dr. Hugh Turral. Hugh led the GIAM project initially and was involved in many initial discussions. He provided useful comments periodically and edited several chapters. Hugh has changed careers and gone into film making, but has continued to provide input for the book project and wrote the first chapter. All of this is much appreciated.

I want to acknowledge the editorial help from Kingsely Kurukulasuriya (Sri Lanka). Kingsely is a very efficient editor and edited all chapters for consistency, grammar, English, and formatting. This was a huge effort from Kingsely and he did an outstanding job by any standards.

In the final phase of this book project, I needed help with different aspects of the book project. Dr. Venkateswarlu Dheeravath and Dr. MuraliKrishna Gumma helped me organize all the chapters and ensured the quality of the figures and tables and proper formatting. Both of them are strong contributors to GIAM work. To them I owe special thanks. Arosha Ranasinghe provided very efficient secretarial support throughout the production of this book. This is much appreciated.

Dr. Tom Loveland (U.S. Geological Survey, USA) was kind enough to write the preface for this book. Tom has always been an inspiration for me, and a true global leader in remote-sensing science. His seminal works on global and continental mapping using remote sensing have provided motivation for me to do something worthwhile. I recommend that everyone reading this book must read his insightful and thought-provoking Foreword. I am very grateful for his kindness, generosity, support, and inspiration. I spent substantial time in editing and finalizing this book at the USGS, Southwest Geographic Science Center (SGSC) at Flagstaff, Arizona. The encouragement and support from Edwin Pfiefer, Team Chief of the SGSC and USGS and Dr. Susan Benjamin, Director, Western Geographic Science Center of USGS are much appreciated and gratefully acknowledged.

I thank the publisher CRC Press/Taylor & Francis Group production team for an outstanding job in producing a high-quality book. Srikanth Gopaalan of Datapage (India) Private Limited edited this book and coordinated with the authors. Srikanth is a professional of the highest order and I owe him a special thanks for the outstanding editing of this book. I also want to thank Kari Budyk, Senior Project Coordinator, Rachael Panthier, Production Editor, Irma Shagla, Acquisitions Editor for Environmental Sciences, and Arlene Kopeloff, Editorial Assistant (all of Taylor & Francis) for managing this book's publication and for their support and coordination.

I would like to give very special thanks to the two most special people to me: Sharmila T. Prasad (wife) and Spandana P. Thenkabail (daughter) for all their understanding, support, and love. Probably when the book comes out they will understand why I had to sit in front of my computer for such long hours—not just working on the book, but innumerable hours spent on GIAM and GMRCA projects. I would like to dedicate this book to my parents whose love, support, and vision gave me an education. They are farmers in a village called Thenkabail (India) and my interest in agriculture comes from them. Applying remote sensing to agriculture has always been one of my topmost priorities and passions because of my background. So, this book on remote sensing of global croplands was both a hobby and a professional endeavour.

Finally, any book project involves many unknown faces, who have worked tirelessly to make it all happen. To them all, my most sincere thanks.

Prasad S. Thenkabail
Editor-in-Chief
Remote Sensing of Global Croplands for Food Security

Editors

Dr. Prasad S. Thenkabail has 25 years experience working as a well-recognized international expert in remote sensing and geographic information systems (RS/GIS) and their applications to land and water resources, agriculture, forestry, sustainable development, and environmental studies. Dr. Prasad has extensive work experience in over 25 countries in Africa, Asia, the Middle East, and North America. Until recently, Dr. Prasad worked as a principal researcher in the Global Research Division and was head of the Remote Sensing and Geographic Information Systems (RS/GIS) at the International Water Management Institute (IWMI), headquartered in Sri Lanka with a network of offices in Asia and Africa. It was during this period that Dr. Prasad led the Global Irrigated Area Mapping (GIAM) that consisted of an energetic, strong, young team. The book is a culmination of GIAM knowledge development and advances. Prior to joining IWMI, Dr. Prasad worked as an associate research scientist (a research faculty position) at the Yale Center for Earth Observation (YCEO) at Yale University, New Haven, Connecticut. In this position, he worked in NASA-funded research in Africa and Asia. He was one of the two principal investigators for the NASA-funded project called "Characterization of Eco-regions in Africa" (CERA). He worked on hyperspectral remote sensing and carbon stock estimations from remote sensing. Prior to this, for nearly five years, he led the remote sensing programs at the International Institute of Tropical Agriculture (IITA), mostly working in West and Central African countries, based in Nigeria, and the International Center for Integrated Mountain Development (ICIMOD), working in the Hindu-Kush Himalayan countries, and based in Katmandu, Nepal. During this period, he led the remote-sensing component of the inland valley wetland characterization and mapping, using Landsat TM and SPOT HRV data. In his early career, Dr. Prasad worked as a scientist with the National Remote Sensing Agency (NRSA), Department of Space, and Government of India. He began his professional career as a lecturer in hydrology, water resources, hydraulics, hydraulics laboratory, and open channel flow in the colleges affiliated with Bangalore and Mysore Universities in India.

The USGS and NASA selected him to be on the Landsat Science Team for a period of five years, starting in 2006 (http://ldcm.usgs.gov/intro.php). Dr. Prasad is one of the editors of *Remote Sensing of Environment*. He is also associate editor-in-chief of the *Journal of Spatial Hydrology* (*JoSH*). In June 2007, Dr. Prasad's team was recognized by the Environmental System Research Institute (ESRI) for "special achievement in GIS" (SAG award) for its tsunami-related work and for its innovative spatial data portals (http://www.iwmidsp.org) and science applications (http://www.iwmigiam.org). In 2008, Dr. Prasad and co-authors were the second place recipients of the 2008 John I. Davidson ASPRS President's Award for practical papers (for their paper on spectral-matching techniques used in mapping global irrigated areas). He won the 1994 Autometric Award of the American Society of Photogrammetric Engineering and Remote Sensing (ASPRS) for superior publication in remote sensing.

Dr. Prasad's publications were selected as one of the best five papers consecutively for three years (2004–2006) in the IWMI's annual research meeting (ARM). His team was also awarded the "best team" at IWMI during ARM 2006. Dr. Prasad was on the scientific advisory board of Rapideye, a private German Earth Resources Satellite Company. He played a pivotal role in recommending the design of wave-bands in the rapideye sensor onboard a constellation of satellites planned for launch during 2007 by Rapideye.

Dr. Prasad has 60+ publications, mostly peer-reviewed international remote-sensing journals. His work on hyperspectral remote sensing of agriculture and vegetation is widely referenced.

Dr. John G. Lyon's research has involved advanced remote sensing and Geographic Information System (GIS) applications to water and wetland resources, agriculture, natural resources, and engineering applications. He is the author of books on wetland landscape characterization, wetland and environmental applications of GIS, and accuracy assessment of GIS and remote-sensing technologies. Dr. Lyon was educated at Reed College in Portland, Oregon, and the University of Michigan in Ann Arbor, and has previously served as a professor of civil engineering and natural resources at The Ohio State University (1981–1999). For approximately eight years, Dr. Lyon was the director (SES) of the U.S. Environmental Protection Agency's Office of Research and Development's (ORD) Environmental Sciences Division, which conducts research on remote sensing and GIS technologies as applied to environmental issues, including landscape characterization and ecology, and hazardous wastes. He currently serves as the senior scientist (ST) in the EPA Office of the Science Advisor in Washington, DC, and is co-lead for work on the Group on Earth Observations and the Global Earth Observation System of Systems, and research on geospatial issues in the Agency.

Dr. Hugh Turral is an irrigation and water resources engineer, and was the leader of the theme on Basin Water Management (Theme 1 of 4) at the International Water Management Institute (IWMI) from 2003 to 2007. He was also the project leader for work on the mapping of irrigated areas using remote sensing.

He is now embarking on a new career as a film-maker back home in Australia, but continues to do some consulting work in water resources management, including the application of remote sensing, most recently for wetland inventory in South Australia.

Chandrashekhar M. Biradar is a senior research scientist, Center for Spatial Analysis, College of Atmospheric and Geographic Science, Stephenson Research and Technology Center, Norman, Oklahoma. Prior to this, he was a research scientist at the Institute for the Study of Earth, Oceans and Space (EOS). Before this, he worked as a post-doctoral researcher at the International Water Management Institute (IWMI). He is one of the key lead researchers in producing the first satellite-sensor-based global irrigated and rain-fed area maps. He developed many innovative methods and techniques for mapping agroecosystems. The global irrigated and rainfed cropland mapping efforts demonstrated, for the first time, the use of quantitative and qualitative spectral-matching techniques, generation of ideal spectra in analyzing long time-series satellite images, and innovative techniques for subpixel area calculations. In addition, extensive use of spatial modelling, decision-tree algorithms, and Google Earth data for accuracy evaluations was investigated. Dr. Biradar has played a key role in the conceptualization and development of the IWMI's spatial data gateway or IWMI Data Storehouse Pathway. His main emphasis is on mapping, modelling, and characterization of agricultural and natural landscape using geospatial data. He was awarded "the best young scientist" award and was also nominated for the "best young scientist" of the Consultative Group of International Agricultural Research (CGIAR). He is the author of more than 20 peer-reviewed scientific papers and book chapters, and co-editor of two books (including this one). His current research interests include global land remote sensing; land use and land cover change; mapping cropping intensity, irrigated and rain-fed areas; applications of remote sensing and GIS in ecosystems science and natural resources; ecosystem modelling at large spatial scales; and climate change and ecology of infectious diseases.

Contributors

Mobin-ud-Din Ahmad
Land and Water Division
Commonwealth Scientific
  and Industrial Research
  Organisation
Canberra ACT, Australia

Ranjith D. Alankara
International Water
  Management Institute
Colombo, Sri Lanka

Chandrashekhar M. Biradar
Center for Spatial Analysis and
  Department of Botany and
  Microbiology
University of Oklahoma
Norman, Oklahoma

Vijendra K. Boken
Department of Geography and Earth
  Science
University of Nebraska at
  Kearney
Kearney, Nebraska

Jesslyn F. Brown
U.S. Geological Survey
  Earth Resources Observation
  and Science Center
Sioux Falls, South Dakota

Michael E. Budde
U.S. Geological Survey
  Earth Resources Observation
  and Science Center
Sioux Falls, South Dakota

Xueliang L. Cai
International Water
  Management Institute
Colombo, Sri Lanka

Francis Canisius
Canada Center for Remote Sensing
Ottawa, Ontario, Canada

G. Chandrakantha
Department of Applied Geology
Kuvempu University
Karnataka, India

Russell G. Congalton
Department of Natural Resources
University of New Hampshire
Durham, New Hampshire

Eddy De Pauw
International Center for Agricultural
  Research in the Dry Areas
Aleppo, Syria

Venkateswarlu Dheeravath
United Nations Joint Logistics Centre
Juba, Sudan

Suchuang Di
Graduate University of Chinese
  Academy of Sciences
Beijing, People's Republic of China

Gregory L. Easson
Department of Geology and
  Geological Engineering
The University of Mississippi
University, Mississippi

Christopher D. Elvidge
Earth Observation Group
NOAA National Geophysics
    Data Center
Boulder, Colorado

M.S.D.N. Gamage
International Water
    Management Institute
Colombo, Sri Lanka

Obi Reddy P. Gangalakunta
National Bureau of Soil Survey and
    Land Use Planning (ICAR)
Nagpur, India

Roland A. Geerken
German Development Cooperation
Eschborn, Germany

MuraliKrishna Gumma
International Water
    Management Institute
ICRISAT Campus
Patancheru, India

Sarath Gunasinghe
International Water
    Management Institute
Colombo, Sri Lanka

Gerrit Hoogenboom
Department of Biological and
    Agricultural Engineering
The University of Georgia
Griffin, Georgia

Asghar Hussain
International Water
    Management Institute
Lahore, Pakistan

Jerry W. Knox
Centre for Water Science
Cranfield University
Bedfordshire, United Kingdom

Maji Amal Kumar
National Bureau of Soil Survey and
    Land Use Planning
Indian Council for Agricultural
    Research
Nagpur, India

Bethany Kurz
Space Studies Department
University of North Dakota
Grand Forks, North Dakota

Y.J. Li
Department of Geography
University of Maryland
College Park, Maryland

Shunbao Liao
Institute of Geographic Sciences and
    Natural Resources Research
Chinese Academy of Sciences
Beijing, People's Republic of China

Thomas R. Loveland
U.S. Geological Survey
    Earth Resources Observation
    and Science Center
Sioux Falls, South Dakota

John G. Lyon
Environmental Protection
    Agency
Washington, District of
    Columbia

Z. Masri
International Center for
    Agricultural Research in the
    Dry Areas
Aleppo, Syria

Susan Maxwell
U.S. Geological Survey
    Earth Resources Observation
    and Science Center
Sioux Falls, South Dakota

Cristina Milesi
Earth System Science and Policy
California State University
Monterey Bay
Seaside, California

Aamir Nazeer
International Water
  Management Institute
Lahore, Pakistan

Ramakrishna R. Nemani
NASA Ames Research Center
Moffett Field, California

Praveen Noojipady
Department of Geography
University of Maryland
College Park, Maryland

Shariar Pervez
Stinger Ghaffarian Technologies Inc.
Greenbelt, Maryland

J.A. Rodriguez-Diaz
Department of Agronomy
University of Cordoba
Cordoba, Spain

James Rowland
U.S. Geological Survey
  Earth Resources Observation
  and Science Center
Sioux Falls, South Dakota

Mitchell A. Schull
Department of Geography and
  Environment
Boston University
Boston, Massachusetts

Santhosh Seelan
Space Studies Department
University of North Dakota
Grand Forks, North Dakota

Gabriel B. Senay
U.S. Geological Survey
  Earth Resources Observation
  and Science Center
Sioux Falls, South Dakota

and

Geographic Information
  Science Center of
  Excellence
South Dakota State
  University
Brookings, South Dakota

S.A.M. Shamal
Centre for Water Science
Cranfield University
Bedfordshire, United Kingdom

Ronald B. Smith
Department of Geology and
  Geophysics
Yale University
New Haven, Connecticut

Wataru Takeuchi
Institute of Industrial
  Science
University of Tokyo
Tokyo, Japan

Prasad S. Thenkabail
U.S. Geological Survey
Flagstaff, Arizona

Hugh Turral
On the Street Productions
Melbourne, Australia

Manohar Velpuri
Geographic Information
  Science Center of Excellence
South Dakota State
  University
Brookings, South Dakota

James P. Verdin
U.S. Geological Survey
    Earth Resources Observation
    and Science Center
Sioux Falls, South Dakota

E.K. Weatherhead
Centre for Water Science
Cranfield University
Bedfordshire, United Kingdom

Xiangming Xiao
Center for Spatial Analysis and
    Department of Botany and
    Microbiology
University of Oklahoma
Norman, Oklahoma

Yoshifumi Yasuoka
National Institute for Environmental
    Studies
Tsukuba, Ibaraki, Japan

Songcai You
Institute of Geographic Sciences
    and Natural Resources Research
Chinese Academy of Sciences
Beijing, People's Republic of China

Ye Yuan
China University of Mining &
    Technology
Beijing, People's Republic of China

Abbreviations and Acronyms

2-D FS BGW	Two-dimensional Feature Space Plot of Brightness–Greenness–Wetness
ACD-crop	Agricultural census-based total cropland area data set
ACF	Apalachicola–Flint–Chattahoochee
ACT	Alabama–Coosa–Tallapoosa
ACZ	Agroclimatic zones
AER	Agroecological region
AET	Actual evapotranspiration
AEZ	Agroecological zones
AIA	Annualized Irrigated Areas
AIC	Akaike Information Criterion
ANIF	Anisotropy Factor
ANIX	Anisotropy Index
AOI	Areas of interest
APC	Agricultural Production Coefficient
AREI	Agricultural resource endowment index
ARPI	Agricultural resource poverty index
ASAR	Advanced Solid State Array SpectroRadiometer
ASM	Army Service Maps
AVHRR	Advanced Very High Resolution Radiometer
BRBD	Bambanwala–Ravi–Bedian–Depalpur
BRDF	Bidirectional reflectance distribution function
BRF	Bidirectional Reflectance Factor
CAAS	Chinese Academy of Agricultural Sciences
CBIP	Central Bureau of Irrigation and Power
CCA	Cultivable Command Area
CCCma	Canadian Center for Climate Modeling and Analysis
CDBPI	Climatically determined biomass productivity index
CIAF	County Irrigated Area Fraction
CPWF	Challenge Program on Water and Food
CRI	Climatic resource index
CRPI	Climatic resources poverty index
CRU	Climate Research Unit
CWANA	Central and West Asia, North Africa and the Horn of Africa
CWRPI	Climatic and water resources poverty index
CWSI	Crop water stress index
DCP	Degree confluence project

DCW	Digital Chart of the World
DEM	Digital elevation model
DES	Directorate of Economics and Statistics
DEFRA	Department for Environment, Food and Rural Affairs
DOQQ	Digital orthophoto quarter quads
DSP	Data Storehouse Pathway
DTED	Digital Terrain Elevation Data
E	Actual evaporation
EA	Environment Agency
EOS	Earth Observing System
EOSDG	Earth Observing System Data Gateway
EROS	Earth Resources Observation and Science
ET	Evapotranspiration
ET_a	Actual evapotranspiration
ETf	ET fraction
ETo	Reference ET
FAO	Food and Agriculture Organization of the United Nations
FEWS NET	Famine Early Warning System Network
FPA	Full Pixel Area
FSI	Feature Space Images
G_0	Soil heat flux
GCM	General Circulation Model
GDAS	Global Data Assimilation Systems
GDD	Growing-degree-days
GEE	Google Earth Estimates
GEVHRI	Google Earth very high-resolution imagery
GIA	Gross Irrigated Area
GIAM	Global Irrigated Area Mapping
GIS	Geographic Information System
GIScCE	Center/Geographic Information Science Center of Excellence
GLC	Global Land Cover
GMIA	Global Map of Irrigated Area
GMRCA	Global map of rain-fed cropland areas
GPS	Global Positioning System
GVP	Gross value of production
H	Sensible heat flux
HRI	High-resolution imagery
IAF	Irrigated area fraction for class
IAMRS	Irrigated Area Mapping and Reporting System
ICAR	Indian Council of Agricultural Research
ICARDA	International Center for Agricultural Research in the Dry Areas
IMW	International Map of World

IPU	Irrigation potential utilized
ISA	Impervious surface area
ISDB	Ideal Spectral Data Bank
IWMI	International Water Management Institute
JERS	Japanese Earth Resources Satellite
JM	Jeffries–Matusita Distance
LAEA	Lambert Azimuthal Equal Area
LCC	Lower Chenab Canal
LISA	Local Indicator of Spatial Association
LPDAAC	Land Processes Distributed Active Archive
LST	Land Surface Temperature
LST/E	Land Surface Temperature/Emissivity
MDFS	Multidimensional Feature Space
METRIC	Mapping Evapotranspiration at High Resolution Using Internalized Calibration
MFDC	Megafile data cube
MFP	Mississippi Flood Plain
MINFAL	Ministry of Food, Agriculture and Livestock
MIR	Midinfrared
MIrAD-US	MODIS Irrigated Agriculture Dataset for the United States
MISR	Multiangle Imaging SpectroRadiometer
MLR	Ministry of Land Resources
MMU	Minimum Mapping Unit
MOA	Ministry of Agriculture
MODIS	Moderate Imaging Spectrometer
MODIS	Moderate Resolution Imaging Spectroradiometer
MoWR	Ministry of Water Resources
MR	Marala–Ravi
MSM	Maximum soil moisture
MVC	Maximum Value Composite
MWR	Ministry of Water Resources
MWREP	Ministry of Water Resources and Electric Power
NALD	National abstraction licensing database
NASS	National Agricultural Statistics Service
NBS	National Bureau of Statistics
NBSS	National Bureau of Soil Survey
NCDC	National Climate Data Center
NDVI	Normalized Difference Vegetation Index
NIA	Net Irrigated Areas
NIR	Near-infrared
NLCD	National Land Cover Database
NOAA	National Oceanic and Atmospheric Administration

NWFP	North-West Frontier Province
PCA	Principle Component Analysis
PET	Potential evapotranspiration
PET_{HG}	Potential evapotranspiration after Hargreaves
PHLC	Doab and Pehur High Level Canal
PMIU	Punjab Monitoring and Implementation Unit
POLDER	POLolarization and Directionality of the Earth's Reflectances
PR	Pattern recognition
RMSE	Root mean square error
R_n	Net radiation
R_{n24}	24-h averaged net radiation
RSD-crop	Remote-sensing-based total cropland data set
SAIC	Science Applications International Corporation
SAR	Synthetic Aperture Radar
SCS	Spectral correlation similarity
SDSS	Spatial Decision Support System
SEBAL	Surface Energy Balance Algorithm for Land
SGT	Stinger Ghaffarian Technologies
SLC	Scan line corrector
SM_{ss}	Steady soil moisture state
SMT	Spectral matching techniques
SOM	Space Oblique Mercator
SPA	Subpixel areas
SPDT	Subpixel Decomposition Techniques
SRPI	Soil resource poverty index
SRTM	Shuttle Radar Topographic Mission
SSEB	Simplified Surface Energy Balance
SSV	Spectral similarity value
STSC	Space Time Spiral Curve
SWIR	Shortwave infrared
TAAI	Total Area Available for Irrigation
TD	Transformed Divergence
TOA	Top of atmosphere
TRI	Topographic Resource Index
TRPI	Topographic resource poverty index
TSNDVI	Time-Series NDVI
UCC	Upper Chenab Canal
UF	University of Frankfurt
UND	University of North Dakota
USAID	United States Agency for International Development
USCC	Upper Swat Canal Command
USDA	United States Department of Agriculture

USGS	United States Geological Survey
UTM	Universal Transverse Mercator
VHRI	Very high-resolution imagery
VNIR	Visible and near-infrared
WP	Water productivity
WPI	Wholesale price index
ZIV	Zoom-in-view
λ	Latent heat of vaporization
λE	Latent heat flux

Section I

Background and History of Global Irrigated Area Maps (GIAM)

1 Context, Need: The Need and Scope for Mapping Global Irrigated and Rain-Fed Areas

Hugh Turral
On the Street Productions

Prasad S. Thenkabail
U.S. Geological Survey

John G. Lyon
Environmental Protection Agency

Chandrashekhar M. Biradar
Univeristy of Oklahoma

CONTENTS

1.1 IRRIGATION, WATER USE, AND CROP PRODUCTION

Water is a topic of significant global public interest, encompassing its provision, quality, sustainable, and equitable use, and its importance in underpinning ecosystem health. As economic development continues and populations continue to grow, all aspects of water management attract closer scrutiny. Competition between different water-using sectors is evident or emerging, especially in arid and semiarid conditions.

Some of the principal goals realized in the development of poorer nations over the last century were the attainment of food security, satisfaction of individual livelihoods, and the reduction of poverty. Agriculture has played a major role in development thinking and progress, particularly as a springboard for economic development in Asia in the latter part of the twentieth century. The Green Revolution, from the 1960s to the late 1980s, resulted in significant increases in the yields of basic staple crops (rice and wheat) through the development of short-season, short-statured varieties that were responsive to increased inputs of nitrogen fertilizer and water. Improvements in other crops followed and were further refined, but did not have quite the same level of impact (FAO, 1996). Over the same period, improvements in agricultural science, agronomy, and husbandry resulted in more than a doubling of yields in developed country agriculture, across a wide range of crops and products.

Irrigation—the provision of water (in addition to rainfall and available soil moisture) to meet crop water demand and ensure secure or enhanced crop production—has played a key role in helping many countries achieve food security following bouts of severe famine in Asia and Africa in the 1960s and 1970s. Massive public investment, through international institutions and from national budgets, saw the rapid expansion of irrigated area and the development of large-scale public canal infrastructure (Jones, 1995). In India, more than one-third of public spending on infrastructure was directed to irrigation development during the 1970s and 1980s.

Such large investments naturally raised questions about the performance of irrigation and the distribution and equity of its benefits (Carruthers and Clark, 1983; Chambers, 1988). It is generally concluded that the economic performance of irrigation has been considerably lower than expected by its proponents, even though there are many cases of positive and even high internal rates of return (Faures et al., 2007). Large-scale surface irrigation has been singled out in particular for cost overruns and protracted development of irrigation service areas, which have lagged well behind the development of construction of dams, diversion structures, and main canals. There has also been concern that smallholders have been bypassed by many irrigation developments, as well as fears that many of the benefits have been captured by larger landowners (Lipton and Longhurst, 1989) (Figure 1.1).

From the early 1980s, there was renewed interest and investment in small-scale (farmer-managed) systems and many initiatives aimed at improving the sociotechnical and management performance of publicly managed systems, through better user participation and modernization. Simultaneously, it became clear that those without access to irrigation might often be considerably disadvantaged, as witnessed by the levels of migration from rain-fed areas to urban and irrigated areas. However, in many countries, landholdings in rain-fed areas may be considerably higher than in irrigated ones, and despite relatively modest improvements in rain-fed crop productivity, incomes and livelihoods can sometimes be better than in irrigated areas (Hussain, 2005).

A second wave of irrigation development occurred spontaneously from the mid-1980s, as farmers invested in increasingly affordable pumping and well-construction technology that allowed them to abstract both shallow and deep groundwater. This resulted in a revolution in groundwater use (Shah, 2008), exemplified by the massive

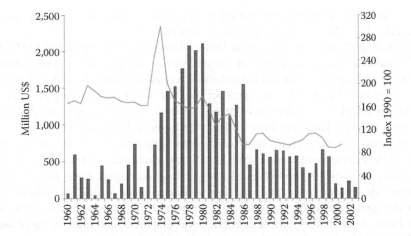

FIGURE 1.1 Investment in irrigation by the World Bank from 1960 to 2002 (bars) and the consumer price index for commodities (line) (From Faures, J.M., Svendsen, M., and Turral, H.N., Irrigation impacts, in *Water for Food, Water for Life: A Comprehensive Assessment of Water Management in Agriculture*, D. Molden (Ed.), Ch. 9, Earthscan, London and International Water Management Institute, Colombo, 2007.).

numbers of private groundwater users in India (> 20 million), China, and many other countries.

Agricultural water use currently accounts for the lion's share of water diverted for human consumption, averaging 80% at a global scale, and ranging from 65% in northern China to close on 90% in India (Molden, 2007). Typically, irrigation accounts for a much lower proportion of water use in industrialized temperate countries (< 30%), but uses similar proportions of diverted water (> 70%) in some larger economies in dry climes—including the western United States, Australia, and Spain.

In many river basins around the world, water resources are fully developed, usually where irrigation is widely practiced. Cities continue to grow rapidly, demanding water for sanitation, drinking, and amenity, as well as to support growing industries. Where water supplies are short, locally or at river basin scale, water moves from agriculture to "higher-value" uses through a variety of formal and informal means (Molle and Berkoff, 2006). Similarly, rising demands for rural drinking-water supply and sanitation are buffered by water transfers from agriculture.

Consumptive use of water, for example through evaporation by irrigated crops, can result in substantial modification of the natural state and health of riverine and aquatic ecosystems (WCD, 2000). The environmental costs of overallocation of water were not foreseen or appreciated by many water developers in the past, but have become increasingly obvious through habitat loss and degradation of water quality (salinity and nonpoint source pollution from agricultural return flows) due to reduced flows and changed flow patterns (MA, 2006). The main driving force for water reform and the specification of environmental flows to maintain or restore river and ecosystem health in basins as far apart as the Murray-Darling in Australia, the Colorado in the United States, and the Yellow River in China, has been the fear that the function of agricultural and other man-made systems will also, in turn, be undermined by a broader damage to ecosystem health.

The Comprehensive Assessment of Water Management for Agriculture (Molden, 2007) makes a broad evaluation of the costs, benefits, alternatives, and the future of irrigation and improved rain-fed agriculture. It concludes that a doubling of agricultural water use efficiency ("water productivity" or the physical output or value per unit of water depleted) is required, using no more water than at present, and preferably less, in order to feed and clothe the world's population in 2050 and thereafter. It notes that the yield and water productivity of rain-fed agriculture must also rise dramatically, as the expansion of rain-fed crop area would have unacceptable environmental consequences, including hydrological ones that would impact water availability in existing rain-fed and irrigated areas. The various chapters of the assessment highlight the interdependence and trade-offs between different sustainable options to meet future food security, enable or enhance development, and restrain water use.

The quantification of actual water use, especially in agriculture, is anything but exact. Climate is naturally variable from season to season and from year to year, and exhibits often surprising spatial differences: the areas of crops planted vary considerably from year to year and, in the wet and semiarid tropics, three or more crops can be grown in a year—with continuous rice being cultivated up to seven times in two years in some parts of Indonesia and Vietnam, subject to water availability. Only a few countries (such as Australia) keep detailed volumetric accounts of water delivered on-farm, and therefore water use is more usually derived from secondary statistics on crop areas and crop patterns, climatic data, and calculated irrigation needs and use. Crop areas are, in turn, often derived from surveys or tax records and this, unsurprisingly, can lead to systematic underestimation or overestimation, depending on the source and socioeconomic context. The estimates of total agricultural production are also usually derived from secondary statistics of crop areas coupled with sample surveys of crop yields.

The situation has become further complicated by the recent surge in planting of biofuel crops, which have displaced less profitable food crops in both developed and developing countries. While many are rain-fed, maize and sugarcane grown for bioethanol are also widely irrigated. The extent and impact of biofuel cropping have taken many by surprise, and have substantially contributed to the first upturn in real food commodity prices for nearly 20 years (The Economist, December 8, 2007), in turn, reigniting fears of food shortages across the globe.

The impacts of climatic change on hydrology are expected to be significant in arid and semiarid regions, with decreased rainfall and increased variability leading to more frequent drought and flood (IPCC, n.d, 2007). This will lead to substantial changes in freshwater runoff, as illustrated in Figure 1.2.

It is uncertain what will happen to evapotranspiration rates, in principle, higher temperatures will increase evaporative loading, but globally there is a significant trend of reduced net solar radiation due to increased levels of particulate matter in the atmosphere, leading to lower potential evapotranspiration (Fischer et al., 2005). The emerging consensus is that the rise in atmospheric carbon dioxide will not offset likely yield decreases due to higher temperatures and poorer water availability (Stern, 2006).

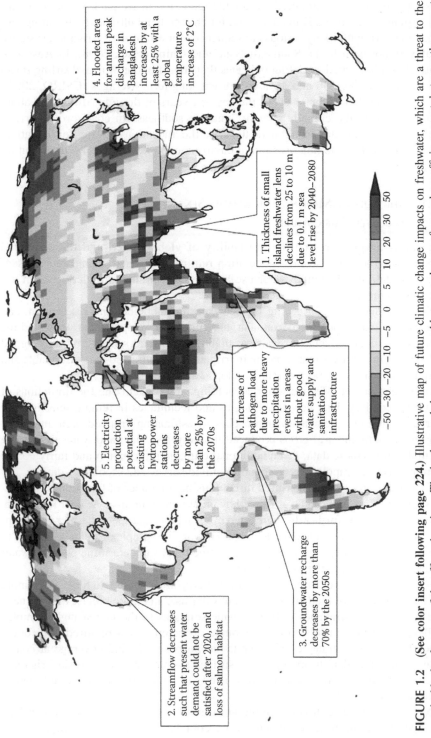

FIGURE 1.2 (**See color insert following page 224.**) Illustrative map of future climatic change impacts on freshwater, which are a threat to the sustainable development of the affected regions. The background shows an ensemble mean change of annual runoff, in percent, between the present (1981–2000) and 2081–2100 for the SRES A1B emissions scenario; blue denotes increased runoff, and red denotes decreased runoff. (From IPCC, *Climate Change Impacts, Adaptation and Vulnerability. Contribution of Working Group II to the Fourth Assessment Report of the Intergovernmental Panel on Climate Change*, Cambridge University Press, Cambridge, 2007.)

Rising temperatures are credited for the reduction in ice-pack in major mountain systems, such as the Himalayas, and for changed and ultimately declining patterns of runoff, with likely serious consequences for established irrigation systems, notably in northern India, Southeast Asia, and China (FAO, 2008). In Australia, relatively sophisticated modeling indicates that runoff in the Murray Darling Basin will decline between 30 and 40% on present long-term averages (CSIRO, 2007), and could even fall to the levels of the past six years of drought, if theories about a step change in the climatic system along the southeastern coast prove correct. Preparations are already being made to revise the allocation of water for irrigation, and to cope with the taxing problem of environmental flow provision in the future (DSE, 2006).

1.1.1 THE ROLE OF MAPPING AND REMOTE SENSING OF IRRIGATED AGRICULTURE

The gathering pressures on the availability of global water, and the continued imperative to feed the world's population require much better (and more disaggregated) data on (1) actual water use and (2) agricultural production, than are available at present.

Strategic decisions concerning further water development and reservation, and allocation of water for the environment, agricultural development strategy, and the impacts of climatic change on water availability and food security require a good understanding of the current situation in irrigated agriculture so that sustainable and profitable alternative options can be developed and implemented.

In the absence of effective water accounting, and recognizing the limitations of existing secondary data, new technologies and techniques in remote sensing offer alternative and complementary approaches to estimating cropped area and cropping intensity (number of crops produced in one year). Yield and water use can be calculated from these data, given additional climatic information and fairly coarse assumptions or extrapolations from more limited survey and secondary data.

Crop water use can be measured directly using a number of techniques of varying sophistication, and can be attributed by pixel or by unit area. At finer resolution, cropped area can be further disaggregated by crop type, and specific water consumption can be determined. Yield models, again of varying complexity, can be applied to extensive cropped areas or to specifically defined crops to determine physical water productivity and to predict total production.

Accurate mapping of irrigated and rain-fed crop areas should, in time, allow more precise assessment of food production and water use, and more precise and nuanced land management strategies to help provide solutions to the increasingly complex challenges outlined in this chapter. A range of techniques is required at different scales for different strategic and operational purposes. Global coverage at coarse scale permits the identification of large surface systems, but small-scale irrigation, and in particular groundwater use, also needs to be identified, monitored, and in some cases regulated.

At river basin scale, it is becoming apparent that the interactions between land use, hydrology, and water development are complex, and not always intuitive. As water resources become fully allocated in a basin, there has been an emerging trend to develop more land in the upper catchments for rain-fed farming, supplemented by water harvesting and groundwater use. This, in turn, changes the nature and patterns of runoff to existing users downstream. Irrigation is often thought to provide differential benefits, further marginalizing the poor who do not have access to irrigation, providing a strong political justification for catchment development and rainwater harvesting, notably in India. Thus, it is becoming increasingly important to establish effective and informed land use planning, with appropriate water accounting, based on a good understanding of landscape processes and hydrology. A better understanding allows more informed trade-offs to be made and the likely consequences to be modeled, leading to more politically and environmentally sustainable policy and outcomes.

Strategies to increase future food production will require intensification of both rain-fed and irrigated agriculture, with similar or reduced levels of water use. At the same time, the possibilities for expansion of cropped area in the semiarid and arid tropics will be limited. Climate change scenarios predict increased rainfall and enhanced growing conditions in the higher latitudes, but the achievement of global food security would then depend on agricultural trade regimes that are substantially different from those working today. The poorest countries will likely continue to seek self-sufficiency in food production, but will be increasingly aware that this cannot be achieved at the price of damaged ecosystems, and even less so in terms of groundwater mining and the desiccation of rivers. Global food production, trade, and trade-off scenarios also benefit from improved data on the potential suitability of lands for development, intensification, extensification, and retirement. The techniques evolving in irrigated area mapping can also be applied to improved land suitability assessment, although some key areas for further research emerge, such as effective detailed soils mapping using remote sensing.

The remote-sensing techniques described in this book contribute to other sorts of trade-off analysis, apart from agricultural development and management, and the management of aquatic ecosystems. For example, there is particular interest in mapping wet rice area for carbon accounting, in a quest to better understand global climate change, since wetlands and paddy rice culture are major contributors of methane.

There are also great possibilities to improve irrigation system operation and management through the use of remote sensing to determine crop area, crop growth stage, and condition, and to monitor water use and the equity of its distribution on a seasonal and even a weekly basis. In general, effective and integrated flow measurement is rare within irrigation systems throughout Asia. Intensification and more equitable water distribution can be improved by understanding the variability of crop water productivity within medium- and large-scale surface irrigation networks. Remote-sensing techniques to derive crop productivity and water use can provide managers with feedback of their performance as well as the performance of the system overall. Modernization strategies should focus on improving performance, where it is needed, to achieve more economically attractive and productive outcomes; again remote-sensing techniques have a role to play through the provision of consistent and repeated monitoring information.

In systems with salinity problems, differential supply strategies can be developed to mitigate degradation in saline areas and enhance production elsewhere. In Australia, rice area is monitored each year, using aerial photography, to help ensure that the hydraulic loading to a largely saline water table is minimized. Zoning and compliance will become increasingly common features of sustainable agricultural management, and remote sensing has a strong role to play in cost-effective and consistent monitoring.

Groundwater use for agriculture in India and China has developed so extensively and rapidly, that statistics may be little more than guesses. Accurate remote sensing at finer resolution should allow unequivocal quantification of groundwater area, and provide a monitoring tool to reliably estimate net groundwater use. Authors such as Tushaar Shah have long commented that conventional approaches to regulation of groundwater are nigh impossible in rural India. Understanding where to target efforts to restrain groundwater use could be improved through remote sensing, allowing a range of local regulatory, power supply, and community-based approaches to be tested.

Techniques developed to map irrigated area, outlined in this book, have broader application in rain-fed crop mapping (also discussed in this volume) and to environmental applications, such as wetland inventory and classification. In time, they have the potential to be developed into strategic and operational monitoring tools that help attack the challenges of the future across a broad front. Remote-sensing techniques will increasingly be integrated with the Geographic Information System (GIS) and modeling frameworks, and better protocols for groundtruth and accuracy assessment will steadily improve their acceptance and use by a broader community of natural resources policy makers and managers. This book is one step on the way.

REFERENCES

Carruthers, I. and Clark, C., *The Economics of Irrigation*, Liverpool University Press, Liverpool, 1983.

Chambers, R., *Managing Canal Irrigation*, Cambridge University Press, Cambridge, 1988.

CSIRO (Commonwealth Scientific and Industrial Research Organisation), *Climate Change in Australia*, CSIRO and Bureau of Meteorology, Canberra, Australia, 2007.

DSE (Department of Sustainability and Environment), *Victoria, Sustainable Water Strategy*, Central Region Discussion Paper, Melbourne, Australia, 2006.

FAO (Food and Agriculture Organization of the United Nations), *World Food Summit Technical Document #6: Lessons from the Green Revolution,* United Nations, New York, 1996, http://www.fao.org.final.e/volume2406-e.htm.

FAO, *Climate Change, Water and Food Security,* High Level Conference on World Food Security – Background Paper HLC/08/BAK/2, 2008, ftp://ftp.fao.org/docrep/fao/meeting/013/ai783e.pdf.

Faures, J.M., Svendsen, M., and Turral, H.N., Irrigation impacts, in *Water for Food, Water for Life: A Comprehensive Assessment of Water Management in Agriculture*, D. Molden (Ed.), Ch. 9, Earthscan, London and International Water Management Institute, Colombo, 2007.

Fischer, G., Shah, M., Tubiello, F., and van Velhuizen, H., Socio-economic and climate impacts on agriculture: An integrated assessment, 1990–2080, *Philosophical Transactions of the Royal Society – Biological Sciences*, 360, 2067–2283, 2005.

Hussain, I., *Pro-poor Intervention Strategies in Irrigated Agriculture in Asia – Poverty in Irrigated Agriculture: Issues, Lessons, Options and Guidelines*, Final Synthesis Report submitted to the Asian Development Bank, International Water Management Institute (IWMI), Colombo, Sri Lanka, 2005.

IPCC, The Fourth Assessment Report of the Intergovernmental Panel on Climate Change, Cambridge University Press, Cambridge, 2007.

IPCC, Summary for policymakers, in *Climate Change Impacts, Adaptation and Vulnerability. Contribution of Working Group II to the Fourth Assessment Report of the Intergovernmental Panel on Climate Change*, M.L. Parry, O.F. Canziani, J.P. Palutikof, P.J. van der Linden, and C.E. Hanson (Eds.), Cambridge University Press, Cambridge, 2007, 7–22.

Jones, W.I., *The World Bank and Irrigation*, A World Bank Operational Evaluation Study, Washington, DC, 1995.

Lipton, M. and Longhurst, R., *New Seeds and Poor People*, Unwin Hyman, London, 1989.

MA (Millennium Ecosystem Assessment), *Ecosystem Services and Human Well-being: Wetlands and Water Synthesis*, World Resources Institute, Washington, DC, 2005.

Molden, D. (Ed.), *Water for Food, Water for Life: A Comprehensive Assessment of Water Management in Agriculture,* Earthscan, London, 2007.

Molle, F. and Berkoff, J., *Cities Versus Agriculture: Revisiting Intersectoral Water Transfers, Potential Gains and Conflicts,* International Water Management Institute, Colombo, 2006.

Shah, T., *Taming the Anarchy: Groundwater Governance in South Asia*, RFF Press, Washington, DC, 2008.

Stern, N., *The Economics of Climate Change. The Stern Review*, Cabinet Office, HM Treasury, Cambridge, 2006.

WCD (World Commission on Dams), *Dams and Development: A New Framework for Decision Making*, Earthscan, London, 2000.

2 A History of Irrigated Areas of the World

Y.J. Li
University of Maryland

Prasad S. Thenkabail
U.S. Geological Survey

Chandrashekhar M. Biradar
University of Oklahoma

Praveen Noojipady
University of Maryland

Venkateswarlu Dheeravath
United Nations Joint Logistics Centre

Manohar Velpuri
South Dakota State University

Obi Reddy P. Gangalakunta
National Bureau of Soil Survey and Land Use Planning (ICAR)

Xueliang L. Cai
International Water Management Institute

CONTENTS

Agriculture is the foundation of a country while irrigation is the spirit of agriculture.

—King of Han Wu, 155–74 BC

2.1   OVERVIEW OF THE EVOLUTION OF GLOBAL IRRIGATED AREAS, THEIR STATISTICS, AND MAPS

Very recent archeological evidence [62] in site Ohalo II in Syria found more than 90,000 plant fragments from 23,000 years ago. This indicates that, settled agriculture began over 10,000 years earlier than previously thought. Previous to this, the settled agriculture was known to have started around 10,000 years back [1–3]. Irrigated agriculture started at Mesopotamia Lowland around 7000–6000 BC [3,4]. An overwhelming proportion of early agriculture took place along the river banks, deltas, and flood plains, where there was plenty of water for agriculture [4]. Notwithstanding that, these may not be strictly defined as irrigation, as they were dependent on nonprecipitation sources for agriculture. Almost all early, great civilizations flourished along the great rivers. With time and increases in populations, humans started using engineering skills to block rivers, create reservoirs and channels, and divert water to areas of interest. Yet, until 1800, irrigation was a modest 8 million hectares (Mha). However, irrigated areas increased swiftly to reach 50 Mha by 1900 and 95 Mha by 1940. From 1940 to 2000, which can be called the irrigation era, irrigated areas increased to around 275 Mha by conventional estimates. During this period, informal irrigation (e.g., groundwater, small reservoirs, minor embankments, and tanks) flourished. After 1950, farmers began tapping groundwater on a large scale, as powerful diesel and electric pumps became available [5]. In the early 1960s, for example, groundwater irrigation was insignificant. In substantial areas of China and India, groundwater levels were falling by 13 m/year [6]. However, by 2000, groundwater irrigation was nearly 60% of the total irrigated areas in India as a result of over 19 million bore wells sunk across the country [7].

The irrigated area statistics of AD 1800 or earlier were very rough estimates based on ancient literature and/or history and/or anecdotal evidence [4,5,8–14]. From 1800 to 1990, irrigated area statistics were primarily made available in many parts of the world as governmental statistics were derived, based mainly on observations and estimates.

Irrigated area maps were a rare phenomenon. In the twentieth century, some irrigated area maps were produced mainly at subnational level [15–17] and some at national level [18,19]. But irrigated areas were still an afterthought, often mapped as one of the many land use/land cover (LULC) classes. At the global level, no map existed until the Food and Agriculture Organization of the United Nations/University of Frankfurt (FAO/UF) effort [20] in the late 1990s, which took the national statistics on irrigated areas and did spatial extrapolation to come up with an irrigated area map of the world.

At the end of the last millennium, the International Water Management Institute (IWMI) commissioned a systematic mapping of the irrigated areas of the world using spatial technologies. The outcome was the first satellite-sensor-based irrigated area map of the world [21,22].

2.2   IRRIGATED AREAS FROM 7000 BC TO AD 2000

Irrigation played an important role in the history of humankind. Early recorded irrigated areas of the world (Figure 2.1) from over 9000 years ago, spread from 7000 BC to AD 1000, were mainly concentrated along the great rivers and great civilizations. Our research found 16 great civilizations with major irrigated areas [4,5,9–14,23,24]. These were, the river basins of the Tigris and Euphrates (7000–6000 BC), the Nile (6000 BC), Indus (5000 BC), Central-east China (4000 BC), and Yellow River (2000 BC). Much later, irrigation became significant in coastal Peru (1300–1000 BC), Japan (660 BC), previous USSR (600–400 BC), Central-west Sri Lanka (400 BC), South Mexico (500–300 BC), Central India (300–200 BC), Italy (AD 100), American southwest (AD 800), and the Lower Mekong River basin (AD 100–900). Most of the ancient irrigated areas were located between 46° N latitude and 10° S latitude (Figure 2.1). Modern-day irrigation has been inspired by historical irrigation development [13]. Today, most of the ancient irrigation systems (Figure 2.1) are the hub of modern irrigation systems and massive expansions. In the following paragraphs, we will look into ancient irrigation development patterns and how they have influenced modern irrigation, by considering some of the leading irrigated area nations.

Global maps on irrigated areas did not exist until 1999 [25]. Such a map described the fraction of each 0.5° cell area that was equipped for irrigation around 1995 [20], and even now, it is being updated [26,27]. Early maps showed agricultural areas that included irrigation: IGBP Land Cover Classification [28,29]; global land cover [30]; Ganges and Indus River basin LULC and irrigated area map [31]; Krishna basin irrigation map [32,33]; Olson map [34,35]. The advent of remote sensing/Geographic Information System (RS/GIS) technologies resulted in LULC maps that, at times, had irrigation as one of the classes. Irrigation maps were mostly limited to small areas, but none for the world [22]. This process of mapping ended with the global irrigated area map [21,22] of FAO/UF and IWMI.

2.3   CHINA

Recently, some archaeologists discovered evidence of some small-scale irrigation in Central-east China that goes back to 4000 BC [9]. So far, it is the earliest activity of irrigation that has been uncovered in China. The history of Chinese in practicing large-scale irrigation started from around 2070 BC (Xia Dynasty).

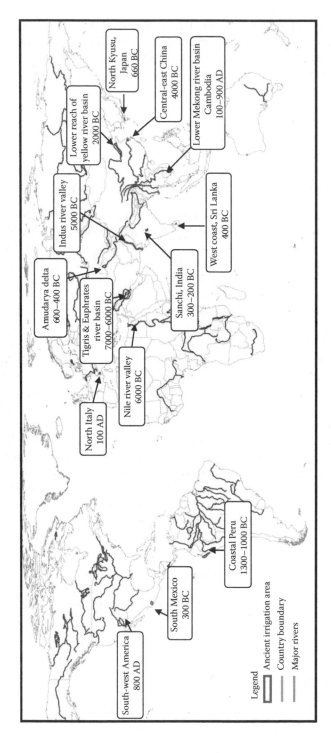

FIGURE 2.1 **(See color insert following page 224.)** Regions of ancient irrigation systems (since 7000 BC).

Organized big projects to control water from flooding and for irrigation purposes from the Yellow River were first led by Yu [10]. However, the recorded history of irrigation in literature is available only since the Qin Han period (600–220 BC). During this period, one of the oldest recorded documents on irrigation could be found (e.g., Figure 2.2). A map shows the irrigated area during that time (Figure 2.3). During the last period of Xi Han, the population of China was about 50 million, with a cultivated area of about 41 million Chinese mu (1 Chinese mu=666.67 m²) [36]. If

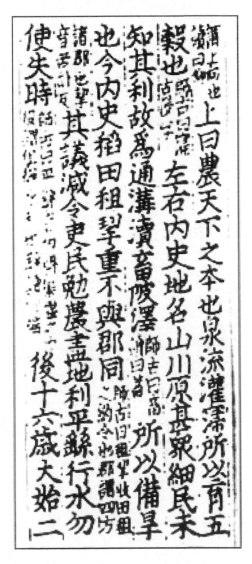

FIGURE 2.2 Early recorded irrigation history of China. This is from one of the oldest history book, called the "Han Book". It was published during the reign of King Han Wu (155–74 BC) the words written there say, "Agriculture is the foundation of a country while irrigation is the spirit of agriculture...."

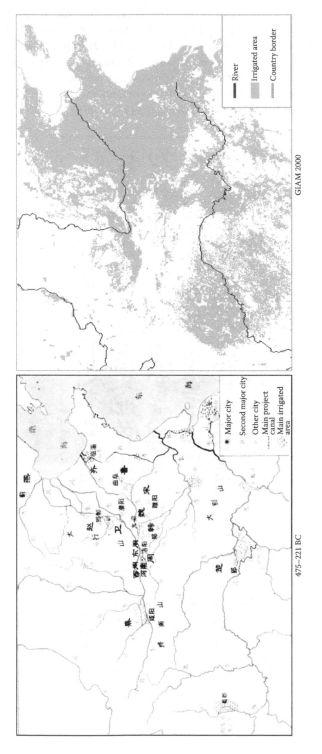

FIGURE 2.3 (See color insert following page 224.) The comparison of irrigated areas in Central-east China.

one-third of the agricultural land was irrigated, there would have been about 0.91M ha during that time.

Figure 2.2 shows the recorded early irrigation history of China. This is from one the oldest history books, called *Han book*, published during the reign of King of Han Wu (155–74 BC). It is written "Agriculture is the foundation of a country while irrigation is the spirit of agriculture. ..."

An early irrigation area map was recovered by Yao [37] for the period of Zhanguo (475–221 BC) (Figure 2.3, left map). The early Chinese irrigation systems were mainly at major river branches and were surrounded by major cities. From figure 2.3, we can see the enlargement of a modern irrigation area developed from the earlier system.

Du Jiang Yan, one of the most famous ancient irrigation projects in China, was constructed around 250 BC. Some other major project constructions for irrigation and flood control were undertaken during China's water control history: An Feng Tang (Huai River, 600 BC), Grand Canal (linking Beijing and Hangzhou, 246–210 BC), and Hsingan Canal (linking Yangtze and Pearl River systems, third century BC). The water wheel was developed and extensively used for low-head lift irrigation during the Han Dynasty (206 BC to AD 220) [13]. During the Tang Dynasty (seventh to thirteenth century), the area downstream of the Yantze River and around Tai Lake, which had systematic irrigation activities, became the major food basket for the whole of China. Later, new irrigation systems were developed in Central and Northwest China, like He Tao plain irrigation and Tai lake TangPu irrigation system.

Irrigation projects in China were developed at greater phases after the Yuan Dynasty (AD 1370). During this period and in later years, more irrigation systems were developed in southeast China, Fujian, HaiNan, Taiwan, in the southwest of China, Yunnan, Guizhou, northwest Xinjiang, Neimenggu, and Ningxia. For sometime after that, irrigation development stopped because of wars and natural disasters, and started blooming again after the establishment of the People's Republic in 1949. In order to match the rapid increase in demand for food from the ballooning population, irrigation attained gigantic development. In 1949, an area of 19.5M ha was irrigated. From 1949 to 1985, the addition of 83,000 reservoirs and 2.3 million wells put an additional 29M ha under irrigation [29,38] (see Figure 2.4).

Traditionally, irrigated area statistics for China were obtained from the government. The investigators go to the villages to collect data by the questionnaire method (Chinese Natural Resources Evironment, economic and population database, http://www.naturalresources.csdb.cn/index.asp [41]). Irrigated area maps were created for some major irrigation systems only [42]. According to national statistics, the total irrigated area during 1999 was 53.4 Mha; quite a contrast to the statistics from Global Irrigated Area Mapping (GIAM) that are shown in Table 2.1.

By comparing the two records, we can see that the GIAM estimate is based more on seasonality. It reports the irrigated area seasonwise and annualwise, and by the availability of total irrigable land. The reason to classify irrigated area by seasons is that the seasonality defines irrigation activity, like effective irrigated area and actual

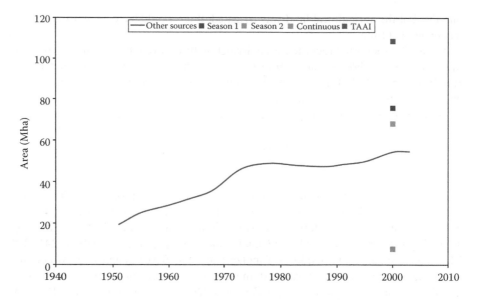

FIGURE 2.4   Irrigated area trends in China between 1940 and 2000. (From Liu, C.M., He, X.W., and Ren, H.Z., *China Water Problem*, Meteorological Press, Beijing, 1996, 160–64.)

TABLE 2.1
GIAM Statistics for China Irrigated Area

Type	GIAM 10-km Area (ha)
Season 1 irrigated area	75,880,320
Season 2 irrigated area	68,233,355
Continuous irrigated area	7,688,411
Annualized irrigated area	151,802,086
Total area available for irrigation	111,988,772

---

effective irrigated area [43]. There are some difficulties in comparing irrigation area data from China:

- Inconsistent area units Mu (varies from 667 to 1500 m^2)
- Benefit driving factors, such as to evade taxes or paying water fees
- Area measurement/estimation errors
- Compilation errors
- Illegal land use change

Figure 2.5 shows the updated irrigated area from GIAM.

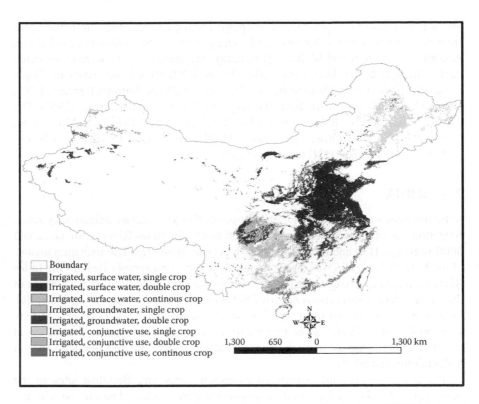

Boundary
Irrigated, surface water, single crop
Irrigated, surface water, double crop
Irrigated, surface water, continous crop
Irrigated, groundwater, single crop
Irrigated, groundwater, double crop
Irrigated, conjunctive use, single crop
Irrigated, conjunctive use, double crop
Irrigated, conjunctive use, continous crop

FIGURE 2.5 **(See color insert following page 224.)** Spatial distribution of GIAM irrigated areas for China at the end of the last millennium, 2000.

2.4 INDUS RIVER BASIN

The Indus River basin is a famous ancient irrigated system. The farmers who migrated, settled in the Indus River Valley around 5500 years ago [5], and there is evidence to show there was irrigation about 5000 years ago [13]. At the beginning, only the narrow strips of land along the river banks were irrigated, but as time went on, irrigation was extended to other nearby areas by reaching the banks or the natural levies of the rivers to bring water to the low-lying field [13]. Canals were built when the demand for food increased as population and migration increased. The first irrigation peak came by 2300 BC, and during that time, water for small canal irrigation was abstracted from the Indus [44]. Another piece of evidence shows the possibility of ancient well irrigation, since the water table at Mohenjo-daro was around 2 m deep between 3000 and 2750 BC [45]. Elaborate drainage channels had existed at Mohenjo-daro, which means buildings with sophisticated underground drainage and irrigation structures existed during the Harappan time [44]. The peak of canal irrigation was between 600 and 700 BC, when there was richness of the area existed in the south of present Multan. But irrigated agriculture declined within the next 200 years due to changes in the course of the Indus. The irrigation era did not come until AD 500. Perennial irrigation started from the seventeenth century and was constructed by the Mughal Emperor Jahangir (1605–1627) to bring water from the right bank of the Ravi to the pleasure gardens of Sheikhupura near Lahore. According to Framji and Mahajan [13], many irrigation projects were constructed during the British rule. They were: Rechna Doab (1892), Chaj Doab (1901), the Triple Canal Project (1915), the Sukkur Barrage Project (1932), the Trimmu Barrage (1939), Kalabagh Barrage (1947), the Kotri Barraqge (1955), and Guddu Barrage (1963). The total irrigated area for Pakistan was 9.44 Mha during 1947–1948, 9.55 Mha during 1958–1959, 11.4 Mha during 1965–1966, 12.9 Mha in 1971, 14.45 Mha in 1979 [13], and 15.976 Mha in 2000 [21] (see Figure 2.6).

2.5 INDIA

India has been known as a country with one of the most famous irrigation systems. Irrigation was practiced even in prehistoric times in the Indus River basin, i.e., about 5000 years ago [13], and the Sanchi dam in central India [46]. The irrigation history of the Indus River basin was discussed in the previous paragraph. Besides the Indus River basin, ancient irrigation systems were found in Central India as well. Some of the earliest dam constructions occurred between the third and second centuries BC, following the chronology of the earliest monuments at Sanchi (see Figure 2.7); they were built to provide irrigation, principally for rice, as a response to the increased population levels suggested by the distribution of habitational and Buddhist sites in Vidisha's hinterland [8].

There was a relationship between irrigation dams and Buddhist sites in the Sanchi area. By the Gupta period, this system appears to have been recast within a Brahmanical framework, as attested by the growing number of Hindu temples on or

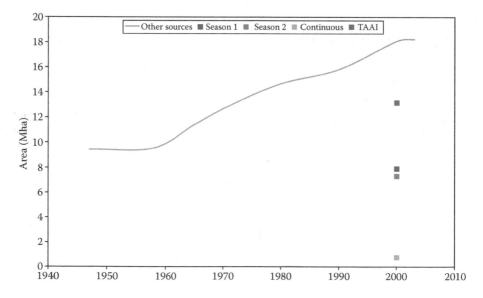

FIGURE 2.6 Historical change of irrigated area in the Indus River basin (Pakistan country statistics) (1927–2000).

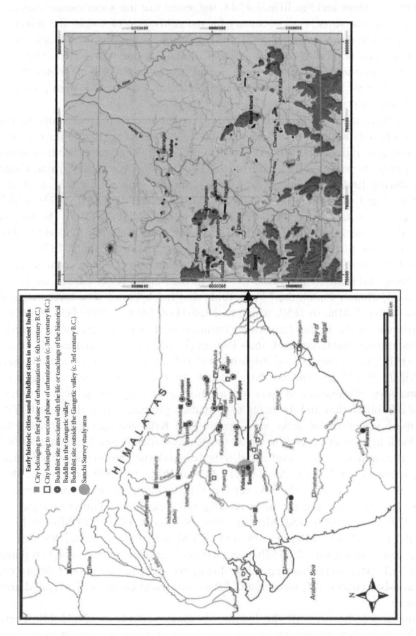

FIGURE 2.7 The location and distribution of the Sanchi dam irrigation system [46].

near the dams [8,46]. These patterns fit with broader developments across India from the Gupta period onward, when we begin to find epigraphic evidence for Brahman land grants and the rise of temple-owned land and water resource structures. In their earlier papers, Shaw and Sutcliffe [8,47,48] suggested that the water storage capacity of the Sanchi reservoirs is not necessary for the cultivation of wheat, the principal local crop today; due to the high moisture storage capacity of the local black cotton soils, wheat can be grown entirely on local rainfall, without irrigation. The total reservoir volume across the study area, based on recent revised estimates [8], is about 19.5×10^6 m^3. With an estimated water requirement for rice of 0.8 m^3, this corresponds to a total irrigated area of 24 km^2 [46].

Most irrigation systems in India, a leading irrigation country, were constructed recently. During 1800–1836, the old Yamuna Canals in the north and the Cauvery delta system (originally built in AD 200) in the south, were remodeled. A series of canal projects was undertaken during 1873–1894. They were: the Sirhind Canal (1882), during 1960–1961 an area of 1,069,177 ha was irrigated by this canal; the Agra Canal and the Lower Ganga Canal (1873); the Betwa Canal (1893), which irrigated 99,553 ha during 1960–1961; the Nira Canal (1881–1894); the Gokak Canal (1882); the Mhaswad Tank (1885); the Periyar Project (1887), which irrigated an area of 61,107 ha during 1960–1961; and the Mutha Canals (1879). From 1866 to 1901, some other small but significant irrigation works were constructed in Tamil Nadu and Bombay Presidencies, such as the Pennar River canals, the Barur Tank, the Hathmati Canal, the Ekruk Tank, and Lakh Canal. The total irrigated area was about 1 Mha in 1850, which reached 11.66 Mha in 1900 [13]. After the establishment of the Indian Irrigation Commission, a large number of new works were undertaken: the Godavari Canals Project (22.662 ha irrigated area); the Sarda Canal (1916), has a capacity of irrigating 445,155 ha; the Pravara Canals Project can irrigate 30,350 ha; and the Nira Canals. In 1921, the Montagu-Chemsford Reforms were introduced and there was a significant change in the policy governing the administration and financing of irrigation projects. From then until 1935, some more important projects were begun—the Krishanarajasagar Project, the north bank high-level canal (known as the Visveswaraya Canal) irrigates 48,562 ha and the right bank low-level canal irrigates 1416 ha; the Nizamsagar Project (1924) irrigates 111,288 ha; the Cauvery Mettur Project (1934) irrigates 121.811 ha; and the Gang Canal or the Bikaner Canal (1927) irrigates a command area of 283,280 ha. The following are the gross irrigated areas in India during the past 56 years: 22.6 Mha in 1951, 25.6 Mha in 1956, 28 Mha in 1961, 30.9 Mha in 1966, 38.2 Mha in 1971, 43.2 Mha in 1976, and 45.3 Mha in 1978 [13]. In 2000, during the first cropping season, 72.66 Mha were irrigated and during the second cropping season, 53.73 Mha were irrigated, while during the same period about 6 Mha were continuously irrigated [21] (see Figure 2.8 for historical changes of irrigated area in India).

China and India are still the leading irrigation countries [21,22,44]. There are other irrigation regions and countries, which are not the leading ones, but have longer irrigation periods. They are Egypt, Peru, and Tigris-Euphrates River basin.

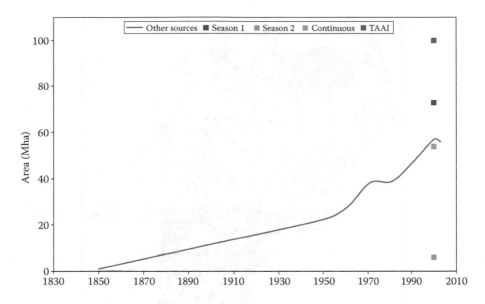

FIGURE 2.8   Historical changes of irrigated area in India (1850–2003).

2.6 EGYPT

Agriculture was introduced to northern Egypt a little before 5000 BC [49]. According to Butzer [50], predynastic settlers in northern Egypt inaugurated irrigation projects, from a base on the floodplain margins, during the fifth and fourth millennia BC, tackling the hydraulic problems of the bottomlands by draining the swamps and cutting down the thickets. A good example of a very early irrigation system in Egypt was an irrigation network inaugurated by the Scorpion King (Figure 2.9) about 3100 BC (Table 2.2).

Artificial irrigation, including deliberate flooding and draining by sluice gates, and water contained by longitudinal and transverse dikes, was established by the first dynasty. Controlled irrigation was easiest among the smaller flood basins of southern Upper Egypt and further north, on the east bank. In these areas, crop cultivation steadily increased at the expense of pastoral activities and the remaining tracts of "natural vegetation." Examination of the impact of strong or deficient floods on such a rudimentary and locally organized irrigation system suggests that the apparent redistributors and managerial functions of the local temples, at least during the new kingdom, would have been of paramount significance in coping with the vagaries of the river flow. The economic history of ancient Egypt was primarily one of continuous ecological readjustments due to the variable water supplies. It was combined with repeated efforts to intensify or expand land use in order to increase productivity. The pioneer, Mohammed Ali, developed deep canals and barrage irrigation systems in 1826. Since then, the irrigated area has increased from 2.02 Mha in 1879 to 2.4 Mha 1900 [13]. The construction of the Aswan Dam in Upper Egypt led to perennial irrigation, a cropping intensity of 2 Mha or more, and an increase in

FIGURE 2.9 The Scorpion King inaugurating an irrigation network, ca. 3100 BC. (From Butzer, K.W., *Early Hydraulic Civilization in Egypt – A Study in Cultural Ecology,* The University of Chicago Press, Chicago, IL, 1976.)

irrigated area to 3 Mha by 1980 and to 3.3 Mha (gross irrigated area) by 2000 [21] (see Figure 2.10).

2.7 TIGRIS AND EUPHRATES RIVER BASIN

The earliest agricultural practice started from approximately 10,000 BC from the Upper Tigris and Euphrates highlands. Lake Zeribar is one of the places of such an area (see Figure 2.11 [1]).

According to Pollock [4] and Maisels [3], irrigated agriculture started around 7000–6000 BC at Mesopotamia Lowlands, since an overwhelming proportion of early agriculture took place along the riverbanks, deltas, and flood plains. The most famous irrigation practice from the Tigris and Euphrates River basin began in 2300 BC with the ancient civilization of Sumerians, Babylonians, and Assyrians. During the Sumerian time (3500–2000 BC), canals were introduced among the first systematic agriculturists. By 3000 BC, there were extensive irrigation systems branching out from the Euphrates River, controlled by a network of dams and channels. At its peak, 10,000 square miles were irrigated [51]. Evidence shows the irrigation water source included both canals and groundwater. The Assyrian king, Sennacherib (705–681 BC), developed a 10-mile long canal in three stages, including 18 freshwater courses from the mountains, two dams, a water diversion, and a chain of canals.

TABLE 2.2

Hypothetical Demographic Developments in Ancient Egypt

Region	4000 BC			3000 BC			2500 BC			1800 BC			1250 BC			150 BC		
	A	B	C	A	B	C	A	B	C	A	B	C	A	B	C	A	B	C
Valley	8,000	(30)	240	8,000	(75)	600	8,000	(130)	1,040	8,000	(140)	1,120	9,000	(180)	1,620	10,000	(240)	2,400
Faiyum	100	(30)	3	100	(60)	6	100	(90)	9	450	(135)	61	400	(180)	72	1,300	(240)	312
Delta	8,000	(10)	80	7,000	(30)	210	9,000	(60)	540	10,000	(75)	750	13,000	(90)	1,170	16,000	(135)	2,160
Desert	25			50			25			25			25			50		
Total (millions)	0.35			0.87			1.6			2.0			2.9			4.9		

Source: Adapted from Butzer, K.W., *Early Hydraulic Civilization in Egypt – A Study in Cultural Ecology*, The University of Chicago Press, Chicago, IL, 1976.

Notes: A=area of cultivable land in km²; B=population density per km²; C=hypothetical population in thousands.

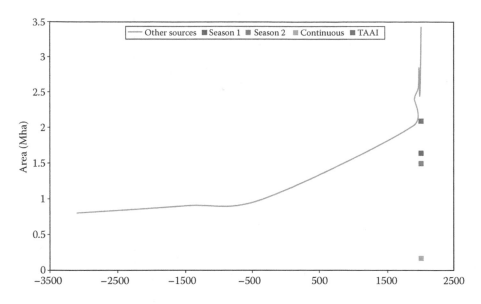

FIGURE 2.10 Historical irrigated area change in Egypt (3100 BC to 2003).

Additionally, traces of wells have been discovered, which suggest that the wheel of buckets technique or Doria was used here to raise the water to the highest point of the terrace. This irrigation era did not last long. By 400 BC, the Persian influence was itself overcome with the invasion of Alexander the Great and later rampages of Arabian nomadic hordes (http://www.theplumber.com/). Another peak time period of settlement and cultivation was under Sassanian rule, from the third to the seventh centuries. By AD 220, the irrigated area in the vast Mesopotamian alluvial plain was around 50,000 km² [52]. In 2000, the irrigated area during the first cropping season was 6.4 Mha, and the irrigated area during the second season was 5.21 Mha [21,22]. The recent statistics for irrigated area in the Tigris and Euphrates basin is 69,683 km² [53], with GIAM 2006.

Figure 2.12 shows the historical change in the irrigated area in the Tigris and Euphrates River Basin.

2.8 PERU

Irrigation started in central coastal Peru around 1800–1000 BC [5]. During this period, small-scale irrigation agriculture was more common. The north central regions of Peru, highland and coast, experienced prosperity and settlements grew to include impressively sized public structures [54]. Irrigation practices were based on floodwater farming in marshlands along the valley floors, and these were based and supplemented progressively by the construction of elaborate gravity-fed irrigation systems during 900 BC. The best-documented irrigation systems on the north coast are those in the Moche Valley [55]. After independence, there were many irrigated projects, like the Joya Irrigation Project; the San Lorenzo Irrigation and

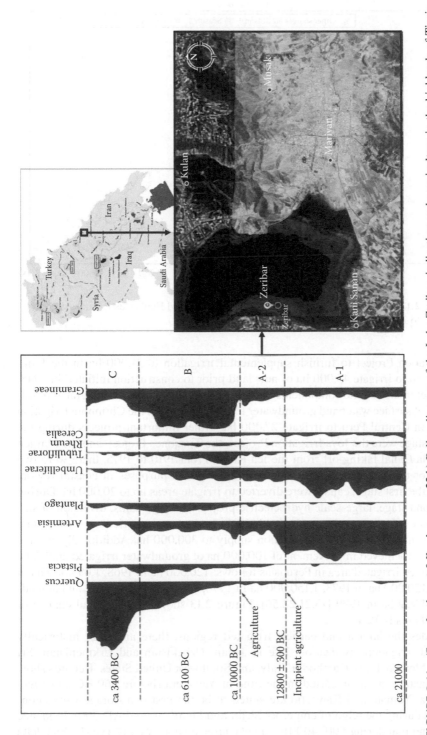

FIGURE 2.11 **(See color insert following page 224.)** Pollen diagram of the Lake Zeribar, indicating early agriculture in the highlands of Tigris and Euphrates River basin. (From Van Zeist, W. and Wright, H.E., Preliminary pollen studies at Lake Zeribar, Zagros mountains, southwestern Iran, Washington, D.C., *Science*, 140, 3562, 65–69, 1963.)

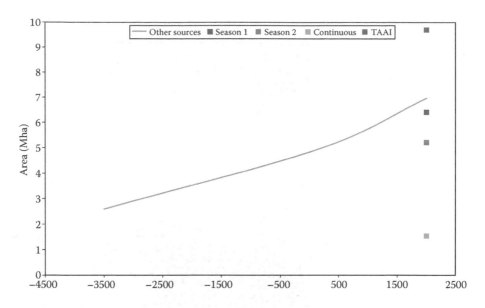

FIGURE 2.12   Irrigated area historical change in Tigris and Euphrates River basin (3500–2000 BC to 2003).

Colonization Project to furnish supplemental irrigation to 31,000 ha in the Piura valley, and to irrigate 45,000 ha of new land prior to construction initiated in 1948, which set in motion the components of the Chira-Piura Scheme for integrated development of surface water and groundwater for 147,000 ha and the Chimbote Irrigation Project in Central Peru to irrigate 27,000 ha. Other important projects during the second stage were the Joya Irrigation Project, the Tinajones Project, and the Margen Izquierda Canal taking off from the Ica River. From 1910 to 1975, there were two stages for utilization of water resources for irrigation purposes in coastal region. During the first stage, canals were diverted to irrigate areas up to 20,000 ha. During the second stage, large-scale hydroelectric projects, storage dams, diversion of surplus water by canals and tunnels to adjoining basins, were carried out by the state government, and it improved irrigation supply to 300,000 ha. Additionally, private enterprises achieved improvement of 100,000 ha of groundwater irrigated land. The documented irrigated area in Peru is as follows: 120,000 ha in 1968, 1,095,000 ha in 1971, 1,120,000 ha in 1973, 1,150,000 ha in 1976, 1,180,000 ha in 1979, and annualized 374,999 ha in 1999 [13,21,22,56]. Figure 2.13 shows the historical change of irrigated area in Peru.

Besides the largest and earliest irrigated regions, there are other historically recorded irrigation civilization, like in Japan, Uzbekistan and Urkmenistan, Sri Lanka, Mexico, Laos, Cambodia, Italy, and Southwest United States. Records show that irrigation in Japan started in the time of Yayoi era (before 660 BC). The reservoirs at Yamato and Kawaetio are stated to be the first two storage works constructed during the reign of emperors Sirjin and Sinjin (around AD 200). Evidence was found that during 600–400 BC, a very large area at Amu Darya (USSR) delta

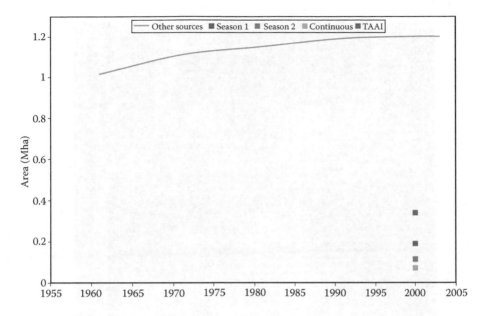

FIGURE 2.13 Irrigated area historical change in Peru (country statistics) (1961–2003).

was irrigated [13]. The delta is located in Uzbekistan and Turkmenistan now. The first irrigation practice in Sri Lanka was recorded about 400 BC. It was written in the Mahawansa (history book of Ceylon) [14]. In Mexico, the earliest irrigation was discovered in a valley named Tehuacan in 300 BC [12]. In south Mekong River basin, where there is a place called Angkorwat, irrigation had started from AD 100 to 900. It was during the time of the earlier kingdoms of Chenla and Funan, that the Khmer of Angkor constructed an extensive and highly advanced system of irrigation canals and huge reservoirs (see Figure 2.14 for the Angkorwat irrigation system) [11]. The Romans practice of irrigation in Italy can be traced back to AD 100, and the importance of irrigation in the Roman economy is indicated by the attention given to irrigation in the famous Justinian Code formulated in AD 500–600 [13]. Irrigation can be traced back to AD 800 in Southwest United States during the Indian settlements [13].

2.8.1 EVOLUTION OF IRRIGATED AREA STATISTICS

By plotting the available irrigated areas together (Figure 2.15), the global irrigated area from 10,000 BC to AD 1,000 was found to be insignificant, but important technological innovations on irrigation slowly developed, especially from 6000 BC.

Figure 2.16 shows the tremendous increase in the irrigated areas of the world. In the regions of the earliest irrigation civilization, during a history of thousands of years, the rate of increase started to slow down during recent years.

A more recent globalwise view shows the world's irrigated area was 8 Mha in 1800 [13]. It was estimated that the total area irrigated in the world by 1900 was about 50 Mha, and the areas irrigated in the major irrigating countries in descending

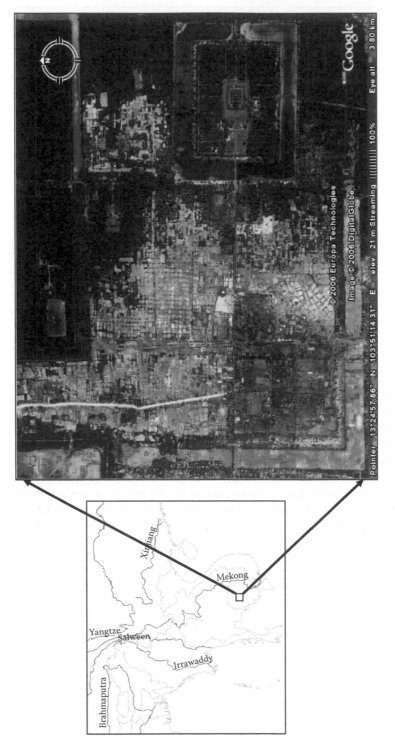

FIGURE 2.14 The zoom-in view from Google earth of Angkorwat ancient irrigation system.

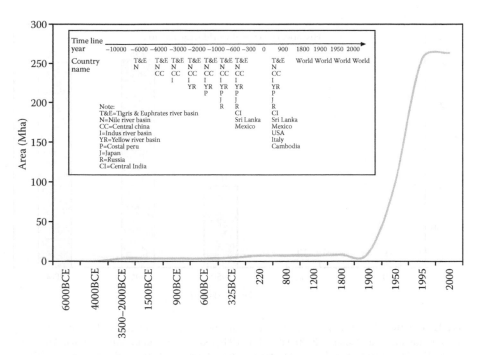

FIGURE 2.15 World irrigated area changes since prehistory. The values used to create this graph were adapted from the shown countries/regions of different literatures.

order were: India (17 Mha), China (15 Mha), previous USSR (4 Mha), United States (3 Mha), Egypt (2.4 Mha), Japan (2 Mha), and Italy (1.6 Mha) [13]. The global irrigated area reached 100 Mha in 1950 [57] and 255 Mha in 1995 [58]. All the records were based on statistical data. The remote-sensing-based GIAM work gives updated statistics—252 Mha for the first cropping season, 174 Mha for the second season, and 41 Mha for continuous cropping, with annualized irrigated areas of 467 Mha during 1999–2000 [21,22].

During 1900–1950, a further 90 Mha of irrigated area appeared, and during 1950–1995, a further 155 Mha of irrigated area appeared. We can see that the world's irrigated area increased rapidly during the twentieth century. The global land devoted to irrigation grew by nearly 3% per year from 1950 to the 1970s, 2% during 1970–1982, and 1.3% during 1982–1994 [5]. The increasing rate is decreasing for irrigated area. About 40% of the world's food now comes from the 17% of irrigated croplands [59]. The rate of abandonment of irrigation systems due to salinization, alkalization, and waterlogging is 15,000–25,000 km^2 per year [60]. Nearly 60% of the world's irrigation systems are less than 50 years old [5], and several decades are typically required for salinity and waterlogging effects to appear. The total rate of loss of irrigation systems is estimated as 1.1–1.5% per year, the anticipated global rate of expansion of irrigated lands is 0.6–1.3% per year [61].

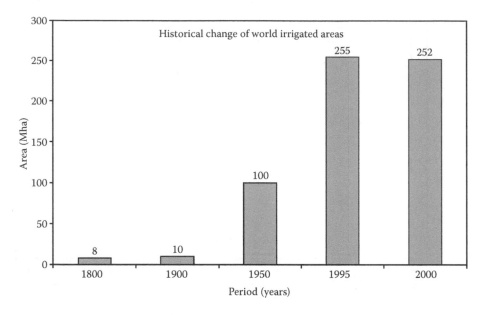

FIGURE 2.16 World irrigated area change during the past 200 years. The first cropping season statistics was used for 2000 data. (From Thenkabail, P.S. et al., *An Irrigated Area Map of the World (1999) Derived from Remote Sensing*, Research Report 105, International Water Management Institute, Colombo, Sri Lanka, 2006. Thenkabail, P.S. et al., Global irrigated area map (GIAM) for the end of the last millennium derived from remote sensing, *International Journal of Remote Sensing*, 2008, (in press).)

2.9 DISCUSSIONS AND CONCLUDING REMARKS

As discussed earlier, the traditional method of making irrigated area maps was a combination of census statistics data with GIS tools [20,26]. Statistics can provide very detailed information, sometimes detail of a very small-scale system, e.g., some irrigated area statistics can be at county scale in China [38]. The problems became more visible as more statistical data were collected and evaluated. First, collecting data individually is expensive and time-consuming. The more details we need, the more difficult it is to get them. Second, it cannot be neglected that statistical data are much affected by collecting methods and the collectors across countries; in other words, the quality of statistical data still needs further validation. The FAO created the first global digital GIS irrigated area map [20,27,28]. FAO's global digital irrigation map was a good example though, that their maps were dependent on the accuracies of the data provided by national governments. Statistics-based maps relate very much to the accuracy of the collected statistical data. Is there any other way we can avoid using uncertain statistical data? The remote-sensing-method has been used to generate global LULC maps [29,30,31,35,36]. But most of them have a resolution no better than 1-km and none of them have focused on irrigated areas [29,30,31,35,36]. In 2006, the first remote-sensing-based irrigated area map was released [21] by the GIAM group. It is the first irrigation-focused, remote-sensing- and GIS-techniques-based map that is backed by extensive groundtruth data and accuracy assessments [22]. The GIAM map used multisensors, multiresolution, and multitemporal remote-sensing data, and generated seasonal, annual, and potential irrigated area maps [21,22]. According to GIAM, the global irrigated areas during 2000 were: (a) 467 Mha annualized with a distribution of 252 Mha during season 1, 174 Mha during season 2, and 41 Mha yearly continuous; and (b) 399 Mha net irrigated areas at any one given time period [22]. Data can be found at www.iwmigiam.org.

With time, new techniques and more detailed remote-sensing imagery will contribute to the future generation of irrigated maps.

REFERENCES

1. Van Zeist, W. and Wright, H.E., Preliminary pollen studies at Lake Zeribar, Zagros mountains, southwestern Iran, Washington, D.C., *Science*, 140, 3562, 65–9, 1963.
2. Hole and McCorriston, as reported in the *New York Times*, 2 April 1991.
3. Maisels, C.K., *The Early Civilization of the Old World*, Routledge, London, 1999.
4. Pollock, S., *Ancient Mesopotamia,* Cambridge University Press, Cambridge, 1999.
5. Postel, S.L., *Pillar of Sand: Can the Irrigation Miracle Last?* W.W. Norton, New York, 1999.
6. Norse, D., Agriculture and environment: Changing pressures, solutions and trade-offs, in *World Agriculture: Towards 2015/2030*, J. Bruinsma (Ed.), UNFAO, Earthscan, London, 331–56, 2003.
7. Shah, T. et al., Sustaining Asia's groundwater boom: An overview of issues and evidence, *Natural Resources Forum*, 27, 130–41, 2003.
8. Shaw, J. and Sutcliffe, J.V., Ancient dams and Buddhist sites in the Sanchi area: New evidence on irrigation, land use, and patronage in central India, *South Asian Studies,* 21, 1–24, 2005.
9. Zou, H.B. et al., *The Discovery of Water Farm of MaJiaBing Culture at CaoXieShan in JiangSu Province*, Cultural Relics Publishing House, Beijing, 2000.
10. Xia Shang Zhou Duan dai project specialists group, *Xia Shang Zhou Duandai Project Mid-term Report 1996–2000*, abridged version, World Book Press, Beijing, 2000.
11. Standen, M., *Passage Through Angkor*, Mark Standen, Bangkhen, 1997.
12. Dolittle, W.E., *Canal Irrigation in Prehistoric Mexico*, University of Texas Press, Austin, TX, 1990.
13. Framji, K.K. and Mahajan, I.K., *Irrigation and Drainage in the World*, Caxton Press, New Delhi, 1981.
14. Brohier, R.L., *Ancient Irrigation Works in Ceylon*, Ceylon Government Press, Colombo, Ceylon, 1934.
15. Huston, D.M. and Titus, S.J., An inventory of irrigated lands within the State of California based on lands and supporting aircraft data, in *Space Sciences Laboratory Series 16*, Issue 56, University of California, Berkeley, CA, 1975, 23.
16. Wall, S.L., California's irrigated lands. Paper presented at the Symposium on Identifying Irrigated Lands Using Remote Sensing Techniques: State of the Art, 15–16 November 1979, at U.S. Geological Survey EROS Data Center, Sioux Falls, SD, 1979.
17. Thiruvengadachari, S., Satellite sensing of irrigation pattern in semiarid areas: An Indian study, *Photogrammetric Engineering and Remote Sensing*, 47, 1493–99, 1981.
18. Marschner, F.J., *Major Land Uses in the United States*, US Department of Agriculture, Agricultural Research Service, Washington, DC, 1950.
19. CNCID (Chinese National Committee on Irrigation and Drainage), *Irrigation and Drainage in China*, China WaterPower Press, Beijing, 2005.

20. Döll, P. and Siebert, S., A digital global map of irrigated areas, *ICID Journal*, 49, 2, 55–66, 2000, http://www.geo.uni-frankfurt.de/.
21. Thenkabail, P.S. et al., *An Irrigated Area Map of the World (1999) Derived from Remote Sensing*, Research Report 105, International Water Management Institute, Colombo, Sri Lanka, 2006.
22. Thenkabail, P.S. et al., Global irrigated area map (GIAM) for the end of the last millennium derived from remote sensing, *International Journal of Remote Sensing*, 2008, (in press).
23. Brander, B., *The River Nile*, 2nd ed., National Geographic Society, Washington, DC, 1996.
24. Hyams, E.S., *Soil and Civilization*, HarperCollins, Ann Arbor, MI, 1976.
25. Döll, P. and Siebert, S., *A Digital Global Map of Irrigated Areas*, Report A9901, Center for Environmental Systems Research, University of Kassel, Germany, 1999.
26. Siebert, S. and Döll, P., *A Digital Global Map of Irrigated Areas – An Update for Latin America ad Europe*, Kassel World Water Series 4, Center for Environmental Systems Research, University of Kassel, Germany, 2001.
27. Siebert, S. et al., Development and validation of the global map of irrigation areas, European Geosciences Union, *Hydrology and Earth System Sciences*, 9, 535–47, 2005.
28. Belward, A.S. (Ed.), *The IGBP-DIS Global 1 km Land Cover Data Set (Discover) – Proposal and Implementation Plans*, IGBP-DIS Working Paper No. 13, IGBP-DIS, Stockholm, Sweden, 1996.
29. Loveland, T.R. et al., Development of a global land cover characteristics database and IGBP DISCover from 1-km AVHRR data, *International Journal of Remote Sensing*, 21, 6/7, 1303–30, 2000.
30. Agrawal, S. et al., Spot-vegetation multi temporal data for classifying vegetation in South Central Asia, Mission of the Joint Research Center (JRC), *Current Science*, 84, 11, 1440–48, 2003, http://www-gvm.jrc.it/glc2000/objectivesGLC2000.htm.
31. Thenkabail, P.S., Schull, M., and Turral, H., Ganges and Indus River Basin land use/land cover (LULC) and irrigated area mapping using continuous streams of MODIS data, *Remote Sensing of Environment*, 95, 3, 317–41, 2005.
32. Biggs, T.W. et al., Vegetation phenology and irrigated area mapping using MODIS time-series, ground surveys, agricultural census data and Landsat TM imagery, Krishna River Basin, India, *International Journal of Remote Sensing*, 27, 19, 4245–66, 2006.
33. Thenkabail, P.S. et al., Spectral matching techniques to determine historical land use/land cover (LULC) and irrigated areas using time-series AVHRR pathfinder datasets in the Krishna River Basin, India, *Photogrammetric Engineering and Remote Sensing*, 73, 9, 1029–40, 2007.
34. Olson, J.S., *Global Ecosystem Framework: Definitions*, USGS EROS Data Center Internal Report, Sioux Falls, SD, 1994.
35. Olson, J.S., *Global Ecosystem Framework: Translation Strategy*, USGS EROS Data Center Internal Report, USGS, Sioux Falls, SD, 1994.
36. Ning, K., *Ningke History Articles*, Guang Ming Daily, History, (Original article in Chinese), Beijing, 1979.
37. Yao, H.Y., *Compendium of China Water Power History*, China WaterPower Press, Beijing, 1987, 44–8.
38. MWREP (Ministry of Water Resources and Electric Power), People's Republic of China, *Irrigation and Drainage in China*, China Water Resources and Electric Power Press, Beijing, 1987.
39. Smil, V., Controlling the Yellow River, *Geographical Review*, 69, 3, 253–72, 1979.
40. Liu, C.M., He, X.W., and Ren, H.Z., *China Water Problem*, Meteorological Press, Beijing, 1996, 160–64.

41. Chinese Natural Resources, Environment, economic and population database, http://www.naturalresources.csdb.cn/index.asp (accessed 21 July 2008).
42. Nanyang Water Power Bureau, He Nan Province, *Irrigation Maps of Nanyang Districts, Xi'an Map*, Nanyang Water Power Bureau, Xi'an, 1998.
43. Thenkabail, P.S. et al., Sub-pixel irrigated area calculation methods, *Sensors Journal (special issue: Remote Sensing of Natural Resources and the Environment)*, 7, 2519–38, 2007.
44. Singh, N.T., *Irrigation and Soil Salinity in the Indian Sub Continent: Past and Present*, Lehigh University Press, Bethlehem, PA, 2005.
45. Marshall, S.J.H., *Asian Educational Services*, (reprint of the 1931 edition), A. Probsthain, London, 1996.
46. Shaw, J. et al., *Asia Perspectives, The Journal of Archaeology for Asia and the Pacific*, 46, 1, University of Hawaii Press, Honolulu, 2007.
47. Shaw, J. and Sutcliffe, J.V., Ancient irrigation works in the Sanchi area: An archaeological and hydrological investigation, *South Asian Studies,* 17, 55–75, 2001.
48. Shaw, J. and Sutcliffe, J.V., Ancient dams, settlement archaeology and Buddhist propagation in central India: The hydrological background, *Hydrological Sciences Journal,* 48, 2, 277–91, 2003.
49. Ralph, E.K., Michael, H.N., and Han, M.C., Radiocarbon dates and reality, *MASCA Newsletter*, 9, 1–20, 1973.
50. Butzer, K.W., *Early Hydraulic Civilization in Egypt – A Study in Cultural Ecology*, The University of Chicago Press, Chicago, IL, 1976.
51. Janick, J., *History of Horticulture*, Department of Horticulture and Landscape Architecture, Purdue University, 2002, http://www.hort.purdue.edu/newcrop/history/default.html (accessed 20 June 2008).
52. Christensen, P., *The Decline of Iranshahr: Irrigation and Environments in the History of the Middle East 500 BC to AD 1500*, Museum Tusculanum, Copenhagen, Denmark, 1993.
53. Water Resource Institute, *eAtlas*, 2003, http://www.iucn.org/themes/wani/eatlas/pdf/eu/eu28.pdf (accessed 21 July 2008).
54. The Metropolitan Museum of Art, South America, 2000–1000 B.C., *Timeline of Art History,* New York, 2000, http://www.metmuseum.org/toah/ht/03/sa/ht03sa.htm (October 2000) (accessed 21 July 2008).
55. Park, C.C., Water resources and irrigation agriculture in pre-Hispanic Peru, *The Geographical Journal*, 149, 2, 153–66, 1983.
56. FAO (Food and Agriculture Organization of the United Nations), *Production Year Book*, Vol. 34, Food and Agriculture Organization of the United Nations, Rome, Italy, 1981.
57. Field, W., World irrigation, *Irrigation and Drainage Systems*, 1, 4, 91–107, 1990.
58. FAO, *FAOSTAT Online Statistical Service,* Rome, FAO, 1997, http://faostat.fao.org.
59. Postel, S., Forging a sustainable water strategy in *State of the World 1996*. A World Watch Institute report on progress toward a sustainable society, W.W. Norton, New York, 40–59, 1996.
60. Gardner, G., *Shrinking Fields: Cropland Loss in a World of Eight Billion*, World Watch Paper, 131, Worldwatch Institute, Washington, DC, 1996.
61. Sundquist, B., *Irrigated lands degradation: A global perspective*, 4th edn., 2005, http://home.alltel.net/bsundquist1/ir0.html.

Section II

Global Irrigated Area Mapping (GIAM) Using Remote Sensing

3 Global Irrigated Area Maps (GIAM) and Statistics Using Remote Sensing

Prasad S. Thenkabail
U.S. Geological Survey

Chandrashekhar M. Biradar
University of Oklahoma

Praveen Noojipady
University of Maryland

Venkateswarlu Dheeravath
United Nations Joint Logistics Centre

MuraliKrishna Gumma
International Water Management Institute

Y.J. Li
University of Maryland

Manohar Velpuri
South Dakota State University

Obi Reddy P. Gangalakunta
National Bureau of Soil Survey and Land Use Planning (ICAR)

CONTENTS

3.1 INTRODUCTION AND BACKGROUND

Irrigation is widely thought to provide 40% of the world's food from around 17% of the cultivated area, and is a factor that has been instrumental in the success of the "Green Revolution" and in meeting food needs, particularly of staple grains, of the ballooning global population over the past 40 years. These successes have come with significant environmental costs through river regulation and land and water degradation from inputs to intensive agriculture.

Following the end of the Second World War, and a period of decolonization, there was a boom in irrigation development that coincided with strongly motivated nation building, particularly in Asia. Irrigated area increased at about 2.6% per annum from a modest 95 Mha in the early 1940s to between 250 and 280 Mha in the early 1990s [1–4]. In this era, a key developmental agenda for many countries was the construction of large and small dams and river diversions to extract and store water for agriculture [3]. Over 40,000 large dams (each higher than 15 m) irrigate about 30–40% of the world's irrigated areas (www.dams.org) and are complemented by an estimated 800,000 smaller dams. The literature on irrigated areas shows that the world's croplands increased from about 265 Mha in 1700 to about 1.4 billion hectares (Bha) in 1990, of which rain-fed croplands alone constituted about 1.2 Bha [5–8].

Since the 1980s, there has been a progressive decline in public and international donor funding for irrigation, which has been replaced in many countries by the private development of groundwater irrigation based on the availability of cheap drilling and pumping technologies. Now, India has an estimated 25 million tube-well irrigators [9,10], accounting for as much as 60% of the irrigated areas, according to some estimates [3]. Similarly, there has been a massive increase in groundwater exploitation in China, the United States, and numerous other countries. This development has allowed food production to keep pace with the rapidly growing global population and the increasing urban world.

3.1.1 GLOBAL POPULATION, FOOD SECURITY, AND IRRIGATED AREAS

As the world's population grew from 2.2 billion in 1950 to 6.1 billion in 2001, irrigation has played a major role in feeding these populations by tripling the world's grain harvest from 640 million tons (mt) to 1855 mt [11]. However, further increases in grain production are being challenged by the growing competition from industries and urbanization for the already strained water resources [12]. In the Northern China Plain, which produces over half of China's wheat and a third of its corn, the declining water tables have resulted in both the loss of irrigated areas and the drop in grain production from its peak of 392 mt in 1998 to 338 mt in 2003, a drop equivalent to Canada's harvest [11]. With the estimated population growth of the world reaching 10 billion by 2050, the pressure on food security without causing environmental damage or unsustainable use of water is a challenge facing the world. The nutritional demands have also risen from 2255 kilocalories per person per day (kcal/cap/day) in 1961 to around 2800 kcal/cap/day in 1998 [13,14]. Yet, the increase in cropland area has been modest, from 1.36 Bha in 1960 to 1.47 Bha in 1990 [15,16]. Based on various estimates, it has remained at around 1.5–1.8 Bha at the end of the last millennium [8], of which about 16–18% is irrigated [17]. Irrigated areas are also increasing at a much slower rate of about 1.3% per year at present compared to 2.6–3% during the peak irrigation era of 1960–80 (e.g., see Ref. [18]). Given these facts, how does the world feed itself? To meet the future food demand of a population of 8 billion by 2025, some estimate that at least another 2000 cubic kilometers (km^3) of water will be needed. This is equivalent to the mean annual flow of 24 additional Nile rivers [18]. Better technology and advances in agronomy and crop breeding (including genetically modified crops) are expected to contribute to increasing cropland area and water productivity. Cultivation intensities (more than one crop a year, especially through irrigation) and increased grain yields from hybrid varieties with shorter growing days have been meeting the world's food demand in recent decades. However, extending the areal extent of the irrigated areas as well as their intensification are increasingly questioned by environmental activists and more ecologically sensitive governments. A key challenge for the irrigation sector lies in using less water to produce more food (better water productivity), while mitigating the negative impacts on the environment, particularly on the aquatic ecosystems.

Farmers currently produce enough to feed the world, although poverty and malnutrition still affect more than a fifth of the global population due to local shortages and inadequate distribution and market systems. Although rates of population increase are now slowing and the world is expected to continue being able to feed itself, there will be continued pressure to expand irrigated area, increase crop and livestock productivity, or substitute intensive irrigation with better and more extensive rain-fed agriculture.

Nevertheless, with the global population increasing at about 90 million annually, alternative demands for water from increasing urban populations, industrialization, environmental flows, and biofuels have put severe stress on available water resources in many parts of the world. This is making the food security for tomorrow a challenge, and it must be confronted today. A large part of the solution to the global and regional food problem is likely to be found through irrigated agriculture. Researchers anticipate cropping intensities to increase in response to the growing demand for food, placing tremendous pressure on land and water resources.

Alternative strategies to feed future populations and decreasing irrigated areas will be to:

1. Grow less water-consuming crops (e.g., more wheat and less rice).
2. Increase water productivity ("more crop per drop") through better water and land management technologies and increasing irrigation efficiency.
3. Educate people to consume less water-consuming food (e.g., more vegetables and grains compared to meat).
4. Emphasize rain-fed crop productivity to reduce stress on water-intensive irrigated croplands.

3.1.2 WATER USE AND IRRIGATED AREAS

Irrigation consumes about 70–80% of all water used by humans. Water use for irrigation varies considerably across the globe. It accounts for 2–4% of diverted water in Canada, Germany, and Poland, but is an impressive 90–95% in Iraq, Pakistan, Bangladesh, Sudan, Kyrgyzstan, and Turkmenistan [19]. Globally, the irrigated landscape remains very dynamic. Although the annual rate of increase of irrigated areas has slowed to about 1%, this still represents an increase of between 2 and 3 Mha each year, which means a significantly greater need for water. There is a smaller corresponding annual loss of irrigated area to salinity and waterlogging as well as to abandonment of uneconomic projects. Countries such as China and India continue to build large multipurpose dam projects that also supply water for irrigation. In sub-Saharan Africa, irrigation is perennially seen as having an unfulfilled potential. Elsewhere in the world, there are moratoria on dam building and even on the decommissioning of dams in western United States. The irrigated landscape of the world will be shaped increasingly by the effects of competition for water from other sectors, notably urban and rural domestic water supply and industrial needs. It is becoming increasingly common for river basins to be overly allocated, with negative downstream effects of competitive upstream development, such as in the Krishna basin in India [20]. Similarly, groundwater is being mined in many places, notably in significant parts of India and in the Ogallala aquifer in the Midwest of the US Reservation, and reallocation of flows for environmental purposes will, in the end, place even greater competing demands in terms of water volume. Climatic change will impose additional challenges that will reshape the irrigated landscape through changes in snowmelt runoff and rainfall.

3.1.3 PREVIOUS ATTEMPTS AT MAPPING IRRIGATED AREAS

Irrigated areas have rarely been mapped over large areas such as subcontinents, continents, and the world. Currently, there are two irrigated area maps of the world. First, the recent work by the International Water Management Institute (IWMI) as described by Thenkabail et al. [3,4] (http://www.iwmigiam.org). The IWMI study is based on time-series remote sensing and secondary data and is reported at nominal 1–10 km resolution; second, the work by the Food and Agriculture Organization of the United Nations and the University of Frankfurt (FAO/UF) as described by Siebert et al. [2] (http://www.fao.org/ag/agl/aglw/aquastat/gis/index.stm). The FAO/UF study [2] is a compilation of the national statistics into a spatial map and has a nominal resolution of 10 km. In other land use land cover (LULC) maps [21,22], irrigated areas are just one of the many classes, not the focus of the study, and there is often a mix of rain-fed and irrigated areas.

Previous efforts to quantify the amount of irrigation in the world have relied on secondary statistics, which in many instances systematically overestimate or underestimate the actual cropped irrigated area. For example, now, some 60% of irrigation in India is practiced using groundwater, most of which is privately developed and not necessarily recorded in government statistics. In the past, irrigated areas of the world have been accounted for primarily by using government statistics and have placed global equipped area for irrigation between 250 and 280 Mha. These estimates use a combination of statistics, existing Geographic Information System (GIS), and other techniques to indicate the distribution of irrigated areas in different parts of the world.

3.1.4   SATELLITE REMOTE SENSING FOR MAPPING IRRIGATED AREAS

Satellite remote sensing offers a consistent, timely, (and an increasingly) free resource to estimate and monitor irrigated areas that also meet high scientific standards. In regions such as Asia where secondary data on cropping intensities are not accurately recorded, it also helps to identify the extent of multiple cropping (planting of two or three crops a year). Remote sensing offers a potential solution. Some of the early remote-sensing applications focused on mapping irrigated croplands [23–26]. In recent years, mapping has continued with enhanced techniques and capabilities for inventorying irrigated areas from different irrigation sources, such as surface water, groundwater, and irrigation tanks [26,27], for assessing crop stress, for discriminating crop types, and for monitoring temporal changes in irrigated areas [27].

Finer-resolution irrigated area mapping using remote sensing is limited to smaller areas such as river basins [20,27,28] or regions [29,30]. With an increase in spatial, spectral, and temporal resolutions of the satellite sensors, medium- to high-resolution satellite data provide valuable and precise information on location, spatial distribution, and extent of irrigated areas in the country for accurate mapping of irrigated areas and to analyze their spatiotemporal changes in remote sensing and GIS environment. As a result, mapping irrigated areas using the Moderate Imaging Spectrometer (MODIS) is attractive. The MODIS data are frequently available (e.g., every 8-day), have global coverage, guarantee high scientific quality, and are free of charge. Further, this frequent availability not only enables monthly time composites that remove an overwhelming proportion of the clouds [28], but also derives a crop calendar and a study of irrigation and nonirrigation dynamics [20].

Irrigated area is also estimated, rather coarsely, in global land use classifications derived from remote sensing, usually focused on other objectives, such as forestry, rangelands, and rain-fed croplands. Examples include the United States Geological Survey (USGS) global land cover [31], Global Land Cover (GLC) 2000 [21], and Global Forest Cover [32,33].

3.1.5   GAPS IN KNOWLEDGE ON IRRIGATED AREAS AND THE IWMI GLOBAL IRRIGATED AREA MAPPING (GIAM) PROJECT

There are a number of motivations for the IWMI's GIAM. First, the considerable uncertainty about the exact extent, area, and cropping intensity of irrigation in different parts of the world, due to the dynamics referred to above and systematic problems of under-reporting and over-reporting of irrigation in different contexts (e.g., groundwater) and countries; second, the absence of a consistent GIAM using remote-sensing approaches; third, the limitations of non-remote-sensing approaches in producing GIAM statistics.

Satellite sensors potentially offer a consistent, continuously updated, timely, and increasingly free resource that meets high scientific standards, such as Advanced Very High Resolution Radiometer (AVHRR) pathfinder 10-km, MODIS 250-m, and SPOT 1-km vegetation data with global coverage every 1–2 days and are processed by science teams to 8-day, 16-day, and monthly composites. These data are backed by numerous high-quality secondary spatial data, such as Shuttle Radar Topographic

Mission (SRTM) digital elevation models (DEMs), Landsat, SPOT, ASTER high-resolution data, and global time-series of precipitation and other climatic variables.

Globally, the irrigated landscape remains very dynamic. Although the annual rate of increase of irrigated areas has slowed to about 1%, this still represents an increase of between 2 and 3 Mha each year. There is a smaller corresponding annual loss of irrigated area to salinity and waterlogging as well as to abandonment of uneconomic projects.

Better technology, advances in agronomy and crop breeding (including genetically modified crops) are expected to contribute to increasing croplands and water productivity. A key challenge for the irrigation sector lies in using less water to produce more food (better water productivity), while mitigating the negative impacts on the environment, particularly on aquatic ecosystems.

3.2   GOAL AND SPECIFIC OBJECTIVES

The main motivation to develop the IWMI map lies in the potential of a wide range of increasingly sophisticated remote-sensed images and techniques to reveal vegetation dynamics that perform the following:

- Define more precisely the actual area and spatial distribution of irrigation in the world.
- Elaborate the extent of multiple cropping over a year, particularly in Asia, where two or three crops may be planted in one year, but cropping intensities are not accurately known or recorded in secondary statistics.
- Account for (a) irrigation intensity, (b) irrigation source, (c) irrigated crop types, and (d) precise location of irrigated areas. Irrigation intensity and irrigation crop types have a huge influence in the quantum of water consumed.
- Develop methods and techniques for a consistent and unbiased estimate of irrigation over space and time for the entire world.
- Discuss issues related to irrigated areas, such as subpixel areas (SPAs), irrigated area fractions for class (IAFs), resolution and areas, and informal irrigation in irrigated area statistics.
- Present irrigated area statistics as maps, statistics, web maps, and Google maps.

3.3   DATA USED IN GIAM

The GIAM products were produced, mostly using freely available, public domain, high-quality satellite-sensor data from multiple sources. The spectral, spatial, radiometric, and temporal characteristics of these data sets are summarized in Table 3.1. The emphasis was in the use of time series in order to capture the temporal characteristics of the irrigated areas and other LULC. Time series potentially allow the often distinct dynamics of irrigated agriculture to stand out from other land uses, but there are many confusing situations: for instance in the tropics, where rice may be mainly rain-fed in the monsoonal season but receives some irrigation, and is followed by one or more dry-season crops that may be completely irrigated. In tropical environments, there is generally a high degree of land cover the whole year-round

TABLE 3.1

Satellite-Sensor Data Characteristics with Multiple Satellite Sensor-Data Characteristics used in GIAM at Various Resolutions

Sensor	Spatial (m)	Spectral (No.)	Radiometric (bit)	Band Range (µm)	Band Width (µm)	Irradiance (W m^{-2} sr^{-1} µm^{-1})	Data Points (No. per hectares)	Frequency of Revisit (days)
A. Primary data used in GIAM								
1. AVHvRR	1000	4	11	0.58–0.68	0.10	1390	0.01	Daily
				0.725–1.1	0.375	1410		
				3.55–3.93	0.38	1510		
				10.30–10.95	0.65	0		
				10.95–11.65	0.7	0		
2. SPOT-1	2.5–20	15	16	0.50–0.59	0.09	1858	1600, 25	3–5
-2				0.61–0.68	0.07	1575		
-3				0.79–0.89	0.1	1047		
-4				1.5–1.75	0.25	234		
				0.51–0.73 (p)	0.22	1773		
3. MODIS	250, 500, 1000	36/7	12	0.62–0.67	0.05	1528.2	0.16, 0.04, 0.01	Daily
				0.84–0.876	0.036	974.3	0.16, 0.04, 0.01	
				0.459–0.479	0.02	2053		
				0.545–0.565	0.02	1719.8		
				1.23–1.25	0.02	447.4		
				1.63–1.65	0.02	227.4		
				2.11–2.16	0.05	86.7		

(Continued)

TABLE 3.1 (Continued)

Sensor	Spatial (m)	Spectral (No.)	Radiometric (bit)	Band Range (μm)	Band Width (μm)	Irradiance (W m⁻² sr⁻¹ μm⁻¹)	Data Points (No. per hectares)	Frequency of Revisit (days)
4. Landsat-7 ETM+	30	8	8	0.45–0.52	0.65	1970	44.4, 11.1	16
				0.52–0.60	0.80	1843		
				0.63–0.69	0.60	1555		
				0.50–0.75	0.150	1047		
				0.75–0.90	0.200	227.1		
				10.0–12.5	2.5	0		
				1.75–1.55	0.2	1368		
				0.52–0.90 (p)	0.38	1352.71		

and everything is "green," making the precise definition of irrigated crops more difficult, especially if relatively coarse-scale imagery is used. The primary and secondary data sets used in GIAM are described below.

3.3.1 PRIMARY REMOTE SENSING DATA SETS AND MASKS

3.3.1.1 AVHRR Pathfinder Data Characteristics

The monthly time-composite National Oceanic and Atmospheric Administration (NOAA) AVHRR 0.1-degree data were obtained from the NASA Goddard DAAC (www.daac.gsfc.gov/data/dataset/AVHRR). The Normalized Difference Vegetation Index (NDVI) monthly Maximum Value Composite (MVC) data from 1981 to 1999 are stored in a single megafile of 239 bands, which was also used to generate animations of NDVI and skin temperature to assist in understanding vegetation dynamics and identify irrigated areas. A subset of 3 years of these data (1997–99) was incorporated into the GIAM megafile data cube (MFDC) (Section 3.8, Table 3.1, Figures 3.1 and 3.2).

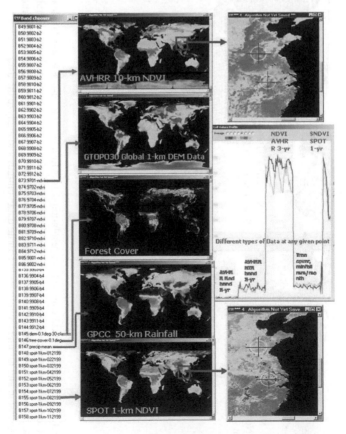

FIGURE 3.1   (See color insert following page 224.) Megafile data cube using time-series satellite-sensor data and secondary data. The time-series satellite-sensor data and secondary data are resampled into a single resolution MFDC of hundreds or even thousands of layers.

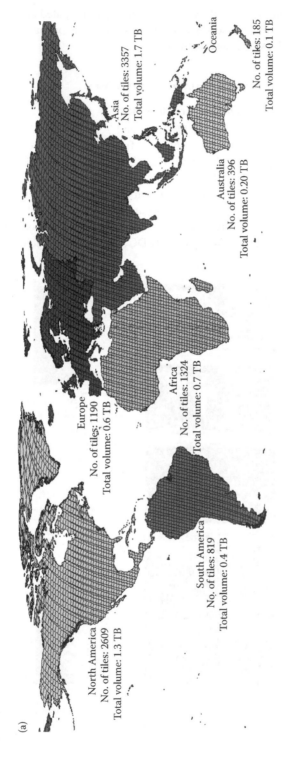

FIGURE 3.2 Megafile data cube using finer-resolution data. Mosaicking finer-resolution data leads to huge data volumes even for a single date. A total of 9770 Landsat tiles were required to cover the entire terrestrial world, requiring about 20 terabytes of disk space (a). When coarser-resolution time series and secondary data are fused, it increases the power of determining irrigated areas and other LULC, but increases the data volume (b), requiring data compression and mining techniques to handle data volumes and analyze the data.

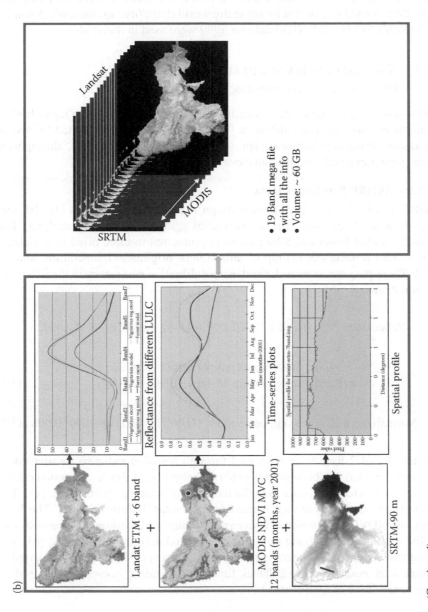

FIGURE 3.2 (Continued)

3.3.1.2 SPOT Vegetation Data Characteristics

The SPOT Végétation (SPOT VGT) 1-km data have four wavebands: blue (0.43–0.47 μm); green (0.61–0.68 μm); near-infrared (NIR) (0.78–0.89 μm); and shortwave infrared (SWIR) (1.58–1.75 μm). There is a 10-day synthesis of SPOT VGT data that can be downloaded free of cost for the entire world (http://free.vgt.vito.be/). A single-year monthly SPOT VGT NDVI data for 1999 were used in this study.

3.3.2 NORMALIZATION LEADING TO AT-SATELLITE SENSOR REFLECTANCE AND ATMOSPHERIC CORRECTION

A critical issue in the long time-series is data normalization. Many factors lead to variations or shifts in data calibration that include, but are not limited to, sensor degradation, changes in sensor design, satellite orbital characteristics, atmospheric effects, topographic effects, and sun elevation [34–36].

3.3.2.1 AVHRR Pathfinder Data

AVHRR Pathfinder data have gone through many processing steps [37–40] and most of these effects were already corrected prior to use in this analysis. The originally scaled 16-bit and 8-bit data were converted to three primary variables: (a) at-ground reflectance, (b) top of atmosphere brightness temperature, and (c) NDVI. These parameters were derived using calibration parameters in the following four equations:

$$\text{Reflectance (\%)} = (\text{Band 1 scaled DN in 16-bit radiance} - 10) * 0.002, \quad (3.1)$$

$$\begin{array}{c}\text{Band 4 brightness temperature (}^\circ \text{ Kelvin)} \\ = (\text{Band 4 scaled DN in 16-bit} + 31990) * 0.005, \end{array} \quad (3.2)$$

$$\text{Normalized difference vegetation index (NDVI)} = (\text{SNDVI} - 128) * 0.008. \quad (3.3)$$

Bands 1 and 2 have been processed through standard radiative transfer equations and corrected for Rayleigh atmospheric scattering. Moisture absorption effects have been corrected using data from the total ozone mapping spectrometer [41]. The resulting reflectances were then normalized for solar illumination [42]. The band values and NDVI distortions due to external forcing (e.g., stratospheric aerosols and satellite orbit degradation) are serious concerns and need to be addressed [43]. For the thermal channel, first the atmospheric radiance was calculated and converted to brightness temperature using a Planck function equivalent look-up table, based on the response curve of each channel.

3.3.2.2 SPOT Vegetation Data

Similar corrections have been made to the SPOT VGT data by the SPOT Image production team, including scattering and moisture absorption, using radiative transfer models. Cloud and snow were detected using a multivariate thresholding technique and neutral networks using data from all four VGT wavebands [44]. Cloud shadows were detected using the geometrical model as described by Lissens et al. [44]. The 10-day synthesis is performed using daily data using MVC.

3.3.3 MASK DATA

Secondary data sets in the megafile are used to segment the world into characteristic regions based on rainfall, elevation, temperature, and known forest cover. For example, in areas where temperatures are less than 280 K, it is unlikely that there is any vegetation and little chance of there being any irrigation.

3.3.4 GTOPO30 1-KM DEM

The GTOPO30 is derived from eight sources consisting of digital terrain elevation data or Digital Terrain Elevation Data (DTED) (50% of global coverage), Digital Chart of the World or DCW (29.9%), USGS 1-degree DEMs (6.7%), Army Service Maps (ASM) at 1:1,000,000 scale (1.1%), the International Map of the World (IMW) at 1:1,100,000 scale (4.7%), the Peru map at 1:1,000,000 scale (0.1%), New Zealand DEM (0.2%), and Antarctic digital database (8.3%) [45–48]. The vertical accuracy of the component DEM data varies significantly from source to source. Accuracies were 30 m for DTED, 160 m for DCW, 30 m for USGS DEM, 250 m for ASM maps, 50 m for IMW maps, 500 m for the Peru map, 15 m for the New Zealand map, and highly variable heights for the Antarctic digital database. In this chapter, GTOPO30 data were used to segment the image based on elevation gradients.

3.3.5 CLIMATE RESEARCH UNIT (CRU) PRECIPITATION AND TEMPERATURE DATA

The 40-year (1961–2000) monthly, 0.5 degree, interpolated rainfall and temperature data were obtained from Dr. Tim Mitchell of the CRU, University of East Anglia, UK [49] (http://www.cru.uea.ac.uk/~timm/index.html). The data have been converted to ESRI GRID format at IWMI and the mean monthly precipitation and temperature for 40 years were computed for each pixel and added to the megafile.

3.3.6 FOREST COVER DATA

Forest cover was derived from the 1992 AVHRR 1-km data by the University of Maryland, which used a continuous fields approach (rather than a discrete number of classes) using a linear mixture model approach [33,34]. This data set was used to mask areas of very high forest cover, which implies that the land is not available for cultivation or irrigation.

3.3.7 JAPANESE EARTH RESOURCES SATELLITE (JERS)-1 SYNTHETIC APERTURE RADAR (SAR)-DERIVED FOREST COVER

Rain forests may contain fragmented irrigation along with shifting cultivation and clearance for livestock production. Mapping irrigated areas in rain forests is more complex than in other parts of the world as a result of forest fragmentation and significant cloud cover. Even the monthly AVHRR and SPOT VGT MVCs contain some cloud cover over rain forests, and irrigated fragments are difficult to discern at 1–10 km. We obtained 100-m resolution JERS-1 SAR tiles (http://southport.jpl.nasa.gov/GRFM/) for South America and Africa, to assist in mapping major rain-forest areas at higher resolution. Unfortunately, well-processed JERS SAR images are not readily available for Asia and hence they could not be used.

The JERS-1 SAR is an L-band (24.5 cm wavelength) imaging radar with an initial full resolution of 18 m that is processed to 100 m, mosaicked and made available for the entire contiguous rain forests of Amazonia and Central Africa. The JERS SAR antenna has a median-look angle of 35 degrees.

The Amazon basin was imaged by JERS-1 during a low flood period from September to December 1995 and in a high flood time from May to August 1996. The Central and West Africa rain-forest images were obtained for January to March 1996 and October to November 1996. Over 20 million km^2 of the rain forests are covered by these images [50]. The JERS SAR image tiles were mosaicked at IWMI into single files for Central Africa and the Amazon using ERMapper 6.5.

The 8-bit JERS SAR data of rain forests were analyzed separately and fused with the overall classification results from other areas.

3.3.8   MEGAFILE DATA CUBE

The MFDC consists of hundreds or even thousands of bands or layers of time series from multiple sensors as well as data from secondary sources all composed into a single file, akin to a hyperspectral data cube.

3.3.8.1   MFDC of Time-series Multispectral Data akin to Hyperspectral Data

The megafile used for the IWMI GIAM consisted of 159 data layers (a few layers are illustrated in Figure 3.1). This MFDC consisted of 144 AVHRR 10-km layers from 3 years (12 layers from 1 band per year * 4 bands including an NDVI band * 3 years), 12 SPOT vegetation 1-km layers from 1 year, and single layers of DEM 1-km, mean rainfall for 40 years at 50-km, and AVHRR-derived forest cover at 1-km. All 159 layers were resampled to a common resolution of 1-km by resampling the coarser-resolution data to 1-km. The longest multitemporal series of remote sensing with global coverage is AVHRR 8-km (reprojected to 10-km). However, since this resolution is coarse, we have combined a 3-year monthly time series of AVHRR 10-km from 1997 to 1999 with a 1-km SPOT Végétation mosaic of the world for 1999. All input data, megafile, and outputs are stored in the IWMI Data Storehouse Pathway (IWMIDSP), an online archive that stores all remote sensing and GIS data collected by IWMI. The site can be accessed at: http://www.iwmidsp.org.

Figures 3.1 and 3.2 illustrate various types of data present in the MFDC. The drop-down menu of bands shows how the layers are ordered. The following sections provide a brief description of each of the data sets, summarized in detail in Table 3.1.

3.3.9   MFDC OF HIGH-RESOLUTION IMAGERY

In finer-spatial-resolution imagery, the megafile is composed using a one-time wall-to-wall coverage of the world due to huge data volumes. Time series of the globe at higher resolution are not feasible due to data volume. For example, a Landsat ETM + mosaic of the world would require 9770 images for a total volume of 20 terabytes (see Figure 3.2a). Nevertheless, the need for time series and secondary data in irrigated area mapping cannot be avoided. For smaller areas, such as a river basin, it is feasible, for example, to fuse finer-resolution Landsat ETM + imagery with coarser-resolution MODIS time-series imagery and secondary data such as SRTM-derived elevation (Figure 3.2b). This still leads to a significant, but manageable, volume of data. However, at the global level, innovative data-reduction or data-mining techniques will be required to handle and analyze the huge volumes in the MFDC that will involve fusing finer-spatial-resolution imagery with time-series coarser-resolution imagery as well as secondary data.

3.3.10   GROUNDTRUTH (GT) DATA

3.3.10.1   GT Data

A precise knowledge of the real situation on the ground is essential to the interpretation of all remote-sensing products for training, class identification and naming, and accuracy assessment. The precise locations of these samples (Figure 3.3) were recorded by Global Positioning System (GPS) in the Universal Transverse Mercator (UTM) and the latitude/longitude coordinate system with a common datum of WGS84. At each location, the following data were recorded (see descriptions in Ref. [28]):

- LULC classes: levels I, II and III of the Anderson approach.
- Land cover types (percentage): trees, shrubs, grasses, built-up area, water, fallow lands, weeds, different crops, sand, snow, rock, and fallow farms.
- Crop types, cropping pattern and cropping calendar for kharif, rabi (winter- or dry-season cropping period from November to March), and interim seasons.
- Water source: rain-fed, full, or supplemental irrigation, surface water, or groundwater.
- Digital photos hot-linked to each GT location.

During the GIAM project, the researchers from IWMI along with its partners such as the Chinese Academy of Agricultural Sciences (CAAS), the Indian Council for Agricultural Research (ICAR), and a number of other international partners collected extensive GT data. Added to this effort was the growing amount of publicly available data, thanks to the Internet source such as the Degree confluence project (DCP) (http://www.confluence.org/). The DCP (http://www.confluence.org/) is an organized sampling of the entire world at every 1-degree latitude and longitude intersection. The confluence points include precise latitude, longitude, and a digital photo of land cover. These were converted to proprietary GIS formats and

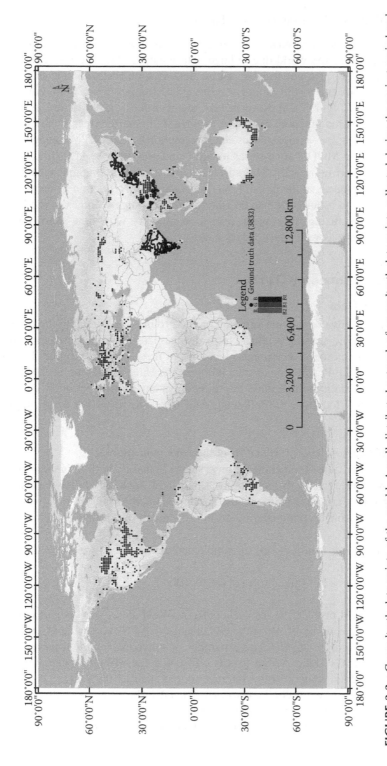

FIGURE 3.3 Groundtruth data points of the world. A well-distributed network of groundtruth data points collected during the project period and sourced from the DCP overlaid on AVHRR pathfinder 10-km data.

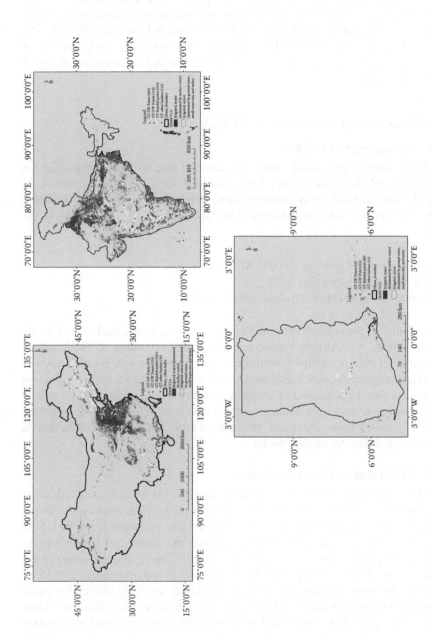

FIGURE 3.4 (See color insert following page 224.) Groundtruth data points of China, India, Ogallala (the United States), and Ghana. Extensive groundtruth data points were collected in a standard format during this project with the goal of using them for (a) ISDB generation, (b) class identification, and (c) accuracy assessment.

added to the DSP in a separate archive to preserve their identity. Figure 3.3 shows the distribution of 3832 of these GT data points, overwhelmingly from the regions of irrigated areas in the world. Of these, 1861 points were used during the class identification process, leaving the other 1971 points for accuracy assessments. Major GT campaigns were conducted in India in 2003–05 and in China in 2007–08. The data for the Ogallala aquifer in the United States were shared by the University of North Dakota. The distribution of these GT data points in four countries is shown in Figure 3.4a, b, and c.

3.3.11 HIGH-RESOLUTION IMAGERY AS GT DATA

3.3.11.1 Google Earth Data Set

Google Earth (http://earth.google.com/) contains an increasingly comprehensive image coverage of the globe at very high-resolution imagery (VHRI) 0.61–4 m, allowing the user to have zoom-in-views (ZIV) into specific areas in great detail, from a base of 30-m resolution data, based on GeoCover 2000. This assists in the following:

- Identification and labeling the GIAM classes
- Area calculations (Section 4.17)
- Accuracy assessment of the classes (Section 4.18)

For every identified class, 20–60 sample locations were cross-checked using Google Earth. The indicative GIAM class name can be updated according to the dominant class identified at high resolution within the sample area. The process also helps to identify mixed classes. Google Earth data are used as a substitute for GT, although images may, in fact, be snapshots of cropping systems taken at different times. The very high-resolution data have some advantage over real GT in that they provide information on much larger areas and, therefore, a more representative area than is normally sampled directly on the ground.

3.3.11.2 ESRI Landsat 150-m GeoCover

The ESRI resampled the 8500 orthorectified Landsat ETM+ "GeoCover" tiles that have been produced by the EarthSat Corporation (http://www.earthsat.com), funded by NASA [45]. The original images can be obtained free of charge from the USGS Earth Resources Observation and Science (EROS) data center and the University of Maryland (http://glcf.umiacs.umd.edu/index.shtml). The resampled images have a pixel resolution of 150 m compared with the original pan-sharpened size of 15 m. GeoCover is the most positionally accurate image set covering the entire globe and shows maximum greenness. These Geocover images offer a detailed "zoom in views" of any part of the world and are used to provide contextual information and pseudo-"groundtruth" by geo-linking to the class maps to identify and label classes.

3.3.12 SECONDARY DATA FOR CLASS VERIFICATION

A number of existing global LULC products were used in the preliminary class identification and labeling process. These included USGS LULC [32], USGS seasonal LULC [32], Global Land Cover 2000 or GLC2000 [21], IGBP [51], and Olson ecoregions of the world [52,53]. These data supplemented/complemented the GT data during the preliminary class identification and labeling processes. The GLC2000 data set was derived using data from SPOT 1-km resolution Végétation Instrument [21,22]. The 10-day synthesis data from November 1, 1999 through December 31, 2000 were used for the classification (http://www.gvm.sai. jrc.it/glc2000/Products/). The GLC characteristics database was developed on a continent-by-continent basis using 1-km, 10-day AVHRR data spanning April 1992 through March 1993 [32]. The same primary data were used in the Global USGS LULC, seasonal USGS LULC, and IGBP LULC (http://edcdaac.usgs.gov/ glcc/globe_int.html). Olson data provides global 94 unique ecosystem classes for the globe [52,53] (http://edcdaac.usgs.gov/glcc/globe_int.html). This approach was developed in the mid-1980s and did not use any remote-sensing information. For convenience, all these land-cover products are made available in standard image-processing formats (e.g., ERDAS Imagine) in IWMIDSP (http://www. iwmidsp.org).

3.4 METHODS

3.4.1 IDEAL SPECTRAL DATA BANK (ISDB) CREATION

Time-series satellite-sensor data enable the creation of ideal spectra for various land use themes, such as irrigated areas, rain-fed areas, classes within irrigated areas, and classes within rain-fed areas.

In this study, knowledge of various classes of irrigated areas was gathered from precise locations established during field visits. Spectral signatures (e.g., NDVI time series or band reflectivity over time) are then generated for these precise locations from the time-series images of AVHRR or SPOT or MODIS. The spectral signatures of similar types of classes (e.g., irrigated, surface water, rice, double crop) are aggregated to form a single composite signature that forms the ISDB for the class, as illustrated for irrigated and rain-fed cropland classes for China (Figure 3.5) and India (Figure 3.6). The ISDB is also illustrated for various LULC classes (Figure 3.7). The characteristic signature of the ideal spectra for any class will be strengthened with a larger number of spectral signatures that are spatially well distributed. However, it must be noted that the same class (e.g., irrigated, surface water, rice, double crop) can have two distinct signatures in two distinct regions as a result of the watering methods or soil type or elevation differences. In the strategy used, all the signatures of a particular class (e.g., irrigated, surface water, rice, double crop) should be plotted and all similar signatures grouped into one category.

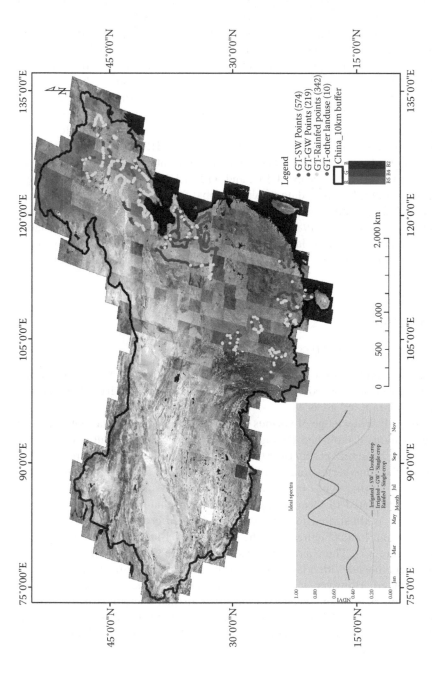

FIGURE 3.5 (See color insert following page 224.) The ISDB for China with illustrations of ISDB for (a) irrigated, surface water, continuous crop; (b) irrigated, groundwater, single crop; and (c) rain-fed, single crop.

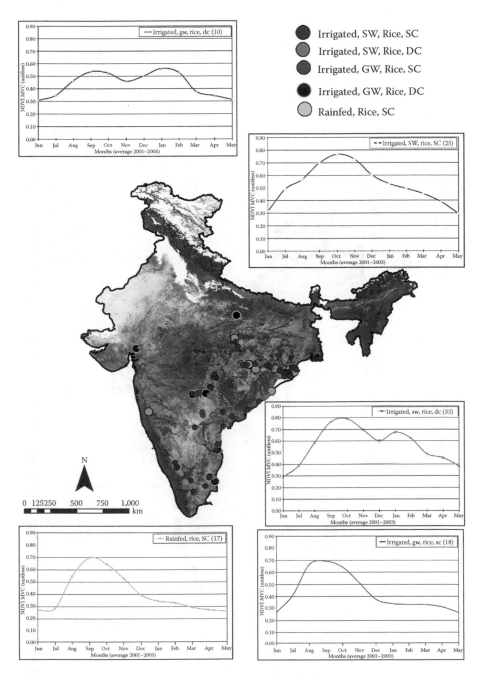

FIGURE 3.6   (See color insert following page 224.) The ISDB for India with illustrations of ISDB for (a) irrigated, surface water, rice, double crop; (b) irrigated, surface water, rice, single crop; (c) irrigated, groundwater, rice, double crop; (d) irrigated, groundwater, rice, single crop; and (e) rain-fed, single crop.

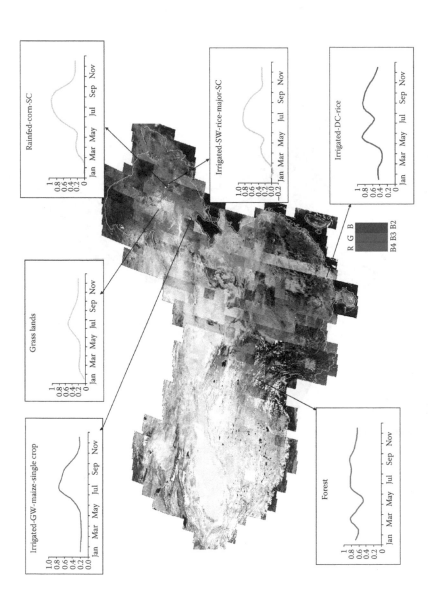

FIGURE 3.7 (See color insert following page 224.) The ISDB for various LULC with illustrations of ISDB for various LULC in China.

3.4.2 IMAGE MASK AND SEGMENTATION

The world's terrestrial area is divided as distinctly masked and segmented, based on elevation, temperature, precipitation, and forest cover. The seven masks are (Figure 3.8):

1. Precipitation less than 360 mm per year segment (PLT360-segment).
2. Precipitation greater than 2400 mm per year segment (PGT2400-segment).
3. Temperature less than 280 Kelvin (K) segment (TLT280-segment).
4. Forest cover greater than 75 mm segment (FGT75-segment).
5. Elevation greater than 1500 mm segment (EGT1500-segment).
6. All other areas of the world segment (AOAW-segment).
7. Rain forest SAR segment (RFSAR-segment).

The above masks were applied on the 159-band MFDC, which led to a 159-band MFDC of each segment.

The segment with less than 360 mm per year identifies areas where any green vegetation has a very high likelihood of being irrigated, since average evaporation rates of 30 mm per month, however distributed in reality, will be considerably less than evaporative demand. This segment will mainly identify arid and semiarid areas and deserts, as shown in Figure 3.8. In contrast, the segment

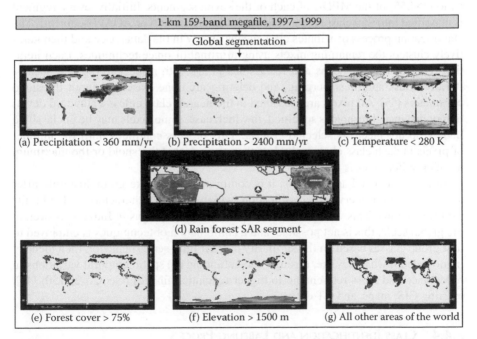

FIGURE 3.8 Masks used to segment the megafile data cube (MFDC). The masks based on elevation, precipitation, and temperature zones are used to mask the areas and segment the images.

with rainfall over 2400 mm per year mainly identifies the rain-forest areas of the world, although there are considerable areas of irrigation within the Southeast Asian lands. Where the temperature is less than 280 K, on average (see Figure 3.8), it is too cold for agriculture, and irrigation is not likely to be practiced there. However, some northern hemispheric areas have a low average temperature but short summer seasons in which supplemental irrigation is actually practiced. If forest density is greater than 75%, it is also rare to find any irrigation, due to high rainfall and limited infrastructure. There is likely to be slash-and-burn agriculture in small fragments. This mask is complemented by a special rain-forest mask derived from the JERS-1 SAR imagery, in order to better identify other land use fragments at higher resolutions within the rain-forest areas, including where there might be irrigation. There is a lower likelihood of irrigation above 1500 m elevation, although there are certainly hill irrigation systems in the Andes, the Himalayas, and the Philippines at higher elevations. The forest is a likely land cover, but should be separable from irrigation and agriculture due to its continuous vegetation signature. Finally, the segment "all other areas of the world" focuses on where there are few biophysical constraints to irrigation and shows where we are most likely to find it in various forms.

3.4.3 CLASS SPECTRA GENERATION THROUGH UNSUPERVISED CLASSIFICATION

Class spectra can be generated through unsupervised ISOCLASS K-means classification [54,55] of the MFDC of each of the seven segments. Initially, every segment is classified into 250 classes (e.g., Figure 3.9 for segment 6 or AOAW-segment). The classification process distributes class means evenly in the data space and then iteratively clusters the remaining pixels using minimal distance techniques. Each iteration recalculates the means and reclassifies pixels with respect to the new means. Iterative class splitting, merging, and deleting are done, based on input threshold parameters [55]. All pixels are classified to the nearest class unless a standard deviation or distance threshold is specified, in which case, some pixels may be unclassified if they do not meet the selected criteria. This process continues until the number of pixels in each class changes by less than the selected threshold or the maximum number of iterations is reached.

In more localized applications, it is common to undertake groundtruthing after a preliminary unsupervised classification, which identifies characteristic land units for investigation. This was done for the IWMI field campaigns in India. However, at the global scale, this is not possible, and a combination of techniques is employed to first group classes based on the similarity of their time-series behavior, then identify in more detail what they are, through understanding the spatiotemporal variations in reflectance and cross-referencing to higher resolution images (GeoCover 150) [45], existing GIS, maps, and GT data.

3.4.4 CLASS IDENTIFICATION AND LABELING PROCESS

The process of class identification and labeling goes through a series of steps that begins with spectral-matching techniques (SMTs) and then goes through the use of

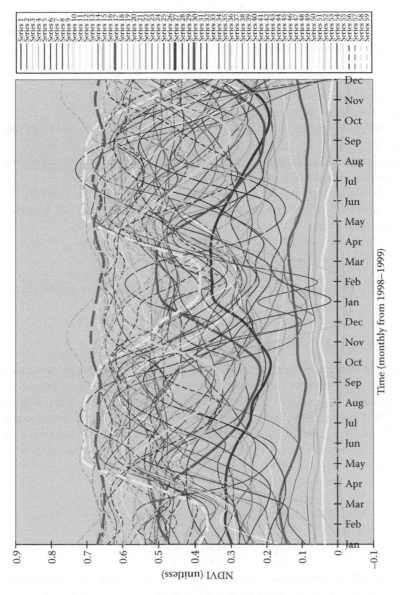

FIGURE 3.9 **(See color insert following page 224.)** Class spectra. The NDVI class spectra were generated for all the 250 classes for one of the segments.

GT data, Google Earth data, and various remote-sensing interpretation techniques discussed in the following paragraphs.

3.4.5 Decision-tree Algorithms

A rule-based decision-tree algorithm (e.g., Figure 3.10) was used to split classes into distinct groups. The approach on applying rule-based decision trees involves the following steps in numerical order:

1. Classify the MFDC into a large number of unsupervised classes.
2. Generate time-series class signatures (e.g., NDVI) for every class.
3. Write decision-tree rules to group classes.
4. Identify the grouped classes, based on the class identification and labeling process.
5. Resolve the mixed classes by involving secondary data (e.g., elevation, precipitation, temperature, zones) in the decision-tree-rule base.
6. If classes still remain mixed after the above applications of rules, the mixed classes in question are masked, the MFDC corresponding to mixed class areas segmented, their class spectra generated through a classification process, and steps 1 to 5 repeated to group classes into distinct categories.

The above six-step iteration will lead to a final set of classes that are all identified and labeled.

3.4.6 Spectral-matching Techniques

The principle in spectral matching is to match the shape, or the magnitude, or (preferably) both to an ideal or target spectrum (commonly known as a pure class or "end-member"). Time series of NDVI or other metrics are analogous to spectra. Earlier, SMTs have mostly been applied to hyperspectral data analyses of minerals [35,36,56–60]. A detailed discussion on SMTs is presented by Thenkabail et al. [61]. The time-series signatures of irrigated crops across the globe can match (the tropics) or be out of phase (the tropics and the Southern Hemisphere).

The SMTs are required to automatically:

- Group classes of similar time-series spectral characteristics among the class spectra.
- Match classes of class spectra with an ISDB.

Since there will be thousands of classes that need to be identified and labeled in a complex global mapping effort such as GIAM, the process can be simplified by SMTs. An example of SMTs is illustrated for five classes in Figure 3.11. The five classes were reduced to three, based on quantitative SMTs described below.

Two quantitative SMTs are widely used in this study. These are as described in Sections 3.4.6.1 and 3.4.6.2.

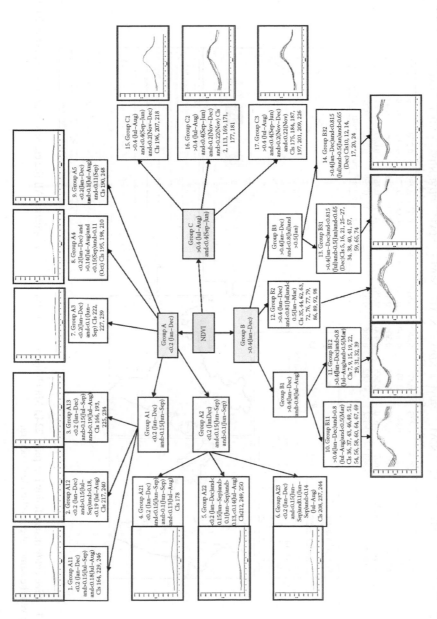

FIGURE 3.10 Decision-tree algorithm. A decision-tree algorithm uses rule base taking the NDVI dynamics of 250 classes generated from Landsat ETM+data for the whole of China. The time-series NDVI were generated for each class using MODIS 250-m data.

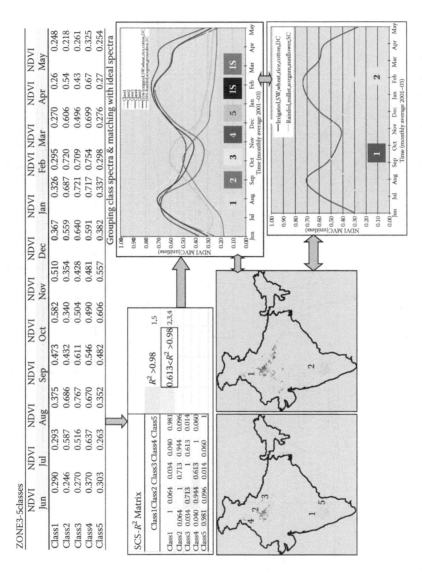

FIGURE 3.11 (**See color insert following page 224.**) Matching class spectra with ideal spectra. All the classes in the class spectra that match ideal spectra are given the same label as the ideal spectra.

3.4.6.1   Spectral Correlation Similarity R^2-value (SCS R^2-value)

The SCS R^2-value was a measure of shape of the spectra and was based on Pearson's Correlation Coefficient applied to an NDVI time series [62]:

$$\rho = \frac{1}{n-1}\left[\frac{\sum_{i=i}^{n}(t_i - \mu_t)(h_i - \mu_h)}{\sigma_t \sigma_h}\right], \qquad (3.4)$$

where:

n = number of samples,

$i = 1$ to n,

t_i = NDVI time series ($i = 1$ to n) of the target class,

μ_t = mean of the NDVI time-series of the target class,

h_i = NDVI time series ($i = 1$ to n) of any other class,

μ_h = mean of the NDVI time series ($i = 1$ to n) of any other class,

σ_t = standard deviation of target class NDVI time series ($i = 1$ to n),

σ_h = standard deviation of NDVI time series of any other class.

The range of Pearson's Correlation Coefficient (ρ) normally lies between -1 and $+1$, but negative values have no meaning in this application. The higher the SCS R^2-value, the greater the similarity between ideal/target spectra and class spectra. The process of identification of classes using the SCS R^2-value is illustrated in Figure 3.12 for a few classes.

3.4.6.2   Spectral Similarity Value (SSV)

The SSV is a measure of the shape as well as of the magnitude of spectra. The SSV is defined as follows [56,60]:

$$SSV = \sqrt{E_d^2 + (1-\rho)^2}, \qquad (3.5)$$

where:

E_d = Euclidian (shortest) distance between two points,

ρ = Pearson's Correlation Coefficient as defined above.

The normal range of SSV is from 0 to 1.415, and the smaller the value, the greater the similarity between ideal/target spectra and class spectra.

In this study, the SCS has been applied to match the shape of any class to the selected target class. The SSV has been used to determine the match of both shape and magnitude and is illustrated in Figure 3.13.

3.4.6.3   Modified Spectral Angle Similarity (MSAS)

The MSAS [29,56,57,59,63] measures the hyperspectral angle between spectra of any two classes or between the target and the sample class spectra. However, the practical implementation of this was troublesome (see also Ref. [4]), often providing uncertain results, and so it is not discussed further.

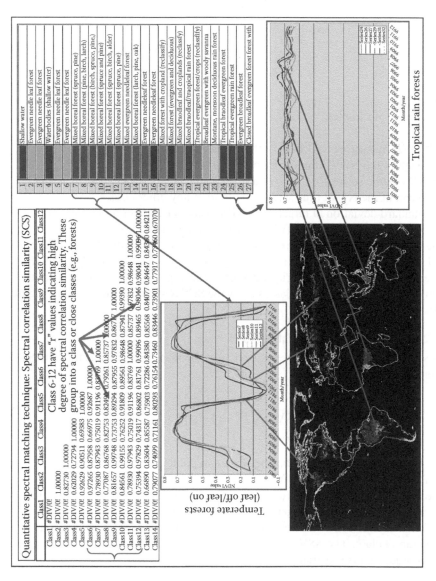

FIGURE 3.12 (See color insert following page 224.) Spectral correlation similarity (SCS). The SCS R^2-value is used to match (a) class spectra with ideal spectra, and (b) classes with similar spectral characteristics in the class spectra. The higher the SCS R^2-value, the greater the match.

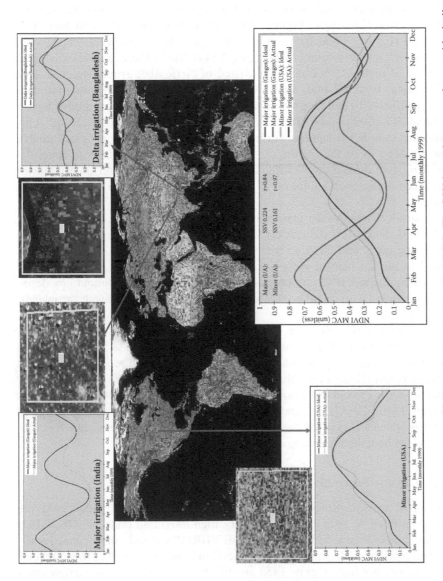

FIGURE 3.13 (See color insert following page 224.) Spectral similarity value (SSV). The SSV is used to group classes with similar spectra in the class spectra. The lower the SSV, the better the match.

3.4.7 GT Data Overlay on Classes

The most accurate means of class labeling requires a good ground-knowledge. We used 3832 GT data points from around the world (Figure 3.3) in the class-labeling process.

The procedure involved overlaying GT data points on a class that needs to be identified and labeled (e.g., Figures 3.14a through 3.15b). First, consider class number 79 (Figure 3.14a) from one of the segments. There were 94 GT data points that fell on class number 79 (Figure 3.14a). Out of these, an overwhelming proportion of them (72 out of 94) consisted of "irrigated, groundwater, double crop" classes (Figure 3.14a). Such overwhelming evidence indicates that class 79 is most likely "irrigated, groundwater, double crop." However, it does have a significant (17 of 94 points) mix of "irrigated, surface water, double crop." This implies that in order to separate surface water and groundwater irrigated double crops, we need to further refine class 79. However, if our goal is to label the class "irrigated, double crop" without reference to surface water or groundwater irrigated areas, then 89 out of 94 points of the class can be very accurately labeled.

In contrast, if we consider class number 60 (Figure 3.14b), a fact that is clearly based on GT data is that the class is "single crop." Twenty-five "irrigated, surface water, single crop" and 32 "rain-fed, single crop" points fall on class 60 (Figure 3.14b). This implies that the class is mixed and needs further refinement before it can be given a name. If we are only concerned with croplands without any desire to learn whether it is irrigated or rain-fed, then the class can be labeled: "croplands, single crop."

The GT data points falling on two classes from a segment are illustrated. First, class 79 (Figure 3.14a): The overwhelming evidence (72 out of 94 points) indicates class 79 as "irrigated, groundwater, double crop." Yet, there are still a significant number of points (17 out of 94) that are "irrigated, surface water, double crop" and a small proportion (5 out of 94 points) that is "rain-fed, double crop." Second, class 60 (Figure 3.14b): The GT data indicate this as a mixed class that cannot distinguish between irrigated and rain-fed classes. However, the class is quite accurate for cropland class (without distinguishing irrigated from rain-fed).

3.4.8 Google Earth Overlay of Very High-resolution ZIV

The irrigated areas identified by the GT in the previous section (Figure 3.14a) are further confirmed by the Google Earth very high-resolution imagery (GE VHRI) ZIV illustrated in Figure 3.15a and b. The GE VHRI ZIV clearly show canals, reservoirs, and croplands. They provide a very good indication of whether the croplands are irrigated or not (e.g., Figure 3.15a and b). Whereas surface water irrigation is fairly easy to detect using GE VHRI, the groundwater irrigation is hard to detect and we will have to rely on the GT knowledge (Figure 3.14a and b). A combination of the GT knowledge (Figure 3.14a and b) and GE VHRI ZIV (Figure 3.15a and b) provides an excellent indication of irrigation vs. nonirrigation and the types of irrigation.

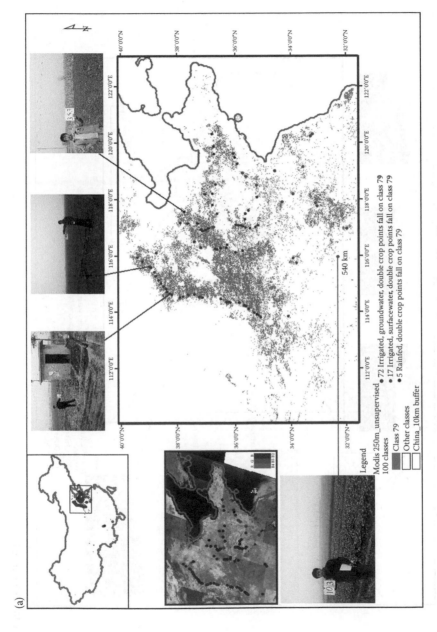

FIGURE 3.14 (See color insert following page 224.) Groundtruth data in the class identification process.

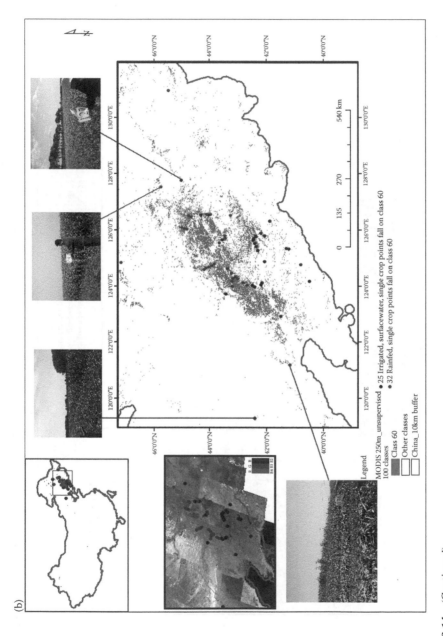

FIGURE 3.14 (Continued)

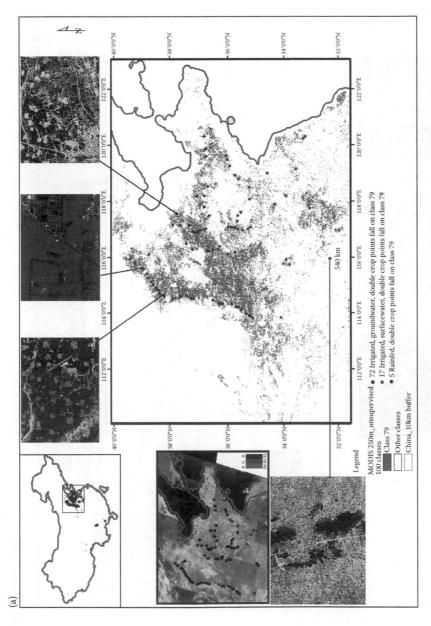

FIGURE 3.15   **(See color insert following page 224.)** Google Earth data in the class identification process. The Google Earth very high-resolution imagery (GE VHRI) data points falling on class 79 (a) and class 60 (b) are illustrated.

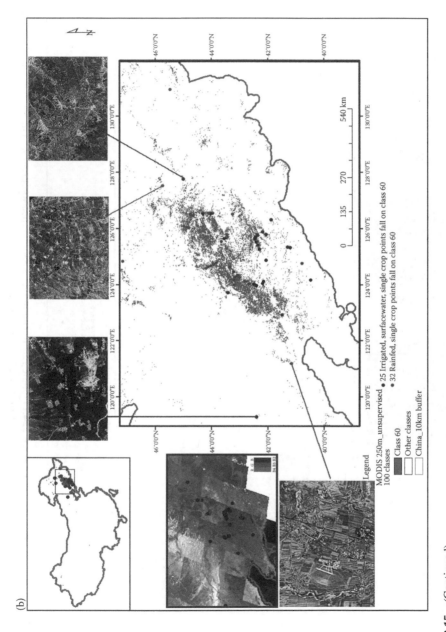

FIGURE 3.15 (Continued)

3.4.9 Two-dimensional Feature Space Plot of Brightness-Greenness-Wetness (2-D FS BGW)

The 2-D FS BGW plots (e.g., Figure 3.16) depict reflectance of classes in red and NIR bands [29,64,65]. These plots are very useful in identifying, grouping, and labeling classes (see Figure 3.16). The 2-D FS BGW plots similar to Figure 3.16 were developed for all seven segments and the class labeling performed. These plots are extremely useful to determine which classes merge together and where they are placed in 2-D FS (e.g., Figure 3.16).

3.4.10 Multidimensional Feature Space (MD FS) Plots

With time, the classes keep moving in 2-D FS, as depicted for three dates in Figure 3.17. Observing such movements will be of value in class identification. We have illustrated such movements for classes, taking class reflectivity in red and NIR during January, May, and September. For example, in Figure 3.17, the movement of class 22 from wetness in January to the relatively neutral zone in May and to the greenness peak area in September indicates this as an agricultural class and most likely irrigated. In contrast, class 17 is mostly moving around in the brightness (drier) area, but sufficiently away from the soil line and has significant vegetation. This class is likely to be rain-fed. Class 12 remains close to the soil line in all three seasons, indicating very sparse vegetation. Even though the answers are not definitive, the strong indications provided by the class movement in distinct seasons will be invaluable in inferring the class name. However, it is important to consider the spatial location of this class in the world. That will help us link the class movement to seasonality.

3.4.11 Space Time Spiral Curve (STSC) Plots

The time-series satellite-sensor data facilitate tracking changes in spectral characteristics of any class (or for that matter any pixel) over time and space. When these changes are plotted in 2-D FS, each class moves around in a "territory" of its own. These movements are referred to as space time spiral curves (STSCs). The STSCs are shown in Figure 3.18 for the Ganges river basin in India. The irrigated and rainfed classes have distinct STSCs. Forests are distinct from irrigated areas during most of the year, but around Julian date 249 the signatures mix. This is because irrigated agriculture is at its peak growth around day 249, resulting in signatures becoming similar to forests.

The STSCs are very useful in studying the class characteristics and in understanding the possible class types (e.g., irrigated or rain-fed). For example, most rainfed classes do not traverse the "territories" of irrigated classes, specifically in the greenness and wetness zones.

3.4.12 Time-Series NDVI (TS NDVI) Plots

The TS NDVI plots (e.g., Figure 3.19) are ideal for understanding the changes that occur (a) within and between seasons, and (b) between classes (e.g., irrigated

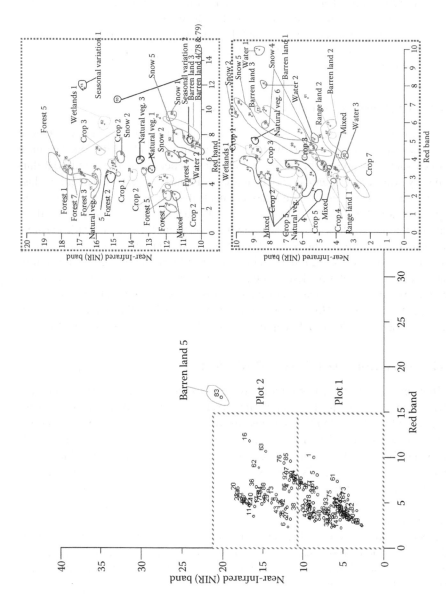

FIGURE 3.16 Bispectral plots for class identification and labeling. The class reflectivity in NIR and red are used for bispectral plots. Classes are identified and labeled based on where they fall in brightness-greenness-wetness zones.

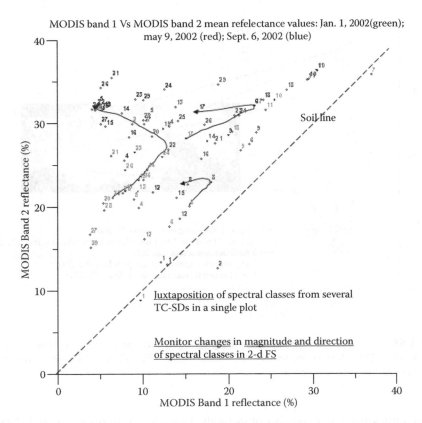

FIGURE 3.17 Multidimensional Feature Space (MD FS) plots. The MD FS classes for understanding and labeling classes.

vs. rain-fed). Figure 3.19 shows the distinct differences between an irrigated vs. a rain-fed class. Irrigated areas have much higher NDVI and are double-cropped (two crops in a calendar year). In contrast, the rain-fed crops have significantly lower NDVI and are limited to one crop. Further, through climatic data, we knew that 2002 was one of the worst drought years. This led to a near-failure of rain-fed crop and the impact was also seen in one season of irrigated crop, where it had a much shortened growing season compared to the normal year of 2001 (Figure 3.19).

The time-series NDVI plots are used to study class characteristics. They (a) provide preliminary class names such as irrigated, rain-fed, and forests; and/or (b) confirm class names established through various subsections in Section 3.4.

3.4.13 RESOLVING THE MIXED CLASSES

In spite of the exhaustive protocols for class separation, identification, and labeling, there are still classes that remain mixed (e.g., Figure 3.14b) because they are not

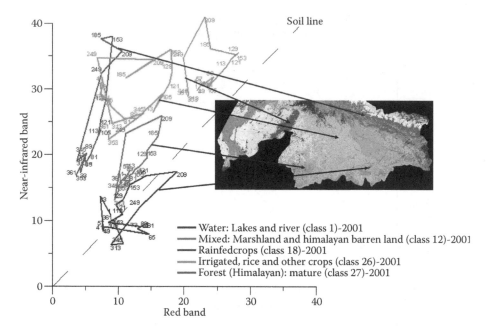

FIGURE 3.18 <u>(See color insert following page 224.)</u> The space time spiral curve (STSC) plots. The STSCs depict time series in 2-D feature space. The STSCs track the movement of the classes over time and space.

spectrally separable. A number of methods were adopted to resolve mixed classes. These include those mentioned in Sections 3.4.13.1 through 3.4.13.3.

3.4.13.1   Spatial-Modeling Approaches Using GIS Data Layers

The above approaches involved taking a mixed class and applying GIS spatial-modeling techniques such as overlay, matrix, recode, and sieve and proximity analysis [55] based on the theory of map algebra and Boolean logic [66–68]. The GIS spatial data layers used included the precipitation zone, elevation zones, Koppen climatic zones, the temperature zone, and tree-cover categories. Any one or a combination of these data layers has often helped separate the mixed classes.

3.4.13.2   Decision-Tree Algorithms to Split
the Mixed Class in Question

Decision-tree algorithms [69] were reapplied for the specific mixed class to resolve it. The decision tree can involve NDVI, waveband reflectivity, thermal temperatures, elevation zones, and so on (e.g., Figure 3.20). The primary goal is to investigate where the mixed class occurs spatially and think of solutions to resolve them through automated processing techniques involving simple decision-tree rule-based algorithms.

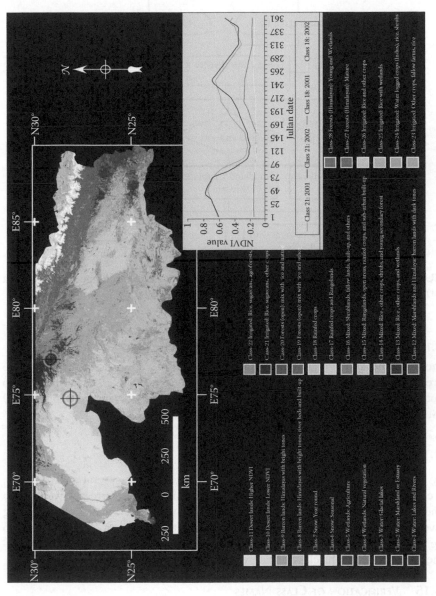

FIGURE 3.19 **(See color insert following page 224.)** The NDVI time-series plots. The time-series NDVI plots for understanding and labeling classes.

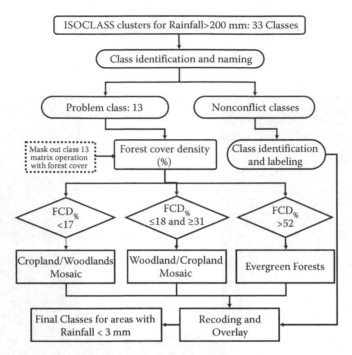

FIGURE 3.20 Resolving the mixed classes through simple decision-tree rules involving information, such as NDVI, temperature, and band reflectivity.

3.4.13.3   Reclassification and Reidentification Process

If none of the above techniques can resolve the mixed classes, then we have to reclassify the mixed classes using the unsupervised classification and go through the process of class identification and labeling discussed exhaustively in Section 3.4.

3.4.14   STANDARDIZED CLASS LABELING

Any large global project such as GIAM is likely to involve several analysts. It is, thereby, important to have a consistent class-naming pattern irrespective of who names a class. In GIAM, the standardized approach used a hierarchical class-naming convention that considered (Figure 3.21) watering method, type of irrigation, crop type, scale, intensity, location, and type of signature. Irrigation intensity is determined directly using the class signature.

3.4.15   VERIFICATION OF CLASS NAMES

In spite of the exhaustive class-naming protocols used, it is useful to verify class names based on further field visits, expert knowledge, and the use of other secondary data [21,22,32].

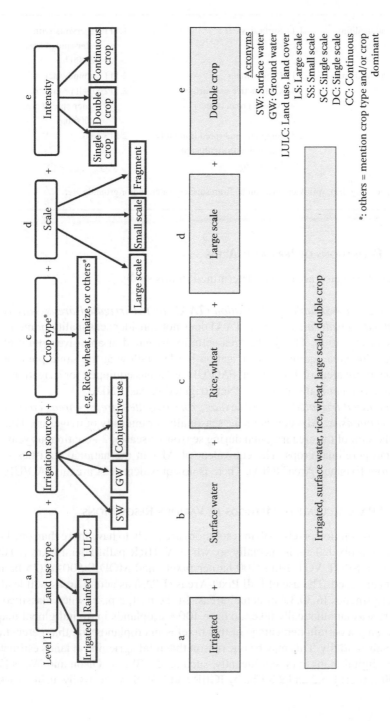

FIGURE 3.21 Standardized class-naming convention. A standardized hierarchical class-naming convention used in GIAM.

The classes can then be grouped and aggregated as follows:

1.1 Major and medium irrigated areas	1.11 Surface water	1.111 Reservoirs with a water-spread area >2,000 ha
2.1 Minor irrigated areas	2.11 Groundwater	2.111 Groundwater
	2.21 Conjunctive use surface water+groundwater)	2.112 Small reservoirs (water-spread area >2000 ha)
	2.31 Supplemental (predominantly rain-fed with significant irrigation)	2.113 Tanks

Note: Irrigation by drip, sprinkler, etc., can be from surface water and/or groundwater.

3.4.16 DEFINITIONS OF IRRIGATED AREAS

Two types of irrigated areas were determined in this study.

A. *Total Area Available for Irrigation (TAAI) or net irrigated areas—areas without considering intensity:* TAAI does not consider seasonality or intensity of irrigation. This is the area actually irrigated at any given point of time plus area equipped for irrigation but left fallow at the same point of time. Equivalent of TAAI in (a) FAO/UFs is "areas equipped for irrigation," and in (b) national statistics is "Net Irrigated Areas" (NIA).

B. *Annualized Irrigated Areas (AIAs) or gross irrigated areas—areas considering intensity:* AIA considers the seasonality or intensity of irrigation. This is the sum of the area irrigated during season 1+season 2+continuous year-round perennial crops. The equivalent of AIA in the national statistics is "Gross Irrigated Area" (GIA). There is no equivalent AIA area in FAO/UF.

3.4.17 SPA CALCULATION METHODS AT VARIOUS RESOLUTIONS

The coarser-resolution pixels (>30 m resolution) are likely to have more than one land cover within a pixel. This is specially so, when AVHRR pathfinder 10-km (10,000 ha per pixel), SPOT VGT 1-km (100 ha per pixel), and MODIS 500-m (25 ha per pixel) data are used. The use of Full Pixel Areas (FPAs) as actual areas will lead to gross overestimates [6,30,31] of actual areas. In the past, a pixel when classified as agriculture was automatically taken to have 100% croplands in digital global maps. In reality, only a certain percentage of this pixel is in croplands and that percentage can vary substantially. This may be one reason that total agricultural lands estimated in various digital maps vary significantly, such as 2.7 Bha by Olson and Watts [70] using a 50-km grid, 3.2 and 2.8 Bha by IGBP and USGS, respectively, using a 1-km grid [51].

Given the above discussion, the precise irrigated areas were calculated as follows (Figure 3.22):

$$SPA_n = FPA_n * IAF_n, \tag{3.6}$$

where:

SPA_n = subpixel area of class n,
FPA_n = full pixel area of class n,
IAF_n = irrigated area fraction for class n.

The FPA is calculated for the six GIAM classes (Figure 3.23a) based on Lambert Azimuthal Equal Area (LAEA) projection. The IAFs are determined using three methods:

1. Google Earth estimate (IAF-GEE)
2. High-resolution imagery (IAF-HRI)
3. Subpixel Decomposition Technique (IAF-SPDT)

The IAFs for TAAI were taken as the average of IAF-GEE and IAF-HRI for one season (fractions include the irrigated cropped area fraction + fallow area fraction, where fallow areas are areas equipped for irrigation but left fallow). The IAFs for AIA were taken as the average of (a) IAF-HRI and IAF-SPDT of season 1, (b) IAF-HRI and IAF-SPDT of season 2, and (c) IAF-HRI and IAF-SPDT of year-round perennial crops. The IAFs were then multiplied with FPAs to obtain SPAs or actual areas. The SPA of TAAI is obtained by multiplying the FPA with the average of IAF-GEE and IAF-HRI. The SPAs of AIA were obtained by summing up SPAs from season 1 + season 2 + year-round perennial crops.

The entire process of SPA computation is illustrated in Figure 3.22. The process involves taking each class, determining the FPAs of the classes, establishing IAFs of the classes, and calculating SPA without intensity (TAAI) and with intensity (AIA). A detailed process of SPA calculation was provided in a recent paper [71].

3.4.18 ACCURACIES, ERRORS, AND UNCERTAINTIES

Accuracies, errors, and uncertainties were assessed using four distinct data sets.

3.4.18.1 Accuracies Based on GT Data

Altogether, 1971 independent GT data points were used in the accuracy and error assessment. The points were gathered from GIAM GT missions as well as sourced from the DCP. Of the 1971 points, 463 represented surface-water irrigated areas, 542 represented groundwater, and the rest represented other LULC classes.

3.4.18.2 Accuracies Based on Google Earth Data

The Google Earth very high-resolution (0.61–4 m) data points (each point representing an area of 10 × 10 km) were randomly generated for the GIAM map, with a higher density of distribution of points where the irrigated area is denser. Altogether, 670

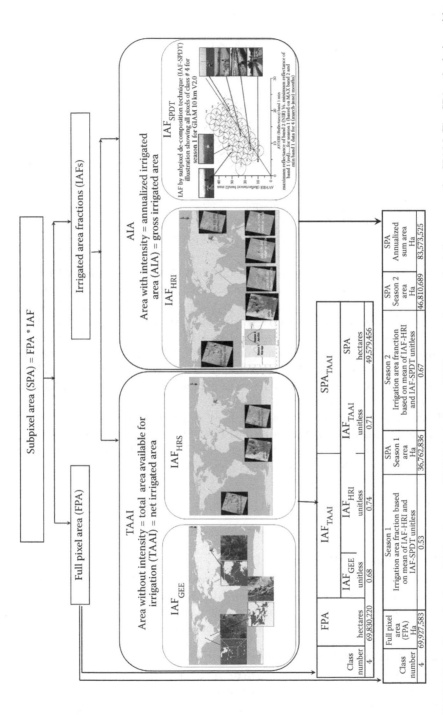

FIGURE 3.22 **(See color insert following page 224.)** Subpixel area (SPA) or actual area calculation process for GIAM. The SPA calculation without considering the intensity (TAAI) and considering the intensity (AIA) is illustrated.

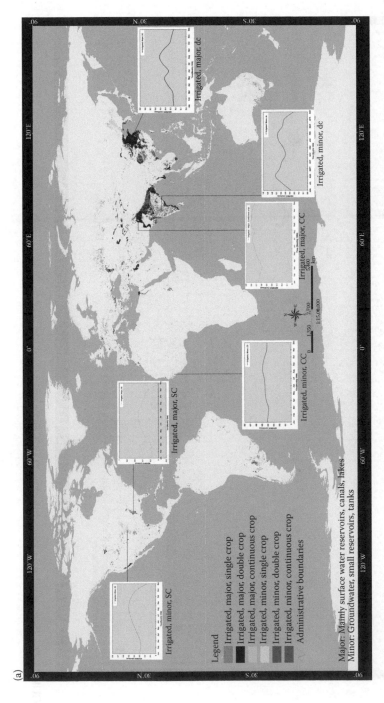

FIGURE 3.23 (See color insert following page 224.) Global irrigated area maps (GIAM). A 6-class GIAM of the world showing the areas where major and minor irrigated, single, double, and continuous (or perennial) crops occur (a). The three major irrigated area classes were combined together and the three minor irrigated area classes were also combined to show a 2-class GIAM (b).

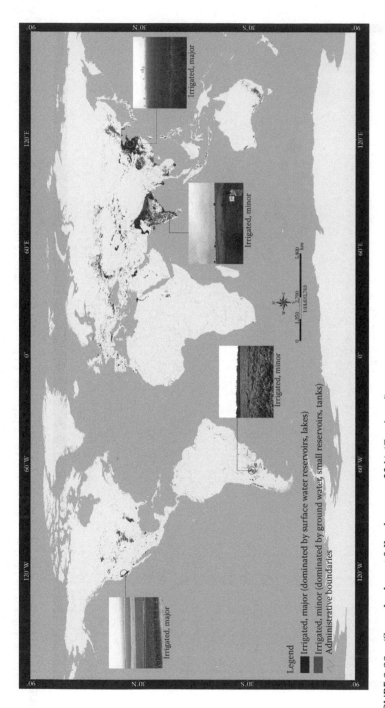

FIGURE 3.23 (See color insert following page 224.) (Continued)

points were generated. Each point was assessed for major irrigation (surface water command areas), minor irrigation (groundwater, small reservoirs, and tanks), and other LULC classes. The major irrigation was easy to establish by canals, reservoirs, tanks, and command areas (supported by secondary data). Minor irrigation was also fairly straightforward for small reservoirs (water-spread areas <2000ha) and tanks, but difficult for groundwater (interpretation was based on secondary data). Of the 670 sample points, 323 were irrigated (220 by surface water and 103 by groundwater).

The accuracies and errors were established for (a) major irrigation, (b) minor irrigation, and (c) a combination of the two. The equations used to determine accuracies and errors for irrigated areas (major+minor irrigated areas) are as follows (the same approach used to determine accuracies and errors for the major and minor irrigated areas):

Accuracy of irrigated area class

$$= \frac{\text{Groundtruthed irrigated points classified as irrigated areas}}{\text{Total number of groundtruthed points for irrigated area class}} \times 100, \quad (3.7)$$

Errors commission for irrigated area class

$$= \frac{\text{Nonirrigated groundtruth points falling on irrigated area class}}{\text{Total number of nonirrigated groundtruthed points}} \times 100, \quad (3.8)$$

Errors of omission for irrigated area class

$$= \frac{\text{Irrigated groundtruth points falling on nonirrigated area class}}{\text{Total number of irrigated area groundtruth points}} \times 100, \quad (3.9)$$

3.4.18.3 Comparisons with FAO/UF Data

The GIAM area results are compared with the FAO/UF [2] computed areas. This comparison is done considering the statistics from 198 countries. It must be kept in mind that the FAI/UF statistics are a collation of the statistics supplied by the countries, extrapolated, and presented in map format.

3.4.18.4 Comparisons with Subnational Census Data

Detailed subnational irrigated area statistics were also available for India from the Ministry of Water Resources [72]. These statistics were compared with the GIAM statistics taking the 28 Indian states and 7 Union territories.

3.5 RESULTS AND DISCUSSIONS

The outcome of this research led to the production of GIAM and the collection of statistics.

3.5.1 GLOBAL IRRIGATED AREA MAPS (GIAMs)

An aggregated GIAM at nominal 10-km resolution consisted of six classes (Figure 3.23a): (a) irrigated, major, single crop; (b) irrigated, major, double crop; (c) irrigated, major, triple crop; (d) irrigated, minor, single crop; (e) irrigated, minor, double crop; and (f) irrigated, minor, triple crop. The major irrigated areas are overwhelmingly surface water command areas or major and medium reservoirs of water-spread areas >2000 ha. Minor irrigation encompasses groundwater, small reservoirs, and tanks. Even when the groundwater irrigated areas encompass large areas, they are still aggregated within the "minor irrigation" category.

Irrigated areas were calculated by combining the six classes. Two types of areas were calculated (Table 3.2): (1) TAAI that does not consider intensity, and (b) AIA that considers intensity. The TAAI and AIA for the end of the last millennium were 399 and 467 Mha, respectively. Irrigated double crop is the most prominent class with about 76% of all global AIA. Of this, 264 Mha were "irrigated, major, double crop" and 93 Mha were "irrigated, minor, double crop" (Table 3.2). The "irrigated, minor, single crop" had 56 Mha (about 12%). The other three classes, combined had the rest of the 12% area. The typical NDVI time-series spectral profiles of the six classes for one year are shown in Figure 3.23a.

Globally, major irrigated areas dominate with 61% (287 Mha) of AIA, while minor irrigated areas occupy the other 39% (183 Mha) (Figure 3.23b, Table 3.3). Figure 3.23b also shows representative photos of the two classes.

3.5.2 IRRIGATED AREAS OF THE CONTINENTS

The Asian continent has 79% (370 Mha) of all global AIA (467 Mha), mostly as irrigated double crop. Europe and North America have about the same (32 Mha each or 7%) global irrigated areas (Table 3.4). Other continents have a very low proportion of global total AIA: South America (4%), Africa (2%), and Australia (1%).

3.5.3 IRRIGATED AREAS OF COUNTRIES

The irrigated areas of the countries (Table 3.5) were computed based on the GIAM classes (Figure 3.23a) and ranked as per the AIAs. China and India constitute 60% of all global irrigated areas (Table 3.5), followed by 11 countries with 1% or higher AIA. These, in order of ranking, are the United States (5.2%), Pakistan (3.4%), Russia (2.4%), Argentina (1.9%), Thailand (1.6%), Bangladesh (1.5%), Kazakhstan (1.4%), Myanmar (1.3%), Australia (1.2%), Uzbekistan (1.1%), and Vietnam (1.06%). There were 13 countries with 0.5–1%. These, in order of ranking, were Brazil, Mexico, Indonesia, Egypt, Spain, Germany, Canada, France, Italy, Iraq, Iran, Japan, and Ukraine. These 25 countries encompass 91% of the global irrigated areas. The TAAI and AIA of 198 countries are listed in Table 3.5, according to ranking based on AIA.

3.5.4 DISTRIBUTION OF IRRIGATED AREAS RELATIVE TO RAIN-FED AREAS

The two GIAM classes mapped in this study (Figure 3.23b) were overlaid on the rain-fed cropland areas reported by Biradar et al. [73] and other LULC classes

TABLE 3.2
GIAM Statistics for the Six Classes with the GIAM Statistics Computed (a) Without Considering Intensity (TAAI) and (b) Considering Intensity (AIAs)

Class No.	Class Names	FPA (ha)	IAF-TAAI (unitless)	TAAI (ha)	IAF-Season 1 (unitless)	Season 1-SPA (ha)	IAF-Season 2 (unitless)	Season 2-SPA (ha)	IAF-Continuous Season (unitless)	Continuous-Season SPA	AIA (ha)
1	Irrigated, major, single crop	31,671,436	0.73	23,275,848	0.53	16,796,945					16,796,945
2	Irrigated, major, double crop	256,982,208	0.67	173,158,856	0.51	131,621,365	0.52	132,584,487			264,205,852
3	Irrigated, major, continuous crop	13,544,336	0.61	8,241,839					0.44	5,914,675	5,914,675
4	Irrigated, minor, single crop	119,008,735	0.70	83,253,497	0.47	56,234,738					56,234,738
5	Irrigated, minor, double crop	102,660,428	0.63	64,911,553	0.49	50,495,221	0.41	42,384,677			92,879,898
6	Irrigated, minor, continuous crop	64,720,961	0.71	45,685,358					0.52	33,852,166	33,852,166

Notes: TAAI = total area available for irrigation; IAF = irrigated area fraction; FPA = full pixel area; SPA = subpixel area; AIA = annualized irrigated areas.

TABLE 3.3
GIAM Statistics for Two Classes with the GIAM Statistics Computed (a) Without Considering Intensity (TAAI) and (b) Considering Intensity (AIAs)

Class No.	Class Names	FPA (ha)	IAF-TAAI (unitless)	TAAI (ha)	IAF-Season 1 (unitless)	Season 1-SPA (ha)	IAF-Season 2 (unitless)	Season 2-SPA (ha)	IAF-Continuous Season (unitless)	Season Continuous-SPA	AIA (ha)
1	Irrigated, major	302,197,980	0.66	204,676,543	0.54	148,418,310	0.51	132,584,487	0.43	5,914,675	286,917,471
2	Irrigated, minor	286,390,124	0.68	193,850,408	0.46	106,729,959	0.47	42,384,677	0.52	33,852,166	182,966,800
	Total	588,588,104		398,526,951		255,148,269		174,969,164		39,766,841	469,884,271

Notes: TAAI=total area available for irrigation; IAF=irrigated area fraction; FPA=full pixel area; SPA=subpixel area; AIA=annualized irrigated areas.

TABLE 3.4

Global Irrigated Areas by Continent and the TAAI and AIA of the Continents

Sl No.	Continent Name	SPA-TAAI[a] (ha)	SPA-HRI/SPDT: IWMI GIAM 10-km V2.0 (actual irrigated area)[b]			AIA (ha)	World Total (%)	FAO/UF V4.0[c] (Area Equipped for Irrigation) (ha)
			Season 1 (ha)	Season 2 (ha)	Continuous (ha)			
1	Africa	8,687,044	5,601,273	3,680,659	1,020,078	10,302,011	2	13,432,285
2	Asia	290,641,673	192,600,664	152,312,096	24,700,163	369,612,923	79	187,600,089
3	Australia	11,865,244	2,991,344	0	2,382,064	5,373,409	1	2,056,580
4	Europe	33,937,745	20,126,797	7,691,113	4,627,243	32,445,154	7	26,770,001
5	North America	35,426,895	22,316,537	6,448,147	3,089,990	31,854,673	7	36,889,071
6	South America	17,842,959	8,055,356	3,363,798	5,608,671	17,027,825	4	11,495,806
7	Oceania	125,390	68,146	58,034	15,505	141,686	0	581,254
	Total	398,526,951	251,760,118	173,553,847	41,443,716	466,757,680	100	278,825,086

[a] Subpixel area from combined coefficients of Google Earth estimate and high-resolution images.

[b] Subpixel area from combined coefficients of high-resolution images and subpixel decomposition technique.

[c] Area irrigated obtained from FAO-Aquastat and Earth trends.

TABLE 3.5
Global Irrigated Areas by Countries

Rank (No.)	Country Name	SPA-TAAI[a] (ha)	SPA-HRI/SPDT: IWMI GIAM 10 km V2.0 (Actual Irrigated Area)[b]				Irrigated Area (%)	FAO/UF V4.0[c,d] (Area Equipped for Irrigation) (ha)
			Season 1 (ha)	Season 2 (ha)	Continuous (ha)	AIA (ha)		
1	China	111,988,772	75,880,320	68,233,355	7,688,411	151,802,086	32.523	53,823,000
2	India	101,234,893	72,612,189	53,685,066	5,956,598	132,253,854	28.335	57,291,407
3	United States	28,045,478	18,182,104	4,006,141	2,120,942	24,309,188	5.208	27,913,872
4	Pakistan	14,036,151	7,895,566	7,302,243	761,533	15,959,342	3.419	14,417,464
5	Russia	13,886,856	8,865,013	2,113,783	224,734	11,203,530	2.4	4,899,900
6	Argentina	9,304,258	3,601,505	1,605,815	3,559,092	8,766,412	1.878	1,767,784
7	Thailand	6,610,586	3,228,550	2,209,523	1,959,295	7,397,368	1.585	4,985,708
8	Bangladesh	5,235,050	3,882,847	3,076,494	206,686	7,166,028	1.535	3,751,045
9	Kazakhstan	7,227,718	4,625,716	1,760,606	83,362	6,469,685	1.386	1,855,200
10	Myanmar (Burma)	4,452,997	3,360,330	2,798,234	148,108	6,306,671	1.351	1,841,320
11	Australia[f]	11,865,244	2,991,344	0	2,382,064	5,373,409	1.151	2,056,580
12	Uzbekistan	3,601,487	2,733,397	2,427,259	134,859	5,295,515	1.135	4,223,000
13	Vietnam	4,384,022	1,865,074	1,419,401	1,665,058	4,949,533	1.06	3,000,000
14	Brazil	4,195,118	2,165,151	869,365	1,051,327	4,085,844	0.875	3,149,217
15	Mexico	3,854,673	1,818,168	916,083	874,479	3,608,730	0.773	6,435,800
16	Indonesia	3,172,879	1,221,384	716,038	1,385,021	3,322,443	0.712	4,459,000
17	Egypt	2,144,099	1,635,323	1,491,605	165,798	3,292,726	0.705	3,422,178
18	Spain	3,421,724	1,516,815	683,698	825,310	3,025,823	0.648	3,575,488
19	Germany	2,197,697	1,642,692	1,318,567	40,415	3,001,674	0.643	496,871
20	Canada	2,658,297	1,727,915	1,124,721	21,616	2,874,252	0.616	785,046
21	France	2,399,518	1,249,368	829,980	607,806	2,687,153	0.576	2,906,081
22	Italy	2,829,523	1,342,442	539,802	761,896	2,644,140	0.566	3,892,202

23	Iraq	2,220,024	1,242,694	1,254,929	128,942	2,626,564	0.563	3,525,000
24	Iran	2,623,336	1,308,727	679,564	500,268	2,488,558	0.533	6,913,800
25	Japan	2,525,096	1,157,850	656,470	654,276	2,468,596	0.529	3,129,000
26	Ukraine	2,995,578	1,631,677	258,515	491,607	2,381,799	0.51	2,395,500
27	Korea, Dem. Rep.	1,467,262	935,934	923,533	194,157	2,053,625	0.44	1,460,000
28	Romania	2,375,239	1,128,692	315,485	605,711	2,049,888	0.439	2,149,903
29	Turkmenistan	1,522,372	994,264	904,352	101,368	1,999,984	0.428	1,744,100
30	Sudan	1,737,118	1,185,252	643,655	101,685	1,930,592	0.414	1,863,000
31	Philippines	1,542,629	1,024,930	589,003	175,175	1,789,108	0.383	1,550,000
32	Turkey	1,753,382	882,867	332,404	362,042	1,577,313	0.338	4,185,910
33	Nepal	1,251,988	681,267	530,989	265,047	1,477,303	0.317	1,168,349
34	Chile	1,514,922	703,120	345,867	396,243	1,445,230	0.31	1,900,000
35	Korea, Rep.	1,192,469	546,413	432,289	335,053	1,313,755	0.281	880,365
36	Morocco	1,045,119	578,582	460,512	114,723	1,153,817	0.247	1,458,160
37	United Kingdom	970,733	810,688	233,603	15,913	1,060,204	0.227	228,950
38	Bulgaria	1,301,804	579,629	62,782	369,652	1,012,064	0.217	545,160
39	Netherlands	870,243	681,847	299,991	29,502	1,011,340	0.217	476,315
40	Denmark	1,164,705	976,705	2,835	0	979,539	0.21	476,000
41	Cambodia	736,318	480,153	329,683	128,606	938,441	0.201	284,172
42	Afghanistan	1,008,138	403,083	218,706	301,701	923,490	0.198	3,199,070
43	South Africa	821,040	574,487	206,929	47,075	828,491	0.177	1,498,000
44	Azerbaijan	835,627	441,335	218,092	162,553	821,980	0.176	1,453,318
45	Sri Lanka	948,029	169,255	111,161	529,164	809,579	0.173	570,000
46	Venezuela	894,880	499,284	93,686	214,109	807,078	0.173	570,219
47	Kyrgyzstan	700,876	447,852	247,134	75,288	770,274	0.165	1,075,040
48	Greece	907,739	271,632	106,151	388,895	766,678	0.164	1,544,530
49	Czech Republic	518,036	380,186	321,296	245	701,727	0.15	50,590

(Continued)

TABLE 3.5 (Continued)

Rank (No.)	Country Name	SPA-TAAI[a] (ha)	SPA-HRI/SPDT: IWMI GIAM 10 km V2.0 (Actual Irrigated Area)[b]				Irrigated Area (%)	FAO/UF V4.0[c,d] (Area Equipped for Irrigation) (ha)
			Season 1 (ha)	Season 2 (ha)	Continuous (ha)	AIA (ha)		
50	Taiwan, Province of China	499,043	282,608	314,359	80,910	677,877	0.145	525,528
51	Cuba	486,898	342,202	269,666	25,291	637,159	0.137	870,319
52	Syria	566,990	302,293	235,219	58,751	596,263	0.128	1,266,900
53	Colombia	546,186	336,538	176,558	79,399	592,495	0.127	900,000
54	Saudi Arabia	678,677	143,187	89,073	318,806	551,066	0.118	1,730,767
55	Belgium	324,796	294,221	204,916	8,293	507,430	0.109	35,170
56	Poland	351,514	268,183	185,150	779	454,111	0.097	134,050
57	Tajikistan	383,243	277,736	156,376	15,040	449,153	0.096	719,200
58	Somalia	372,476	162,324	117,817	123,434	403,574	0.086	200,000
59	Mongolia	422,332	265,966	110,413	0	376,378	0.081	57,300
60	Peru	355,956	189,766	113,945	71,243	374,954	0.08	1,729,069
61	Uruguay	381,403	311,863	25,602	22,591	360,055	0.077	217,593
62	Guinea	302,633	153,448	95,459	71,442	320,350	0.069	94,914
63	Portugal	358,865	133,115	54,464	126,330	313,908	0.067	792,008
64	Senegal	211,416	148,318	129,202	13,052	290,572	0.062	119,680
65	Ecuador	288,581	127,918	85,157	68,091	281,166	0.06	863,370
66	Malaysia	258,766	123,739	66,638	84,189	274,565	0.059	362,600
67	Serbia	171,939	140,266	92,171	1,910	234,348	0.05	163,311
68	Moldova	294,070	161,373	20,311	47,749	229,433	0.049	307,000
69	Albania	223,777	117,469	55,223	53,172	225,864	0.048	340,000
70	Nigeria	197,909	103,154	61,884	51,115	216,154	0.046	293,117
71	Libya	230,656	67,173	60,076	82,773	210,022	0.045	470,000

72	Hungary	241,714	166,069	14,990	5,162	186,221	0.04	292,147
73	Bolivia	214,091	28,854	9,777	124,404	163,036	0.035	128,240
74	Ethiopia	184,239	62,157	25,604	75,047	162,808	0.035	289,530
75	Guinea Bissau	108,042	84,650	66,770	3,969	155,389	0.033	22,558
76	Georgia	128,538	96,950	46,285	2,907	146,141	0.031	300,000
77	New Zealand	125,390	68,146	58,034	15,505	141,686	0.03	577,882
78	Algeria	144,349	90,667	34,731	11,548	136,946	0.029	569,418
79	Macedonia	169,843	113,105	9,610	8,905	131,620	0.028	127,800
80	Armenia	106,695	73,185	37,092	8,047	118,324	0.025	286,027
81	Laos	105,585	78,350	21,795	7,589	107,734	0.023	295,535
82	Israel	99,806	39,883	37,020	27,639	104,542	0.022	183,408
83	Kenya	85,401	53,025	37,354	14,148	104,527	0.022	103,203
84	Guyana	96,276	61,736	30,935	10,259	102,930	0.022	150,134
85	Cote d'Ivoire	95,138	79,392	20,756	1,742	101,890	0.022	72,750
86	Tunisia	109,144	30,355	23,663	46,628	100,647	0.022	394,063
87	Austria	116,456	69,017	19,025	10,509	98,551	0.021	97,480
88	Swaziland	149,274	97,004	0	0	97,004	0.021	49,860
89	Guatemala	69,373	47,776	40,864	2,673	91,313	0.02	129,803
90	Dominican Republic	70,876	45,462	25,851	8,335	79,648	0.017	269,710
91	Yemen	91,688	42,912	16,073	20,203	79,188	0.017	388,000
92	Honduras	70,584	51,034	21,071	5,623	77,729	0.017	73,210
93	Slovakia	109,904	71,826	1,044	2,618	75,488	0.016	15,643
94	Madagascar	72,359	41,627	19,039	14,490	75,156	0.016	1,086,291
95	Finland	125,307	71,961	0	0	71,961	0.015	103,800
96	Ghana	60,647	28,411	24,173	19,181	71,764	0.015	30,900
97	Sweden	83,918	69,968	1,140	0	71,108	0.015	188,470

(Continued)

TABLE 3.5 (Continued)

Rank (No.)	Country Name	SPA-TAAI[a] (ha)	SPA-HRI/SPDT: IWMI GIAM 10 km V2.0 (Actual Irrigated Area)[b]				Irrigated Area (%)	FAO/UF V4.0[c,d] (Area Equipped for Irrigation) (ha)
			Season 1 (ha)	Season 2 (ha)	Continuous (ha)	AIA (ha)		
98	United Arab Emirates	93,810	10,249	4,867	55,487	70,603	0.015	280,341
99	Mali	56,355	38,220	26,100	1,559	65,879	0.014	235,791
100	Rwanda	80,067	64,806	0	0	64,806	0.014	8,500
101	Thegambia	39,872	34,993	28,422	0	63,415	0.014	0
102	Belarus	84,088	60,731	195	0	60,926	0.013	115,000
103	Mozambique	56,415	39,402	16,753	4,587	60,742	0.013	118,120
104	Haiti	50,848	29,974	15,438	8,490	53,903	0.012	91,502
105	Jordan	72,717	574	568	51,399	52,541	0.011	76,912
106	Cameroon	52,694	35,415	5,861	10,852	52,128	0.011	25,654
107	Tanzania	47,022	33,678	7,852	5,467	46,998	0.01	184,330
108	Panama	49,069	21,997	6,477	16,574	45,048	0.01	34,626
109	Croatia	35,202	28,102	15,511	1,018	44,630	0.01	5,790
110	Lithuania	57,272	41,591	0	0	41,591	0.009	4,416
111	Switzerland	29,523	21,079	15,897	0	36,976	0.008	40,000
112	Angola	23,316	16,671	14,371	3,116	34,158	0.007	80,000
113	Uganda	30,017	26,957	3,447	183	30,586	0.007	9,150
114	Oman	17,853	15,247	14,898	0	30,145	0.006	72,630
115	Sierra Leone	21,807	16,343	12,481	213	29,037	0.006	29,360
116	Chad	25,234	15,932	8,020	3,747	27,698	0.006	30,273
117	Qatar	38,509	0	0	27,596	27,596	0.006	12,520
118	Kuwait	37,333	0	0	26,753	26,753	0.006	6,968
119	Lebanon	24,747	11,240	8,170	5,859	25,268	0.005	117,113

120	Paraguay	28,582	12,913	1,670	10,445	25,029	0.005	67,000
121	Togo	21,727	9,624	7,433	6,786	23,843	0.005	7,300
122	Nicaragua	16,439	12,165	9,941	614	22,720	0.005	61,365
123	Suriname	19,845	14,491	5,070	1,213	20,774	0.004	51,180
124	Congo, Dem. Rep.	21,833	19,326	191	857	20,375	0.004	10,500
125	Mauritania	15,124	9,814	10,007	214	20,036	0.004	45,012
126	Costa Rica	12,628	9,730	5,448	613	15,791	0.003	103,084
127	Benin	15,173	4,383	3,797	7,235	15,415	0.003	12,258
128	Burkina Faso	15,663	4,539	4,420	5,702	14,660	0.003	25,000
129	Estonia	24,637	14,476	0	0	14,476	0.003	1,363
130	Bosnia and Herzegovina	10,766	6,696	5,445	2,062	14,203	0.003	4,630
131	Montenegro	10,331	6,940	5,604	1,364	13,908	0.003	0
132	Eritrea	17,017	11,467	2,309	0	13,776	0.003	21,590
133	Puerto Rico	11,964	7,082	1,582	2,588	11,253	0.002	37,079
134	El Salvador	11,592	7,839	2,508	54	10,401	0.002	44,993
135	Namibia	10,526	7,508	1,795	0	9,303	0.002	7,573
136	Burundi	11,793	534	36	7,921	8,490	0.002	21,430
137	Latvia	12,683	7,260	65	0	7,325	0.002	1,150
138	Gaza Strip	5,909	3,192	3,223	375	6,790	0.001	0
139	Cyprus	7,099	2,751	129	1,983	4,863	0.001	55,813
140	Jamaica	4,881	3,058	492	1,006	4,556	0.001	25,214
141	Niger	4,129	3,121	1,196	0	4,317	0.001	73,663
142	Botswana	5,417	3,687	590	0	4,278	0.001	1,439
143	East Timor	3,800	3,257	804	0	4,061	0.001	14,000
144	Mauritius	5,312	2,381	0	1,528	3,910	0.001	21,222
145	Lesotho	5,675	3,681	0	0	3,681	0.001	2,638

(Continued)

TABLE 3.5 (Continued)

Rank (No.)	Country Name	SPA-TAAI[a] (ha)	SPA-HRI/SPDT: IWMI GIAM 10 km V2.0 (Actual Irrigated Area)[b]					FAO/UF V4.0[c,d] (Area Equipped for Irrigation) (ha)
			Season 1 (ha)	Season 2 (ha)	Continuous (ha)	AIA (ha)	Irrigated Area (%)	
146	Zimbabwe	4,744	3,234	299	0	3,533	0.001	173,513
147	Belize	3,887	2,919	306	286	3,510	0.001	3,000
148	French Guyana	2,860	2,217	351	254	2,822	0.001	2,000
149	Malawi	3,293	2,794	0	0	2,794	0.001	56,390
150	Equatorial guinea	2,812	2,644	0	0	2,644	0.001	0
151	Antigua and Barbuda	2,270	1,378	706	384	2,468	0.001	130
152	Guadeloupe	1,894	1,498	342	183	2,022	0	2,000
153	Trinidad and Tobago	1,859	1,672	0	48	1,720	0	3,600
154	West Bank	1,612	538	533	471	1,542	0	0
155	Norway	2,072	1,323	130	0	1,453	0	134,396
156	St. Kitts and Nevis	1,650	1,314	84	48	1,445	0	18
157	Bhutan	997	796	600	0	1,396	0	38,734
158	Central African Republic	1,155	1,086	0	0	1,086	0	135
159	Virgin Islands	827	563	361	91	1,015	0	185
160	Brunei	799	481	369	152	1,002	0	1,000
161	Reunion	651	517	329	0	846	0	13,000
162	San Marino	1,102	0	0	797	797	0	0
163	Djibouti	905	587	0	0	587	0	1,012
164	Zambia	779	0	0	536	536	0	155,912
165	Slovenia	439	293	217	0	510	0	0

166	Comoros	241	218	199	0	417	130
167	Anguilla	489	404	0	0	404	0
168	Liberia	237	201	100	0	300	2,100
169	Turks and Caicos Islands	214	117	0	53	170	0
170	Montserrat	69	51	65	0	115	0
171	St. Pierre and Miquelon	70	59	0	0	59	0
172	Cayman Islands	66	55	0	0	55	0
173	Monaco	73	0	0	53	53	0
174	Seychelles	66	44	0	0	44	260
175	Andorra	0	0	0	0	0	150
176	Bahrain	0	0	0	0	0	4,060
177	Barbados	0	0	0	0	0	1,000
178	Cape Verde	0	0	0	0	0	3,109
179	Congo	0	0	0	0	0	2,000
180	Fiji	0	0	0	0	0	3,000
181	Gabon	0	0	0	0	0	4,450
182	Gambia	0	0	0	0	0	2,149
183	Grenada	0	0	0	0	0	219
184	Guam	0	0	0	0	0	312
185	Ireland	0	0	0	0	0	1,100
186	Liechtenstein	0	0	0	0	0	0
187	Luxembourg	0	0	0	0	0	27
188	Malta	0	0	0	0	0	2,300
189	Martinique	0	0	0	0	0	3,000

(Continued)

TABLE 3.5 (Continued)

Rank (No.)	Country Name	SPA-TAAI[a] (ha)	SPA-HRI/SPDT: IWMI GIAM 10 km V2.0 (Actual Irrigated Area)[b]				Irrigated Area (%)	FAO/UF V4.0[c,d] (Area Equipped for Irrigation) (ha)
			Season 1 (ha)	Season 2 (ha)	Continuous (ha)	AIA (ha)		
190	Northern Marianna Islands	0	0	0	0	0	0	60
191	Palestine	0	0	0	0	0	0	19,466
192	Papua New Guinea	0	0	0	0	0	0	0
193	Pitcairn Islands	0	0	0	0	0	0	0
194	Sao Tome and Principe	0	0	0	0	0	0	9,700
195	Singapore	0	0	0	0	0	0	225,310
196	St. Lucia	0	0	0	0	0	0	297
197	St. Vincent and the Grenadines	0	0	0	0	0	0	0
198	Vatican city	0	0	0	0	0	0	0
	Total	398,526,951	251,760,118	173,553,847	41,443,716	466,757,680[e]	100	278,825,086

Note: The global irrigated area by country is ranked according the AIAs.

[a] Subpixel area from combined coefficients of Google Earth estimate and high-resolution images.

[b] Subpixel area from combined coefficients of high-resolution images and subpixel decomposition technique.

[c] Area equipped for irrigation from Food and Agricultural Organization and University of Frankfurt Global Map of Irrigated Area. Version 3.0 (based on national statistics).

[d] Area irrigated obtained from FAO-Aquastat and Earth trends (http://faostat.fao.org/faostat /http://earthtrends.wri.org/country_profiles/).

[e] World total was computed as per table 3.41 (480 m ha).

[f] Australian irrigated area was computed using the procedure described in annex 2 of Thenkabail et al. (2006).

mapped during this project. This led to a composite map consisting of the global irrigated areas, rain-fed areas, and other LULC classes (Figure 3.24). The irrigated areas (Tables 3.2 through 3.5) and their spatial distributions (Figure 3.23a and b, and Figure 3.24) show a number of interesting facts:

1. Irrigated areas are mainly distributed in China, India, the United States, Pakistan, and Central Asian countries (Figure 3.23a).
2. China and India, combined, have about 60% of the global AIA, mainly as irrigated double crop and constitute the main pillar of food security, feeding a population of about 2.6 billion (Figure 3.23b).
3. North America and Europe are dominated by rain-fed cropland areas (see Ref. [73]), which together support a population of about 1.3 billion and export food to numerous countries in the world (Figure 3.5).
4. Africa with a population of about 900 million has less than 2% (10 Mha) of global irrigated areas (Figure 3.23a).
5. South America with a population nearing 400 million has 17 Mha of irrigated area.

The distributions of irrigated areas are expected to change significantly in the coming years. First, as a result of increasing populations and the need to feed them, areas may increase or rain-fed areas may come under supplemental irrigation to increase productivity. Second, increasing urbanization is consuming large amounts of agricultural lands. Third, cultivation of biofuels is eating up agricultural lands and taking up water previously used for irrigation. Fourth, change in consumer habits (e.g., more vegetables and fruits in place of grain-only diet in many parts of the world) will lead to changes in the cropping pattern and water requirements. Fifth, the concept of "virtual water," if put into practice, can rapidly change the spatial distribution of irrigated areas.

3.5.5 ACCURACIES AND ERRORS

Based on GT data, the irrigated areas were mapped with an accuracy of 79% with errors of omission and commission less than 23% (Table 3.6). With Google Earth data, these accuracies increase to 91% and errors of omission and commission decrease to less than 16%. The major irrigated areas are significantly better mapped than minor irrigated areas, with better accuracies and lower levels of errors of omission and commission (Table 3.6).

3.5.6 COMPARISONS BETWEEN GIAM AND FAO/UF

The country-by-country GIAM TAAI was compared with the FAO/UF-reported [2] irrigated areas. A 1:1 line considering all 198 countries showed the relationship GIAM TAAI (Mha)=0.5385 * FAO/UF "areas equipped for irrigation" (R^2=0.93). The high R^2 value of 0.93 indicates good correlation. However, the slope of 0.5385 indicates higher values in GIAM by a factor of 2.

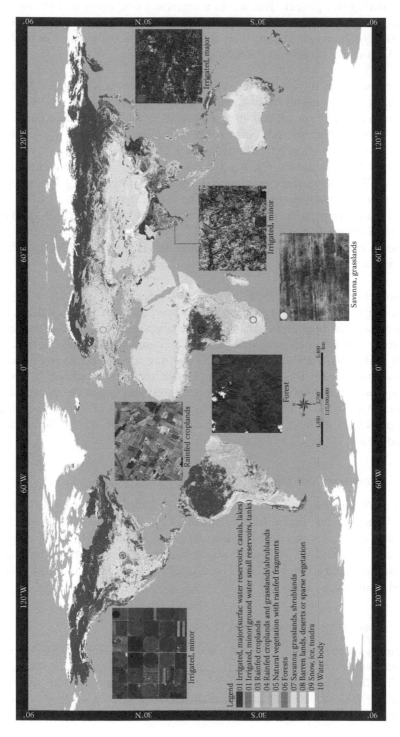

FIGURE 3.24 (See color insert following page 224.) GIAM with global rain-fed and other LULC.

TABLE 3.6
Accuracies and Errors of GIAM

	Total Groundtruth Sample Size (No.)	Correctly Classified Groundtruth (No.)	Accuracy of Irrigated Area Classes (%)	Errors of Omissions (%)	Errors of Commissions (%)
I. Accuracy based on independent groundtruth data points					
A. Major and minor irrigated areas	1005	793	79	21	23
B. Major irrigated areas (surface water reservoirs, command areas)	463	347	75	25	25
C. Minor irrigated areas (groundwater, small reservoirs, tanks)	542	385	71	29	26
II. Accuracy based on Independent Google Earth data points					
A. Major and minor irrigated areas	323	295	91	9	16
B. Major irrigated areas (surface water reservoirs, command areas)	220	187	85	15	17
C. Minor irrigated areas (groundwater, small reservoirs, tanks)	103	79	77	23	36

Note: The accuracies and errors based on groundtruth data.

Then, we considered 154 countries that have 10 Mha or less of irrigated areas; leaving out 44 countries. Of these 44 countries, 37 have zero or near-zero irrigated areas in GIAM TAAI and/or FAO/UF. In seven other countries (China, India, Russia, Australia, Argentina, Kazakhstan, and Iraq) the differences between GIAM and FAO/UF were very high. The 1:1 line considering the 154 countries showed the relationship (Figure 3.25) GIAM TAAI (Mha)=0.952 * FAO/UF "areas equipped for irrigation" (R^2=0.94). This has both a high correlation (R^2=0.94) and a near-perfect slope (0.952). However, it must be noted that the GIAM TAAI estimate of 399 Mha for the end of the last millennium is significantly higher than the FAO/UF estimate of 278 Mha. The two huge differences occur in China and India. For China, GIAM TAAI is 112 Mha compared to the FAO/UF estimate of 54 Mha. For India, GIAM TAAI is 101 Mha whereas the FAO/UF estimate is 57 Mha. The main cause of these differences is the inadequate accounting of minor irrigation (irrigation from groundwater, small reservoirs, tanks) in the FAO/UF, as their statistics are taken directly off the nationally reported statistics. Detailed discussions on the issues related to differences are presented in Section 3.5.

3.5.7 COMPARISONS WITH THE INDIAN NATIONAL STATISTICS

The state-by-state census-based irrigation potential utilized (IPU) from the Ministry of Water Resources [72] of India was related to its equivalent GIAM AIA statistics: (a) considering all 28 states (Figure 3.26a), and (b) dropping two "outlier" states (Figure 3.26b), resulting in the following relationships:

GIAM AIA (Mha)=1.3422 * MoWR IPU (Mha) (R^2=0.76) for all 28 states

GIAM AIA (Mha)=1.3499 * MoWR IPU (Mha) (R^2=0.89), leaving out two outlier states.

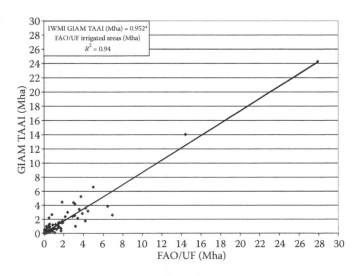

FIGURE 3.25 GIAM vs. FAO/UF. The GIAM statistics of the 154 countries that have less than 10 Mha of irrigated areas in both GIAM and FAO/UF are compared.

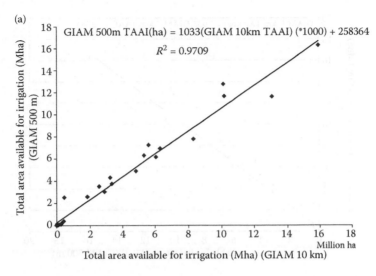

(a)

GIAM 500m TAAI(ha) = 1033(GIAM 10km TAAI) (*1000) + 258364

$R^2 = 0.9709$

Total area available for irrigation (Mha) (GIAM 500 m)

Total area available for irrigation (Mha) (GIAM 10 km)

Million ha

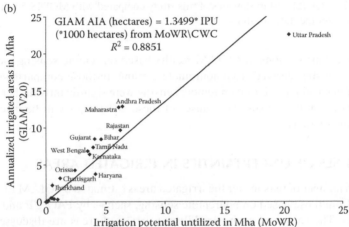

(b)

GIAM AIA (hectares) = 1.3499* IPU (*1000 hectares) from MoWR\CWC

$R^2 = 0.8851$

Annualized irrigated areas in Mha (GIAM V2.0)

Irrigation potential untilized in Mha (MoWR)

FIGURE 3.26 GIAM vs. Indian National Statistics. The GIAM statistics of the Indian states were compared with the corresponding values provided by India's Ministry of Water Resources (MoWR) for the 28 states (a) and the 26 states after removing two "outlier" states (b).

3.5.8 COMPARISONS WITH THE HIGHER RESOLUTION IMAGERY

The GIAM 10-km TAAI and AIA derived from this study were compared with the TAAI and AIA derived from MODIS 500-m data for the 28 states in India [74]. The results showed an excellent correlation between the two resolutions with R^2 values of 0.97 for the 1:1 plot (Figure 3.27a and b). These results demonstrate that approaches based on remote sensing provide similar results even when there are significant differences in resolution.

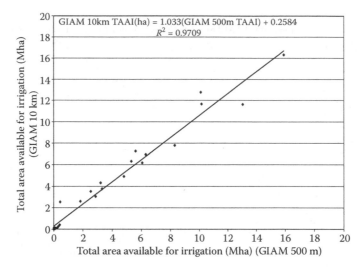

FIGURE 3.27 GIAM 10-km statistics of this study compared with MODIS 500-m derived irrigated areas for the states in India.

In the previous sections, the GIAM results based on remote sensing were compared with statistics derived from nonremote sensing. In those comparisons, it was made clear that estimates based on remote sensing were significantly higher than the statistics based on the census. The causes of these differences will be discussed in the following sections.

3.6 CAUSES OF UNCERTAINTIES IN IRRIGATED AREAS

There are a number of reasons for the irrigated areas estimated by GIAM to vary significantly from those based on nonremote sensing, such as by FAO/UF and national census data. The causes of such variability in irrigated areas are discussed below (Sections 3.6.1 through 3.6.7).

3.6.1 INADEQUATE ACCOUNTING OF MINOR IRRIGATED AREAS IN THE CENSUS STATISTICS

The overwhelming proportion of the irrigated area statistics is dependent on the command areas sourced by major and medium irrigation schemes. The statistics on minor irrigation from informal sources (groundwater, small reservoirs, tanks) are either not fully accounted for or, at times, missing in the statistics on irrigated area. For example, the number of groundwater wells in India increased from a meager 100,000 in the early 1960s to 19–26 million [9], most of them used for irrigation purposes. However, it is not clear how well these informal irrigation sources are accounted for in the statistics on irrigation. Similarly, there are numerous small reservoirs and tanks in private control whose presence may never be noted in the government census. For example, a recent study by Velpuri et al. [75] showed that

6100 small reservoirs and tanks, using Landsat 30-m data in the Krishna river basin, irrigated as much area as that by the 24 major and medium irrigation schemes. But, it is unclear what proportion of the 6100 minor reservoirs and tanks is accounted for in the census data in determining areas irrigated by them. As a result of these facts, there is significant uncertainty in accounting for the areas irrigated using non-remote-sensing census data and remote sensing approaches.

3.6.2 DEFINITION ISSUES

In GIAM, any significant amount of water used by artificial means to sustain crop growth is considered an amount irrigated. Typically, if there is more than one irrigation turn during a crop-growing season, then the area is considered irrigated. There are instances in the irrigated area maps and statistics, where the areas irrigated by groundwater, small reservoirs, and tanks are often not considered as irrigated. Many maps show informal irrigated areas under rain-fed conditions. We determined this by putting the GT data points on a number of global maps [2,21,32–34,69,76], where the GT shows the location as irrigated while the maps indicate the location as rain-fed, or even as other classes.

There are a number of cultivated areas with significant water application during the growing season to supplement rainfall. Depending on the definition, these areas are either mapped as rain-fed or irrigated. In GIAM, all supplemental irrigated areas are mapped as irrigated.

3.6.3 IRRIGATED AREA FRACTIONS FOR CLASS (IAFs)

The SPAs provide actual areas in any coarse-resolution (>30 m) remote sensing. The SPAs are computed by multiplying FPAs with IAFs. So, determining IAFs accurately is key to estimating irrigated areas accurately. The strategy to increase the accuracy of IAFs is to have a large number of samples of GT data establishing IAFs for every class.

3.6.4 RESOLUTION OR SCALE

Velpuri et al. [75] established that the finer the spatial resolution, the greater the area. At finer spatial resolution, the IAFs can be determined better. At coarser resolution, they are dependent on how well the IAFs are determined. Also, fragmented irrigated areas can miss out completely in coarser-resolution remote sensing.

3.6.5 MINIMUM MAPPING UNIT (MMU)

When irrigated areas are fragmented, they will not be mapped when the scale is too large. This is specifically a problem in many developing countries where fragmented irrigated areas are widespread and contiguous irrigated areas are limited.

3.6.6 REMOTE SENSING VS. CONVENTIONAL DATA-GATHERING APPROACHES

Remote sensing data provide a consistent approach to mapping irrigated areas, as demonstrated in various chapters of this book. Conventional data gathering involves eye estimates of irrigated areas, recording them, and aggregating them at various administrative units. The chances of making errors due to subjective estimates are quite high. Further, the precise spatial component of the areas is missing.

3.6.7 OTHER CAUSES OF UNCERTAINTY IN IRRIGATED AREAS

There will be many other causes for uncertainty in irrigated areas in the coming decades. First, rain-fed croplands are identified as areas for productivity increases in the new millennium [77], and may yet have an impact on limiting irrigated areas in the coming decades. Second, if serious advances are made in using less water to produce more food (better water productivity), the distribution of irrigated areas may drastically change. Third, the spatial distribution of irrigated areas may also change if the concept of "virtual water trade" (where countries with surplus water grow food and export to water-deficit countries for other trade benefits) takes hold. Fourth, irrigated croplands are being converted significantly, to biofuel farms in certain parts of the world. Fifth, genetic engineering may yet help increase yields, but is increasingly questioned by environmental activists and more ecologically sensitive governments. Sixth, the irrigated landscape of the world will be shaped increasingly by the effects of competition for water from other sectors, notably urban and rural domestic water supply and industrial needs. Seventh, groundwater overdraft may ultimately exhaust and/or substantially reduce irrigated areas in the Ogallala aquifer in the Midwest United States, Northeast China, and most of India. Eighth, reallocation or overallocation of much needed flows for environmental and health purposes, as in the case of the Krishna basin in India [20], will, in the end, place even greater competing demands in terms of water volume. Last, climatic change is likely to impose additional challenges, probably the biggest challenge of all, which will reshape the irrigated landscape through changes in snowmelt runoff and rainfall.

3.7   GIAM PRODUCTS AND DISSEMINATION

The GIAM products are disseminated through a dedicated web portal (http://www.iwmigiam.org). The data and products include: (a) irrigated area statistics (http://www.iwmigiam.org/stats), web maps (http://giam.iwmi.org/mapper.asp), and GIAM on the Google kmz file.

3.8   CONCLUSIONS

The GIAM and statistics were produced at the end of the last millennium at nominal 10-km resolution using remotely sensed data. Three GIAM products were produced. First, a 6-class GIAM consisting of (a) irrigated, major, single crop; (b) irrigated, major, double crop; (c) irrigated, major, continuous crop; (d) irrigated, minor, single crop; (e) irrigated, minor, double crop; and (f) irrigated, minor, continuous crop. Second, a 2-class GIAM map by aggregating the 6-classes that led to irrigated major and minor. Third, a composite map that consisted of global irrigated areas, rain-fed areas, and other LULC.

The GIAM statistics were derived for the 198 countries in the world in the form of (a) TAAI that does not consider the intensity of cropping, and (b) AIAs that do consider the intensity of irrigation. At the end of the last millennium, the TAAI was 399 Mha while the AIA was 467 Mha. Of this, 61% are irrigated, major, and the balance 39% are irrigated, minor.

Asia has 79% (370 Mha) of the AIA (467 Mha) with other continents falling way behind with Europe and North America (7%), South America (4%), Africa (2%), and Australia (1%). China (32.5%) and India (28.5%) together encompass about 60% of all global AIAs, followed by the United States (5.2%) and Pakistan (3.4%). Twenty-five leading irrigated area countries have 90% of all global AIAs. The irrigated areas are dominated by double crops in Asia.

Accuracies of mapping irrigated areas varied between 79% and 91%, with errors of omission and commission of 16–23%. The irrigated areas estimated by GIAM using remote sensing correlated well with the national census data, but generally provided significantly higher estimates. The causes for higher estimates were (a) inadequate accounting of informal irrigation (groundwater, small reservoirs, tanks) in the national statistics; (b) definition issues; (c) IAF estimates; (d) resolution or scale; (e) MMUs; (f) type of data used (e.g., remote sensing vs. nonremote sensing); and (g) methodologies used.

The GIAM data and products are made available through the GIAM web portal (http://www.iwmigiam.org).

ACKNOWLEDGMENTS

The authors thank Professor Frank Rijsberman, former Director General, IWMI (currently with Google.org), for his support and encouragement throughout the GIAM project. No project of this magnitude or complexity would have been feasible without the sustained support of visionary leaders like Frank. There are others we should thank: Professor Mei Xurong and Dr. Hai Weiping from the Chinese Academy of Agricultural Sciences, Dr. Songcai You from the Chinese Academy of Sciences, Dr. Mangala Rai, Dr. J.S. Samra, Dr. A.K. Maji, and Dr. Obi Reddy of the Indian Council for Agricultural Research, Dr. Bharat Sharma of IWMI Delhi office, and Professor Seelan Santhosh Kumar and Bethany Kurz of the University of North Dakota. Kingsley Kurukulasuriya did very insightful and perfect editing of this book and his contribution is very gratefully acknowledged. Many of the authors of other chapters of this book have provided insights that have been invaluable. We would like to thank them all for their support and consideration.

REFERENCES

1. Van Schilfgaade, J., Irrigation – A blessing or a curse? *Agricultural Water Management*, 25, 203–19, 1994.
2. Siebert, S., Hoogeveen, J., and Frenken, K., *Irrigation in Africa, Europe and Latin America – Update of the digital global map of irrigation areas to version 4*, Frankfurt Hydrology Paper 05, Institute of Physical Geography, University of Frankfurt, Frankfurt am Main, Germany and Food and Agriculture Organization of the United Nations, Rome, Italy, 2006.

3. Thenkabail, P.S. et al., A global irrigated area map (GIAM) using remote sensing at the end of the last millennium, *International Journal of Remote Sensing*, 2009, (accepted in press).

4. Thenkabail, P.S. et al., *An irrigated area map of the world (1999) derived from remote sensing*, Research Report 105, 65 International Water Management Institute, Colombo, Sri Lanka, 2006. Also, see under documents in: http://www.iwmigiam.org.

5. Richards, J.F., Land transformation, in *The Earth as Transformed by Human Action*, B.L. Turner, Ed., Cambridge with Clark University, New York, 1990, 163–78.

6. Cramer, W.P. and Soloman, A.M., Climatic classification and future global redistribution of agricultural land, *Climate Research*, 3, 97–110, 1993.

7. Grubler, A., Technology, in *Changes in Land Use and Land Cover: A Global Perspective*, W.B. Meyer and B.L. Turner II, Eds., Cambridge University Press, New York, 1994, 287–328.

8. World Resources 1992–1999, *A Guide to the Global Environment*, World Resources, Oxford, 1992.

9. Endersbee, L., *A Voyage of Discovery: A History of Ideas About the Earth, With a New Understanding of the Global Resources of Water and Petroleum, and the Problems of Climate Change*, Monash University Bookshop, Frankston, Victoria, Australia, 2005.

10. Shah, T. et al., Sustaining Asia's groundwater boom: An overview of issues and evidence, *Natural Resources Forum*, 27, 130–40, 2003.

11. Brown, L.R., *Plan B: Rescuing a Planet Under Stress and Civilization in Trouble*, Earth Policy Institute, Washington, DC, 2003.

12. Flavin, C. and Gardner, G., China, India, and the new world order, in *State of the World 2006 Special Focus: China and India*. The World Watch Institute, Washington, DC, 2006, 3–23.

13. Webb, P., *Cultivated capital: Agriculture, food systems and sustainable development*, Food Policy and Applied Nutrition Program, Tufts Nutrition Discussion Paper No. 15, The Gerald J. and Dorothy R. Friedman School of Nutrition Science and Policy, Boston, MA, 2002.

14. Webb, P., Land degradation in developing countries: What is the problem? *International Journal of Agricultural Resources, Governance and Ecology*, 1, 2, 124–36, 2001.

15. Ramankutty, N. and Foley, J.A., Estimating historical changes in global land cover: Croplands from 1700 to 1992, *Global Biogeochemical Cycles*, 13, 997–1027, 1999.

16. Ramankutty, N. and Foley. J., Characterizing patterns of global land use: An analysis of global croplands data, *Global Biogeochemical Cycles*, 12, 4, 667–85, 1998.

17. FAO (Food and Agriculture Organization of the United Nations), *FAOSTAT Database*, http://apps.fao.org.

18. Postel, S., *Pillars of Sand: Can the Irrigation Miracle Last?* W.W. Norton, New York, 1999.

19. Merrett, S., *Water for Agriculture: Irrigation Agriculture in International Perspectives (Spon's Environmental Science and Engineering Series)*, Taylor & Francis, New York, 2002.

20. Biggs, T. et al., Vegetation phenology and irrigated area mapping using combined MODIS time-series, ground surveys, and agricultural census data in Krishna River Basin, India, *International Journal of Remote Sensing*, 27, 19, 4245–66, 2006.

21. Bartholomé, E. and Belward, A.S., GLC2000: A new approach to global land cover mapping from earth observation data, *International Journal of Remote Sensing*, 26, 9, 1959–77, 2005.

22. Agarwal, S. et al., SPOT vegetation multi temporal data for classifying vegetation in south central Asia, *Current Science,* 84, 11, 1440–48, 2003.

23. Huston, D.M. and Titus, S.J., An inventory of irrigated lands within the state of California based on lands and supporting aircraft data, *Space Sciences Laboratory Series* 16, 56, University of California, Berkeley, CA, 1975.

24. Draeger, W.C., *Monitoring irrigated level acreage using Landsat imagery: An application example*, U.S. Geological Survey Open-File Report No. 76-630, Sioux Falls, SD, 1976 (Duplicated).
25. Wall, S.L., California's irrigated lands. Paper presented at the Symposium on Identifying Irrigated Lands Using Remote Sensing Techniques: State of the Art, 15–16 November 1979, at U.S. Geological Survey EROS Data Center, Sioux Falls, SD, 1979.
26. Thiruvengadachari, S., Satellite sensing of irrigation pattern in semiarid areas: An Indian study, *Photogrammetric Engineering and Remote Sensing*, 47, 1493–99, 1981.
27. Thiruvengadachari, S. and Sakthivadivel, R., *Satellite remote sensing techniques to aid irrigation system performance assessment: A case study in India*, Research Report 9, International Water Management Institute, Colombo, Sri Lanka, 1997.
28. Thenkabail, P.S. et al., A global map of irrigated area at the end of the last millennium using multi-sensor, time-series satellite sensor data, Draft for International Water Management Institute, Colombo, Sri Lanka, 2005, http://www.iwmigmia.org.
29. Xiao, X. et al., Uncertainties in estimates of cropland area in China: A comparison between an AVHRR-derived dataset and a Landsat TM-derived dataset, *Global and Planetary Change*, 37, 297–306, 2003.
30. Xiao, X. et al., Mapping paddy rice agriculture in South and Southeast Asia using multi-temporal MODIS images, *Remote Sensing of Environment*, 100, 95–113, 2006.
31. Loveland, T.R. et al., Development of a global land cover characteristics database and IGBP DISCover from 1-km AVHRR data, *International Journal of Remote Sensing*, 21, 6/7, 1303–30, 2000.
32. DeFries, R. et al., Global land cover classifications at 8 km spatial resolution: The use of training data from Landsat imagery in decision tree classifiers, *International Journal of Remote Sensing*, 19, 16, 3141–3168, 1998.
33. DeFries, R. et al., A new global 1 km data set of percent tree cover derived from remote sensing, *Global Change Biology*, 6, 247–55, 2000.
34. Thenkabail, P.S. et al., Hyperion, IKONOS, ALI, and ETM+sensors in the study of African rainforests, *Remote Sensing of Environment*, 90, 23–43, 2004.
35. Thenkabail, P.S. et al., Accuracy assessments of hyperspectral waveband performance for vegetation analysis applications, *Remote Sensing of Environment*, 91, 2–3, 354–76, 2004.
36. Thenkabail, P.S., Inter-sensor relationships between IKONOS and Landsat-7 ETM+NDVI data in three ecoregions of Africa, *International Journal of Remote Sensing*, 25, 2, 389–408, 2004.
37. Smith, P.M. et al., The NOAA/NASA Pathfinder AVHRR 8-km land data set, *Photogrammetric Engineering and Remote Sensing*, 63, 12–31, 1997.
38. Rao, C.R.N., *Nonlinearity corrections for the thermal infrared channels of the advanced very high resolution radiometer: Assessment and recommendations*, NOAA Technical Report NESDIS-69, NOAA/NESDIS, Washington, DC, 1993.
39. Rao, C.R.N., *Degradation of the visible and near-infrared channels of the advanced very high resolution radiometer on the NOAAP9 spacecraft: Assessment and recommendations for corrections*, NOAA Technical Report NESDIS-70, NOAA/NESDIS, Washington, DC, 1993.
40. Kidwell, K., *NOAA Polar Orbiter Data User's Guide*, NCDC/SDSD, National Climatic Data Center, Washington, DC, 1991.
41. Fleig, A.J. et al., User's guide for the solar backscattered ultraviolet (SBUV) and the total ozone mapping spectrometer (TOMS) RUT-S and RUT-T data sets: October 31, 1978 to November 1980, *NASA Reference Publication No. 1112*, 1984.
42. NGDC (National Geophysical Data Center), 5 Minute gridded world elevation, *NGDC Data Announcement DA*, 93-MGG-01, Boulder, CA, 1984.

43. Kogan, F.N. and Zhu, X., Evolution of long-term errors in NDVI time-series: 1985–1999, *Advances in Space Research*, 28, 1, 149–54, 2001.

44. Lissens, G. et al., Development of cloud, snow, and shadow masking algorithm for VEGETATION imagery, in *Proceedings of IGARSS 2000*, Hawaii, 2000.

45. Tucker, C.J., Grant, D.M., and Dykstra, J.D., NASA's global orthorectified Landsat data set, *Photogrammetric Engineering & Remote Sensing*, 70, 313–22, 2005.

46. USGS (U.S. Geological Survey), *Digital Elevation Models, Data User Guide 5*, Reston, VI, 1993.

47. Verdin, K.L. and Greenlee, S.K., Development of continental scale digital elevation models and extraction of hydrographic features, in *Proceedings of the Third International Conference/Workshop on Integrating GIS and Environmental Modeling*, National Center for Geographic Information and Analysis, Santa Barbara, CA, 1996.

48. Verdin, K. and Jenson, S., Development of continental scale DEMs and extraction of hydrographic features, *Proceedings of the Third Conference on GIS and Environmental Modeling*, Santa Fe, NM, NCGIA, 1996.

49. Mitchell, T.D. et al., A comprehensive set of high-resolution grids of monthly climate for Europe and the globe: The observed record (1901–2000) and 16 scenarios (2001–2100), 2003, *Journal of Climate* (submitted).

50. Saatchi, S.S. et al., Mapping land cover types in the Amazon Basin using 1 km JERS-1, *International Journal of Remote Sensing*, 21, 1201–35, 2000.

51. Loveland, T.R. et al., An analysis of the IGBP global land-cover characterization process, *Photogrammetric Engineering and Remote Sensing*, 65, 9, 1021–32, 1999.

52. Olson, J.S., *Global ecosystem framework: Definitions*, USGS EROS Data Center Internal Report, Sioux Falls, SD, 1994.

53. Olson, J.S., *Global ecosystem framework: Translation strategy*, USGS EROS Data Center Internal Report, 1994.

54. Tou, J.T. and Gonzalez, R.C., *Pattern Recognition Principles*, Addison-Wesley, Reading, MA, 1975.

55. Leica. *Earth Resources Data Analysis System (ERDAS Imagine 9.2)*, Leica Geosystems A.G. Heinrich-Wild-Strasse, Heerbrugg, Switzerland, 2007.

56. Homayouni, S. and Roux, M., *Material Mapping from Hyperspectral Images Using Spectral Matching in Urban Area*, 2003i. Submitted to IEEE Workshop in Honor of Professor Landgrebe, Washington, DC, October, 2003.

57. Shippert, P., *Spectral and Hyperspectral Analysis with ENVI. ASPRS ENVI User's Group Notes,* Annual Meeting of the American Society of Photogrammetry and Remote Sensing, St. Louis, MI, April, 22–28, 2001.

58. Bing, Z. et al., Study on the classification of hyperspectral data in urban area, *SPIE*, 3502, 169–72, 1998.

59. Farrand, W.H. and Harsanyi, J.C., Mapping the distribution of mine tailings in the Coeur d'Alene River Valley, Idaho, through the use of a constrained energy minimization technique, *Remote Sensing of Environment*, 59, 64–76, 1997.

60. Granahan, J.C. and Sweet, J.N., An evaluation of atmospheric correction techniques using the spectral similarity scale, *IEEE 2001 International Geosciences and Remote Sensing Symposium,* 5, 2022–24, 2001.

61. Thenkabail, P.S. et al., Spectral matching techniques to determine historical land use/ land cover (LULC) and irrigated areas using time-series AVHRR pathfinder datasets in the Krishna river basin, India, *Photogrammetric Engineering and Remote Sensing*, 73, 9, 1029–40, 2007.

62. SAS Institute, *SAS/STAT User's Guide and Software Release, Version 6.12*, SAS Institute, Inc., Ed., Cary, NC, 2007.

63. Schwarz, J. and Staenz, K., Adaptive threshold for spectral matching of hyperspectral data, *Canadian Journal of Remote Sensing*, 27, 3, 216–24, 2001.
64. Kauth, R.J. and Thomas, G.S., The tasseled cap – A graphic description of the spectral-temporal development of agricultural crops as seen by Landsat, in *Proceedings of the Symposium on Machine Processing of Remotely Sensed Data*, Purdue University, West Layfayette, IN, 1976, 4B-41–4B-50.
65. Crist, E.P. and Cicone, R.C., A physically-based transformation of the Thematic Mapper data – the tassled cap, *IEEE Transactions on Geoscience and Remote Sensing*, GE-23, 256–63, 1984.
66. Tomlin, D., *Geographic Information Systems and Cartographic Modeling*, Prentice Hall College Div., Canada, 1990.
67. Tomlinson, R., *Thinking about Geographic Information Systems Planning for Managers*, ESRI Press, San Diego, CA, 2003.
68. Peuquet, D.J. and Marble, D.F., *Introductory Readings in Geographic Information Systems,* Taylor & Francis, New York, 1993.
69. DeFries, R. et al., Global land covers classifications at 8 km resolution: The use of training data derived from Landsat imagery in decision tree classifiers, *International Journal of Remote Sensing*, 19, 3141–68, 1998.
70. Olson, J.S. and Watts, J.A., *Major World Ecosystem Complex Map,* Oak Ridge National Laboratory, Oak Ridge, TN, 1982.
71. Thenkabail, P.S. et al., Sub-pixel irrigated area calculation methods, *Sensors Journal (Special Issue: Remote Sensing of Natural Resources and the Environment [Remote Sensing Sensors*, A.M. Melesse, Ed.]), 7, 2519–38, 2007. http://www.mdpi.org/sensors/papers/s7112519.pdf.
72. MoWR (Ministry of Water Resources), *3rd Census of Minor Irrigation Schemes (2000–01)*, Ministry of Water Resources, Government of India, New Delhi, 2005. CD and http://mowr.gov.in/micensus/mi3census/index.htm.
73. Biradar, C.M. et al., A global map of rainfed cropland areas (GMRCA) at the end of last millennium using remote sensing. *International Journal of Applied Earth Observation and Geoinformation*. doi10.1016/j.jag.2008.11.002. January, 2009.
74. Dheeravath, V. et al., Irrigated areas of India derived from MODIS 500 m data of years 2001–2003, *Photogrammetric Engineering and Remote Sensing,* 2008, (submitted).
75. Velpuri, M. et al., Methods for mapping irrigated areas using Landsat ETM + 30 meter, SRTM 90 meter, and MODIS 500 meter time-series data taking Krishna river basin India, *Photogrammetric Engineering and Remote Sensing*, 2008, (submitted).
76. DeFries, R.S. and Townshend, J.R.G., NDVI-derived land cover classifications at a global scale, *International Journal of Remote Sensing*, 15, 3567–86, 1995.
77. CA (Comprehensive Assessment of Water Management in Agriculture), Water for food and water for life: A comprehensive assessment of water management in agriculture, in *Comprehensive Assessment of Water Management in Agriculture,* D. Molden, Ed., Earthscan, London, and International Water Management Institute, Colombo, 2007.

Section III

GIAM Mapping Section for Selected Global Regions

4 Uncertainty of Estimating Irrigated Areas in China

Songcai You and Shunbao Liao
Chinese Academy of Sciences

Suchuang Di
Graduate University of Chinese Academy of Sciences

Ye Yuan
China University of Mining & Technology

CONTENTS

4.1 INTRODUCTION, BACKGROUND, AND RATIONALE GOALS

4.1.1 REVIEW OF IRRIGATION DEVELOPMENT IN CHINA

Agriculture has long been considered the basis of the national economy in China. Historically, China's agricultural production and cultural development were closely related to the development of irrigation. The irrigation history in China can be traced back to 4000 years. The world-renowned Dujiangyan irrigation system was built 2250 years ago and still irrigates over 330,000 hectares (ha) of rice [1].

With the rapid economic development, China is facing water shortages and environmental deterioration, and hence China's irrigation development faces serious challenges. Although the total irrigation water supply in relation to the total water resource is decreasing, agriculture is the top water consumer in all sectors. In recent years, China has rebuilt 208 large irrigation facilities focused on water storage. Irrigation has played, and will continue to play, an important role in China's social and economic development.

In 1949, China's irrigated area was only 15.9 million hectares (Mha). Since then, irrigation has rapidly developed. China's farmland irrigation development can be roughly divided into three stages (Figure 4.1) [2,3].

- Phase I: 1949–1980
  China focused mainly on the construction of irrigation facilities, such as reservoirs, water diversion projects, and wells. Its irrigated area increased from 15.93 Mha in 1949 to 48.89 Mha in 1980.

- Phase II: 1981–1990
  Institutional reform was carried out in the rural areas, including reform of the agricultural structure and investment mechanism, with inputs to national irrigation development and facility maintenance heavily reduced. Meanwhile, as there was a tremendous increase in water demand in the industrial and urban-living sectors, irrigated area decreased from 48.89 Mha in 1980 to 48.39 Mha in 1990, with an annual decrease of 50,000 ha per year.

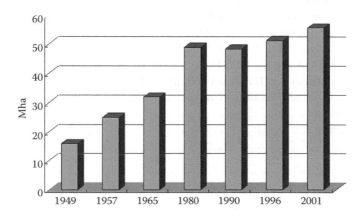

FIGURE 4.1   Irrigated area changes in China.

- Phase III: 1991 to date

  China mainly focuses on the development of water-saving irrigation, improved irrigation management, revising management institutions, and renovation of large-scale irrigation facilities. Investment in the irrigation sector has increased, and the irrigated area increased from 48.39 Mha in 1990 to 55.53 Mha in 2001, at 690,000 ha per year.

  China has 85,136 reservoirs with a total water storage capacity of 528 Bm^3. The large irrigation facilities provide distribution in China as shown in Figure 4.2. In 2000, the total consumption in China was 549.8 Bm^3, of which 360 Bm^3 were for irrigation [4].

  Located in the Asian monsoonal area, China is strongly impacted by the monsoonal climate. The temporal and spatial distribution of rainfall is uneven. The mean annual precipitation decreases from 1600 mm in the Southeast to less than 200 mm in the Northwest. Over 80% of the rainfall occurs from June to September. The spatial distribution of water resources is not the same as that of land resources. South China, with 38% of the total area, possesses 80% of the total water resources, while northern China, with 62% of the total area, possesses only 20% of the corresponding resources. Because of the uneven nature of the temporal and spatial distribution of rainfall, irrigation is essential in China's agricultural production. According to water demand, China's irrigation can be broadly divided into three regions: perennial irrigation district, supplementary irrigation district, and rice irrigation district (Figure 4.3).

- Perennial irrigation district

  The perennial irrigation district includes the North China Plain and parts of Northeast China and Northwest China. The mean annual rainfall in this district is less than 400 mm. The amount of annual rainfall and its temporal

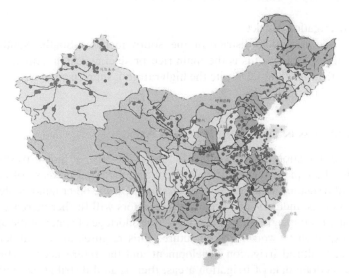

FIGURE 4.2 (See color insert following page 224.) Distribution of large irrigation area in China.

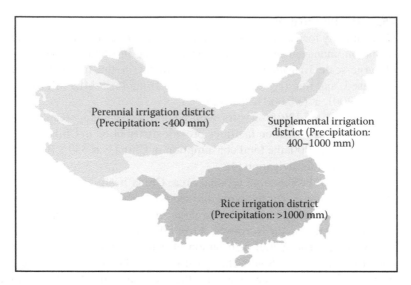

FIGURE 4.3 Precipitation and irrigation in China.

distribution do not meet the demand of crops growth. There would be no agricultural production without irrigation in this region.

- Supplemental irrigation district
  The supplementary irrigation district includes the North China Plain and most parts of Northeast China. The annual rainfall in this district varies from 400 to 1,000 mm. Rainfall distribution is very uneven due to the strong monsoonal climate. Irrigation water requirements change a lot in different years. Irrigation is necessary to achieve the high crop yield.

- Rice irrigation district
  The rice irrigation district in the south includes South, Southeast, and Southwest China. This is the main rice-producing region. Annual rainfall is more than 1000 mm. Despite the high rainfall, supplemental irrigation is necessary during droughts.

4.1.2 Problems in Estimating Irrigated Areas

Agricultural irrigation water accounts for about 70–80% of global freshwater use. Along with the rapid economic development and the rapid increase of population, agricultural irrigation water demand is growing. On the other hand, water demand from ecosystems, industry, and urban-living sectors will further increase the diversion of water used in agriculture. Therefore, the shortage of water in the agriculture sector is expected to worsen in the future. This requires us to reunderstand the status of agricultural irrigation development and the future needs, which means that more accurate data of irrigation areas, their spatial distribution, and dynamic changes are needed to assess different types of irrigation and regional characteristics, and to schedule irrigation planning in a more rational way.

Traditional methods of collecting data on an irrigation area are bottom-up in China. This time-consuming work needs a massive input of manpower. Updating of irrigation statistical data takes a long time. Application of remote-sensing (RS) technology will help us assess the distribution and dynamic changes of the irrigated area.

In recent years, the International Water Management Institute (IWMI), along with other international organizations, carried out a global irrigated area study. According to the IWMI's Global Irrigated Area Mapping (GIAM), at a resolution of 10 km, the total irrigated area at the end of the last century in China, taking into consideration its cropping intensity, was 151 Mha. This accounted for 31.5% of the world's irrigated area. The irrigated areas for the first crop and the second crop in the growing season are 76 and 68 Mha, respectively, and the irrigated area for the continuous crop is 7 Mha. The effective irrigated area or net irrigated area is 108 Mha. These results are inconsistent with current data in China.

4.2 METHODS

4.2.1 STUDY AREA AND METHODS

The study area is China; however, there are no data for the Taiwan province, Hong Kong, and Macau. The nominal year 2000 was chosen as the baseline year to compare the data from different sources and to identify the problems in estimating irrigated areas in China.

4.2.2 DATA SOURCES

The analysis uses different sources of data, so it is necessary to give an introduction on data sources and the role of different organizations in China.

1. Data from the National Bureau of Statistics (NBS)
   Data published by the NBS are always considered as official data. However, the NBS does not collect all data and the specific data are provided by related ministries and organizations.
   Since 1997, the NBS has not published information on available arable land in its yearly China Statistics Yearbook; therefore, data from *Data of China Agricultural Statistics* [5,6] were used to analyze the reliability of the NBS data.

2. Data from the Ministry of Agriculture (MOA)
   The MOA has its own data-collecting system. In general, information on the irrigated area published by the MOA is similar to that published by the NBS, but because the NBS has its own survey system of sampling, data on the irrigated area published by the MOA and the NBS are slightly different in some years or some regions.

3. Data from the Ministry of Land Resources (MLR)
   The MLR has its own survey system to collect data on the irrigated area, mainly focused on land resources. The MLR's definition of arable land is slightly different from other definitions.

4. Data from the Ministry of Water Resources (MWR)
   As not all irrigation facilities were constructed by the MWR, there exists the potential omission of data on some irrigated areas. For example, small irrigation projects constructed by local farmers may not be surveyed by the MWR, but they are reported to the local agricultural administration.

5. Data from the Chinese researchers' reports
   These data are published by researchers rather than by organizations. Their contribution is on the methodology of estimating arable land area and to clarify reasons for the informational gap among four government organizations.

4.2.3 SOME DEFINITIONS USED IN CHINA

Some terms related to irrigated areas were defined in China and should be clearly introduced, as a common language is important in comparing the data obtained from different sources.

1. Effective irrigated area
   An effective irrigated area refers to the flatness of land with a water supply facility. Irrigation can be guaranteed in mean climatic conditions. Under normal circumstances, the effective irrigated area should be equal to irrigated paddy field area plus irrigated non-paddy field area. This is the general definition, but with the following revisions [7]:
   (a) An arable plot of land, with an irrigation facility, is not irrigated due to abundant water supply from rainfall, or planted crops do not needt to be irrigated; in either case, land is still accounted for as an effective irrigated area.
   (b) An arable plot of land is not accounted for as an effective irrigated area when key components of the irrigation system are missing, e.g., wells without pumps or motors and reservoirs without irrigation canals.
   (c) An arable plot of land irrigated by gravity flow water should not be counted as an effective irrigated area because the gravity flow is not considered as a facility. This is quite normal in northern China, especially along the Yellow river in the Henan Province.
   (d) An arable plot of land, without an irrigation facility, but that fully depends on rainfall, should not be counted as an effective irrigated area, and this is quite normal in South China.
   (e) An arable plot of land, without irrigation facilities, but irrigated temporarily for planting, should not be accounted for as an effective irrigated area. For example, even when pumps are used to withdraw water from the river to irrigate land for only planting, this land is not counted as an irrigated area.
   (f) An arable plot of land should not be counted as an effective irrigated area when irrigation facilities are damaged.

2. Design irrigation area and guaranteed irrigation area
   According to *Technical Terminology for Irrigation and Drainage*, published by the MWR, the design irrigation area is the capability of an irrigation facility to provide irrigation at a given guaranteed rate; however, the
   guaranteed rate is not presented. The guaranteed irrigation area takes into
   consideration the guaranteed rate and cropping intensity. As the cropping
   intensity changes along with the changes in the cropping system, so the
   guaranteed irrigation area should not be a constant. It should be less than
   the design irrigation area.

3. Arable land
   From the above definition, the important key phrase about irrigation area is
   "flatness of land" with "irrigation facility." However, the definition of arable
   land is not clear, as two conflicting definitions have been used.
   (a) Definition 1: According to *Specification of Land Use Survey* [8], arable
        land includes yearly cultivated land, newly opened wasteland, cultivated
        land fallowed for less than 3 years, crop-fruit or crop-forest intercrop-
        ping land, and shoal, but excludes land for mulberry trees, tea plants,
        orchards, forests, reeds, and natural grassland, etc.
   (b) Definition 2: According to the MLR [9], arable land includes croplands,
        orchards, forests, grazing lands, and others.
   Clearly, there are some conflicts between the above two definitions. The first is
   defined by the National Committee for Agricultural Regionalization, which is
   led by the Ministry of Agriculture, and the second is defined by the MLR.

4.3 RESULTS AND DISCUSSION

4.3.1 NBS DATA

The NBS data clearly show that the area of arable land has risen dramatically
(Table 4.1) at the country level, with an increase of 36.2%. At provincial level, except
for Beijing, arable land area in all provinces has increased; the highest was in the

TABLE 4.1
Changes in Arable Land Area between 1996 and 2001 ('000 ha)

No.	Province	1996	2001	Changes (%)
1	Anhui	4,280	5,972	39.5
2	Beijing	396	344	−13.1
3	Fujian	1,196	1,435	19.9
4	Gansu	3,486	5,025	44.1
5	Guangdong	2,304	3,272	42
6	Guangxi	2,632	4,408	67.5
7	Guizhou	1,839	4,904	166.6
8	Hainan	429	762	77.5
9	Hebei	6,499	6,883	5.9

(Continued)

TABLE 4.1   (Continued)

No.	Province	1996	2001	Changes (%)
10	Henan	6,786	8,110	19.5
11	Heilongjiang	9,175	11,773	28.3
12	Hubei	3,349	4,950	47.8
13	Hunan	3,239	3,953	22
14	Jilin	3,959	5,578	40.9
15	Jiangsu	4,435	5,062	14.1
16	Jiangxi	2,302	2,993	30
17	Liaoning	3,384	4,175	23.4
18	Inner Mongolia	5,924	8,201	38.4
19	Ningxia	814	1,269	56
20	Qinghai	590	688	16.6
21	Shandong	6,679	7,689	15.1
22	Shanxi	3,624	4,589	26.6
23	Shaanxi	3,359	5,141	53
24	Shanghai	287	315	9.7
25	Sichuan	6,165	9,169	48.7
26	Tianjin	426	486	14
27	Tibet	228	363	59.1
28	Xinjiang	3,176	3,986	25.5
29	Yunnan	2,889	6,422	122.2
30	Zhejiang	1,614	2,125	31.7
31	Taiwan	N/A	N/A	–
32	Hong Kong	N/A	N/A	–
33	Macao	N/A	N/A	–
	National/Total	95,467	130,039	36.2

Source: Ministry of Agriculture of China, *Data of China Agricultural Statistics* (in Chinese), China Agricultural Press, Beijing, 1996 and 2001.

Guizhou Province, which increased by 166.6%, followed by the Yunnan Province, which increased by 122.2% (Figure 4.4). Such a change does not mean that although actual land increases, it reflects the change of survey methodology and the failure of previous survey methodologies. It also implies that the statistical data on arable land published before 1996 are no longer suitable for comparison of studies.

4.3.2 CHINESE RESEARCHERS' REPORT

Liu and Buhe [10] made an analysis of land-use change in China, using RS data platforms and Geographic Information System (GIS) tools, in order to explore and develop a methodology for studying land-use change based on modern techniques. They estimated significantly larger areas (Table 4.2) amounting to a total of 138 Mha, about 6.2% and 44.6% more than MOA values for 2001 and 1996, respectively. The estimates are derived from the interpretation of thematic mapper (TM) images and both methodology and basic data are believed to be reliable.

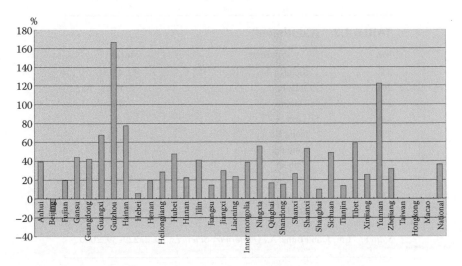

FIGURE 4.4 Changes in arable land between 1996 and 2001 by statistical data.

TABLE 4.2
Arable Land Area ('000 ha)

No.	Province	Total
1	Anhui	6,221
2	Beijing	523
3	Fujian	1,488
4	Gansu	5,342
5	Guangdong	3,475
6	Guangxi	4,175
7	Guizhou	4,148
8	Hainan	740
9	Hebei	6,609
10	Henan	8,180
11	Heilongjiang	13,522
12	Hubei	4,783
13	Hunan	4,314
14	Jilin	6,020
15	Jiangsu	5,556
16	Jiangxi	2,958
17	Liaoning	5,426
18	Inner Mongolia	7,833
19	Ningxia	1,327
20	Qinghai	605
21	Shandong	8,170
22	Shanxi	4,633
23	Shaanxi	5,521
24	Shanghai	366

(Continued)

TABLE 4.2 (Continued)

No.	Province	Total
25	Sichuan	11,441
26	Taiwan	735
27	Tianjin	550
28	Tibet	361
29	Hong Kong	0
30	Xinjiang	5,703
31	Yunnan	4,823
32	Zhejiang	2,505
33	Macao	0
	Total	138,050

Source: Liu, J. and Buhe, A., *Quaternary Sciences*, 20, 2, 229–39, 2000.

4.3.3 FOOD AND AGRICULTURE ORGANIZATION OF THE UNITED NATIONS (FAO) DATA

The FAO reported the total irrigated area [11,12] to be 53,820,300 ha (Table 4.3) in 2000. Comparing it with statistical data provided by the NBS, the highest difference was 8% for the Heilongjiang Province, followed by Qinghai (7.8%), Inner Mongolia (6.4%), and Ningxia (5.9%). For other provinces, the differences were less than 2%. At country level, the difference was less than 1.1%. It may be deduced from these differences that the FAO's data were mainly dependent on statistical data, though it used the 1:1 M land-use atlas [13] to revise the statistical data. Therefore, the reliability of the FAO data depends on that of the NBS data, which is now dubious.

Some comments on this atlas [13] are that it was based on satellite images and large-scale aerial photographs taken in the 1980s and that there still exists mismatching of resolution of the basic data. Also, the different sheets of the atlas were edited by different researchers from different research organizations, which potentially include interpreter errors.

4.3.4 ASSESSMENT TO ARABLE LAND AREA PROVIDED BY THE MLR AND THE MOA

As there are several sources of data, different research organizations may use different data sets provided by different organizations, which leads to difficulty in evaluating different studies. As access to the data is difficult, the MLR [14] and MOA statistical data were used to assess their consistency, although the two data sets have different reference years. Table 4.4 shows that there was only a 0.6% difference between them. But at the provincial level, data on arable land provided by the MOA were 5.1% and 4% more than those provided by the MLR in Inner Mongolia and Shanghai, while in Xinjiang, the situation was the reverse, and the difference was 4%. These are not significant differences and indicate that the same or similar basic data were used with slight modifications, as the same terminology has different definitions by the MLR and the MOA.

TABLE 4.3
Comparison of FAO Data and State Statistical Data ('000 ha) for Irrigated Land Areas

| | | Total Area Available for Irrigation | | |
No.	Province	FAO	Mean Value of NBS Data from 1998–2000	Difference (%) Baseline: NBS Data
1	Anhui	3,197	3,150	1.5
2	Beijing and Tianjin	681	668	2
3	Fujian	940	935	0.6
4	Gansu	982	973	0.9
5	Guangdong	1,479	1,498	−1.3
6	Guangxi	1,502	1,483	1.3
7	Guizhou	653	645	1.3
8	Hainan	180	176	2.1
9	Hebei	4,482	4,438	1
10	Henan	4,725	4,629	2.1
11	Heilongjiang	2,032	1,882	8
12	Hubei	2,073	2,122	−2.3
13	Hunan	2,678	2,673	0.2
14	Jilin	1,315	1,286	2.2
15	Jiangsu	3,901	3,881	0.5
16	Jiangxi	1,903	1,901	0.2
17	Liaoning	1,441	1,429	0.9
18	Inner Mongolia	2,372	2,229	6.4
19	Ningxia	399	377	5.9
20	Qinghai	211	196	7.8
21	Shandong	4,825	4,953	−2.6
22	Shanxi	1,105	1,089	1.5
23	Shaanxi	1,308	1,307	0.1
24	Shanghai	286	280	2.3
25	Sichuan	3,094	3,049	1.5
26	Taiwan	No data	No data	−
27	Tibet	157	155	1.6
28	Hong Kong	2	No data	
29	Xinjiang	3,095	3,048	1.5
30	Yunnan	1,403	1,376	2
31	Zhejiang	1,403	1,402	0.1
32	Macao	No data	No data	−
	Total	53,823	53,226	1.1

Note: FAO data from http://www.fao.org/ag/agl/aglw/aquastat/irrigationmap/cn/index.stm.

TABLE 4.4

Comparison of MLR and MOA for Arable Land Areas ('000 ha)

No.	Provinces	MLR 1999	MOA 2001	Difference (%) Baseline: MOA
1	Anhui	5,961	5,972	0.2
2	Beijing	340	344	1.3
3	Fujian	1,405	1,435	2.1
4	Gansu	5,027	5,025	0
5	Guangdong	3,223	3,272	1.5
6	Guangxi	4,409	4,408	0
7	Guizhou	4,795	4,904	2.2
8	Hainan	764	762	−0.2
9	Hebei	6,849	6,883	0.5
10	Henan	8,096	8,110	0.2
11	Heilongjiang	11,768	11,773	0
12	Hubei	4,931	4,950	0.4
13	Hunan	3,932	3,953	0.5
14	Jilin	5,580	5,578	0
15	Jiangsu	5,033	5,062	0.6
16	Jiangxi	2,974	2,993	0.7
17	Liaoning	4,170	4,175	0.1
18	Inner Mongolia	7,785	8,201	5.1
19	Ningxia	1,274	1,269	−0.4
20	Qinghai	688	688	0.1
21	Shandong	7,667	7,689	0.3
22	Shanxi	4,563	4,589	0.6
23	Shaanxi	5,044	5,141	1.9
24	Shanghai	303	315	4
25	Sichuan[a]	9,120	9,169	0.5
26	Tianjin	485	486	0.1
27	Tibet	367	363	−1.1
28	Xinjiang	4,144	3,986	−4
29	Yunnan	6,405	6,422	0.3
30	Zhejiang	2,106	2,125	0.9
31	Hong Kong	N/A	N/A	−
32	Macao	N/A	N/A	−
33	Taiwan	N/A	N/A	−
	Total	129,205	130,039	0.6

[a] The value of Chongqing is added to that of Sichuan.

4.3.5 IRRIGATED LAND AREA COMPARISON BETWEEN GIAM AND THE NBS DATA

For this analysis, the version "2.0 release" (update May 10, 2007) of IWMI's GIAM and associated products and data were used.

The GIAM products are produced using time series consisting of (a) Advanced Very High Resolution Radiometer (AVHRR) 10-km monthly from 1997 to 1999, (b) Système Pour l'Observation de la Terre (SPOT) 1-km monthly for 1999, (c) GTOPO30 1-km elevation, (d) Climatic Research Unit (CRU) 50-km grid monthly precipitation from 1961 to 2000, and (e) AVHRR-derived 1-km forest cover.

The Total Area Available for Irrigation (TAAI) from GIAM (version 2) for China is 112,471,300 ha. One can see that there are large differences in provincial estimates between the GIAM output and the MLR/MOA data. In some provinces, the TAAI from GIAM is even bigger than the arable land area. The conclusion is that GIAM needs more improvement in interpreting RS data, with better information on the cropping system, crop calendar, and geographical elements.

As stated above, traditional methods of collecting statistical data were bottom-up, and potentially there is inconsistency in data, and the potential for misunderstanding by each surveyor involved. Therefore, application of high-resolution RS data, combined with local knowledge and data on local geographical elements, under support of GIS technology, may be a good method in mapping irrigated area. Here, we take output by Liu and Buhe [10] as baselines to analyze the accuracy of GIAM. Table 4.5 shows the comparison of irrigated land areas at provincial level for China ('000 ha).

TABLE 4.5
Comparison of Irrigated Land Areas at Provincial Level for China ('000 ha)

No.	Province	GIAM A	Difference (%)[a] (A–B)/B×100	(A–C)/C×100	(A–D)/D×100
1	Anhui	7617	27.8	27.5	22.4
2	Beijing	532	56.6	54.6	1.7
3	Fujian	352	−74.9	−75.4	−76.3
4	Gansu	1248	−75.2	−75.2	−76.6
5	Guangdong	1661	−48.5	−49.2	−52.2
6	Guangxi	5079	15.2	15.2	21.7
7	Guizhou	4371	−8.9	−10.9	5.4
8	Hainan	307	−59.8	−59.7	−58.5
9	Hebei	6903	0.8	0.3	4.4
10	Henan	9728	20.2	19.9	18.9
11	Heilongjiang	6274	−46.7	−46.7	−53.6
12	Hubei	6118	24.1	23.6	27.9
13	Hunan	6120	55.6	54.8	41.9
14	Jilin	4460	−20.1	−20.1	−25.9
15	Jiangsu	6736	33.8	33.1	21.2
16	Jiangxi	3133	5.4	4.7	5.9
17	Liaoning	3185	−23.6	−23.7	−41.3
18	Macao	No data	–	–	–
19	Inner Mongolia	3451	−55.7	−57.9	−55.9
20	Ningxia	394	−69.1	−68.9	−70.3
21	Qinghai	153	−77.8	−77.8	−74.7

(*Continued*)

TABLE 4.5　(Continued)

No.	Province	GIAM A	Difference (%)[a] (A–B)/B×100	(A–C)/C×100	(A–D)/D×100
22	Shandong	9910	29.3	28.9	21.3
23	Shanxi	3672	–19.5	–20	–20.7
24	Shaanxi	2615	–48.2	–49.1	–52.6
25	Shanghai	374	23.5	18.6	2.2
26	Sichuan	3559	–23.1	–23.5	–68.9
27	Taiwan	491	–	–	–33.2
28	Tianjin	605	24.8	24.6	10
29	Tibet	3	–99.3	–99.3	–99.2
30	Hong Kong	1	–	–	–
31	Xinjiang	3826	–7.7	–4	–32.9
32	Yunnan	1018	–84.1	–84.1	–78.9
33	Zhejiang	1870	–11.2	–12	–25.3

[a] Data from MLR and MOA in Table 4.4 taken as B and C, respectively. Independent data from Liu and Buhe, 2000, taken as D.

TABLE 4.6
Comparison between the Data of the Chinese Academy of Sciences and Statistical Data

No.	Province	Land Type	RS-based Estimated Area ('000 ha)	MOA[a] ('000 ha)
1	Fujian	Paddy land	1,443.50	1,434.70
		Dryland	328.6	
		Total	1,772.10	
2	Hebei	Paddy land	253	6,883.30
		Dryland	10,086.90	
		Total	10,339.90	

[a] From MOA, *Data of China Agricultural Statistics* (in Chinese), China Agricultural Press, Beijing, 2001

4.3.6　COMPARISON BETWEEN THE STATISTICAL DATA WITH THE DATA OF THE CHINESE ACADEMY OF SCIENCES

The Data Center for Resources and Environmental Sciences, Chinese Academy of Sciences, made an RS data-based land-use map in 2000. According to their experiences, RS data-based maps have large errors in hilly and mountainous areas and have fewer errors in flat areas. As these data cannot be fully accessed freely, only data from two provinces (Fujian and Hebei) were provided. Fujian and Hebei were taken as samples of typical hilly and mountainous region and flatness regions, respectively. The former is located in South China and the latter in the North China Plain. Table 4.6

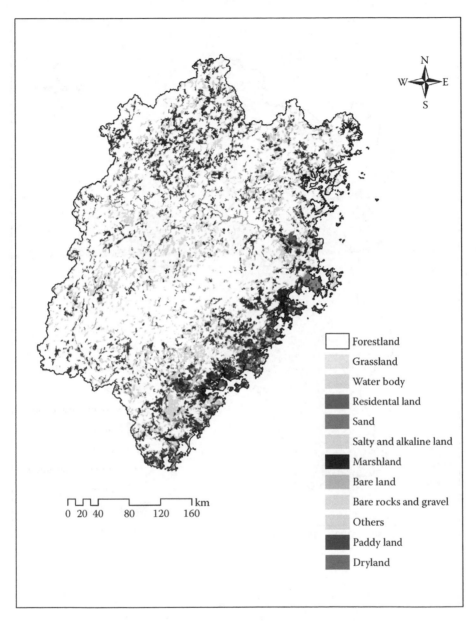

FIGURE 4.5   **(See color insert following page 224.)** Land-use map at scale 1:1 M of Fujian.

shows that RS-based estimates were higher than statistical data. Figures 4.5 and 4.6 shows the land-use map at a scale 1:1,000,000 of Fujian and Hebei, respectively.

4.4   CONCLUSIONS

Based on a comparison between state statistical data, GIAM data, FAO data, and data from Chinese researchers' reports, some preliminary conclusions can be made:

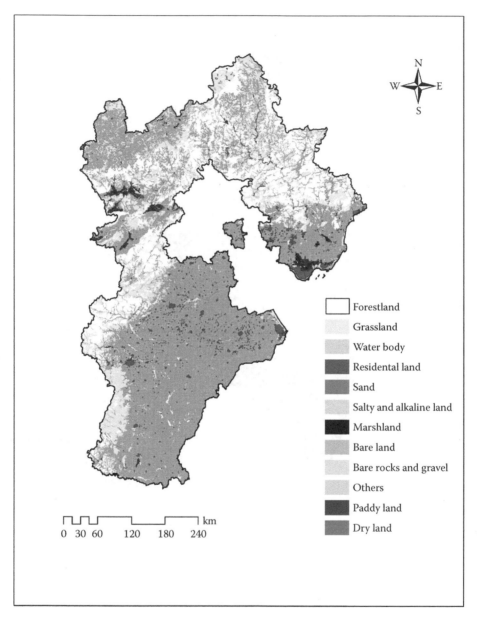

FIGURE 4.6   (See color insert following page 224.) Land-use map at scale 1:1 M of Hebei.

1. The statistical data reflect the historical tendency, but they are not reliable due to conditions mentioned in the text.
2. FAO data were found to be of the same level of reliability as the NBS data.
3. Data from different official organizations are inconsistent to some extent. It is difficult to say which one is better than another. These data must be used

very carefully due to differences in the collection methods, definitions, and in other details.

4. The same terminology may have different meanings for staff in different organization.

ACKNOWLEDGMENTS

This study was jointly supported by the project (Grant No: 40671118, 1997–1999) funded by the National Natural Science Foundation of China and Knowledge Innovation Program (III) of the Chinese Academy of Sciences. The authors appreciate comments from Professor Gaohuang Liu, Professor Dafang Zhuang and data support from Zehui Li. Thanks are also expressed to RESDC (http://www.resdc.cn) for providing land-use map (2000) for illustration of Hebei and Fujian Provinces' estimates from remote sensing.

REFERENCES

1. Yao, H., *History of Water Conservancy in China* (in Chinese), IAHS Publication, Beijing, 1987.
2. Division of Rural Water of Ministry of Water Resources, *Development History of Farmland Irrigation in New China* (in Chinese), IAHS Publication, Beijing, 1999.
3. Xu, P., Development history of farmland irrigation in China, *Journal of Henan Vocation-Technical Teachers College* (in Chinese), 29, 2, 29–32 2001.
4. Gao, Z., The development and role of irrigation in China, *Water Conservancy Economics* (in Chinese), 24, 1, 36–39, 2006.
5. MOA (Ministry of Agriculture), *Data of China Agricultural Statistics* (in Chinese), China Agricultural Press, Beijing, 1996.
6. MOA, *Data of China Agricultural Statistics* (in Chinese), China Agricultural Press, Beijing, 2001.
7. Liu, C.X., Liu, K., and Jin, Z., Chapter 14: Agriculture – (2) explanations to main statistical index. In: Liu et al. (eds.), *How to Use Statistical Yearbook* (in Chinese), China Statistical Yearbook Press, Beijing, 2000, 153–55.
8. National Committee for Agricultural Regionalization, Specification of Land Use Survey (in Chinese), 1984, http://gtt.xinjiang.gov.cn/10010/10010/00004/00001/00004/00001/2005/11052.htm
9. MLR (Ministry of Land Resources of China), *Notes to Land Classification System* (in Chinese), Beijing, 2001. http://www.dtgtzy.gov.cn/zcfg/ShowArticle.asp?ArticleID=1803
10. Liu, J. and Buhe, A., Study on spatial-temporal feature of modern land-use change in China: Using remote-sensing techniques (in Chinese), *Quaternary Sciences*, 20, 2, 229–39, 2000.
11. Siebert, S. et al., Development and validation of the global map of irrigation areas, *Hydrology and Earth System Sciences*, 9, 535–47, 2005.
12. Siebert, S. et al., The digital global map of irrigation areas – development and validation of map version 4, Conference on International Agricultural Research for Development, Tropentag 2006, University of Bonn, October 11–13, 2006.
13. Wu, C., (Ed.), *Land-Use Map of China (1:1,000,000 scale)* (in Chinese), Science Press, Beijing, 1990.
14. MLR, *China Land and Natural Resources Yearbook 2000* (in Chinese), The Geological Publishing House, Beijing, 2000.

5 Irrigated Areas of India Derived from Satellite Sensors and National Statistics: A Way Forward from GIAM Experience

Obi Reddy P. Gangalakunta
National Bureau of Soil Survey and Land Use Planning (ICAR)

Venkateswarlu Dheeravath
United Nations Joint Logistics Centre

Prasad S. Thenkabail
U.S. Geological Survey

G. Chandrakantha
Kuvempu University

Chandrashekhar M. Biradar
University of Oklahoma

Praveen Noojipady
University of Maryland

Manohar Velpuri
South Dakota State University

Maji Amal Kumar
National Bureau of Soil Survey and Land Use Planning

CONTENTS

5.1   INTRODUCTION

India continues to be a predominantly agrarian economy with the majority of its population depending on agriculture for their livelihood. The contribution of the agriculture sector is about 25% of the gross domestic product (GDP). India, with a geographical area of about 328 Mha, experiences diversified climatic conditions. Spatiotemporal variability in rainfall and the availability of groundwater resources have a direct impact on the irrigation system of the country. In the last 50 years, there has been a phenomenal expansion in irrigation development in India, resulting in an increase in gross irrigation area from 22.5 Mha in 1951 to about 76.3 Mha by the end of the year 2001 [1]. At the same time, the Net Irrigated Areas (NIA) in the country has increased from 21.2 Mha to about 57.2 Mha. India has about 27.5% of the global Gross Irrigated Area (GIA), which is second only to China with a corresponding ratio of 31.5% [2,3]. Many experts opined that, to meet the ever-increasing demand of the food grains in the country, a major contribution to the increase in production is attributable to the expansion of the irrigation facility. As per the estimates of the Ministry of Agriculture (MoA), a 1% increase in the irrigated area would increase production by 4 million tons (if the area and production technology and fertilizer use increase at the same rate as in the recent past). Keeping in view the importance of irrigated agriculture in the national food production, the dynamics of irrigated areas needs to be monitored periodically in order to map the irrigated areas for their effective management. The complexities are more in mapping irrigated areas in India, where subsistence agriculture, diversity in crop types and calendar, and small landholdings prevail.

Spatial data on the distribution of irrigated areas and their dynamics are a prerequisite for effective planning, management, and monitoring at the local, regional, and national level for overall agricultural development in the country. In order to accomplish this, it is first necessary to obtain reliable data on spatiotemporal patterns of irrigated areas. As a part of agricultural statistics, the Directorate of Economics and Statistics (DES), under the MoA, reports the irrigated area statistics every year. Traditionally, village *patwari* (grassroots-level revenue department officials) report irrigation statistics as a part of agricultural statistics that are, in turn, compiled at different levels, like the village, tehsil, district, state, and national. It is a rigorous, time-consuming, and resources-intensive process. The major limitation in the existing statistical data lies in the visualization of the spatial pattern of irrigated areas of any given administrative unit. The other limitations are the lack of representation of irrigation intensity and information on both the irrigated crop types and the precise location of irrigated areas. Given the huge implications of irrigated areas on water use, food production, and population growth, the availability of precise irrigated area maps and statistics and information on their spatial distribution are invaluable in the planning and management of irrigated agriculture in the country.

In view of the existing status of water resources and increasing demands for water to meet the requirements of the growing population of the country, a holistic and well-planned, long-term strategy is needed for sustainable water resources management in the irrigation sector in India. The availability of spatiotemporal data on sourcewise irrigated areas will be of immense help in monitoring and planning the different irrigation sources in the country. Recent studies indicate that the number of tube wells in India has increased from about 100,000 in the early 1960s to anywhere between 19 and 26 million by early 2000 [4]. This has resulted in an increase in the area irrigated by tube wells alone to be about 42% of all irrigated areas in India. Sinha [5] reported that the irrigated areas in India had reached 100 Mha by the year 2000. Shah et al. [6,7] have observed about the massive overexploitation of groundwater in most parts of India, and have also mentioned that most of the groundwater potential is already exploited.

Recently, there have been two major efforts in mapping irrigated areas at national and global scales. One is Global Irrigated Area Mapping (GIAM) by the International Water Management Institute (IWMI) [3,8] and the other is the Global Map of Irrigated Area (GMIA) by the Food and Agriculture Organization of the United Nations and the University of Frankfurt (FAO/UF) [9,10]. The IWMI GIAM (http://www.iwmigiam.org) effort is overwhelmingly remote-sensing-based and the FAO/UF GMIA (http://www.fao.org/AG/agL/aglw/aquastat/main/index.stm) effort is based on national statistics combined with GIS techniques. In the present study, an attempt has been made to analyze the irrigated areas of India, derived from national statistics and based on satellite sensors from the GIAM project. It helps to diagnose their similarities and dissimilarities to strategize the way forward for systematic mapping of irrigated areas in the country.

5.2 BACKGROUND AND RATIONALE

Keeping in view the diversified agroclimatic conditions, rainfall patterns, available land, and water resources, it is necessary to accurately map irrigated areas in India in order to properly understand its contribution to food production and to estimate its water use, as competition for water increases with rising urban and industrial needs. Though the green revolution has brought additional areas under irrigation to meet the food demands of the increasing population, the challenges of maintaining the momentum in the growth of food production continues to receive serious attention from all concerned with agriculture. In view of the available limited land and water resources, it is time we knew the accurate irrigated areas and their spatial distribution, dynamics, and irrigation sources, at regular intervals, for their effective management and periodical monitoring. Even though irrigated area statistics are available at different administrative levels, the major limitation lies in visualizing the irrigated areas spatially in any given administrative unit. To overcome this limitation, satellite remote sensing offers a relatively cheap, replicable, and accurate technology to generate spatiotemporal irrigated area maps and statistics. Realizing the utility of remote-sensing data applications in agriculture, the National Remote Sensing Agency (NRSA) and the Space Application Centre (SAC), under the Department of Space, have initiated some of the pilot projects at national level, under a joint Indian Space Research Organization (ISRO)-Indian Council of Agricultural Research (ICAR) collaboration. However, hardly any effort has been made to map exclusively irrigated areas under different irrigation sources at the national scale.

In the recent past, IWMI initiated the GIAM project under the Comprehensive Assessment (CA) of Water Management in Agriculture, to map the irrigated areas of the globe, using the latest time-series National Oceanic and Atmospheric Administration Advanced Very High Resolution Radiometer (NOAA AVHRR) satellite sensor and selected groundtruth data sets [11,12]. As a part of a global initiative under the GIAM project, IWMI mapped the irrigated areas of India using time-series NOAA AVHRR satellite sensor and selected groundtruth data sets. The irrigated areas were also mapped using time-series satellite-sensor data of Moderate Resolution Imaging Spectroradiometer (MODIS) 500-m for the Indian subcontinent and LANDSAT 30-m for selected benchmark river basins (e.g., Krishna basin in India). As per the GIAM estimates, China and India are occupying nearly 31.49% and 27.53%, respectively, of the world's irrigated areas [2]. The GIAM experience shows that the available satellite-sensor data at different resolutions offer new perspectives for inventorying and monitoring irrigated areas. Technological advancement in spatial, spectral, and temporal resolutions of the satellite sensors provide valuable information on location, spatial distribution, and extent of irrigated areas for accurate mapping and for analyzing their spatiotemporal changes in the GIS environment. The other advantages of remote-sensing data are easy acquisition on a weekly/monthly basis during the crop seasons at different scales and resolutions. Keeping in view the methodological and procedural deficiencies in manual data collection on irrigated areas, India needs precise satellite sensor-based irrigated area maps and statistics on a periodical basis, to effectively monitor and manage irrigated areas in the country, for sustainable food production.

5.3　GOALS

The main goals of the present study are the following:

- To analyze the irrigated areas of the country and their trends based on the available national statistics.
- To analyze the irrigated areas of the country derived from satellite sensors at different resolutions under the GIAM project.
- To compare the irrigated areas derived from national statistics and the GIAM project.
- To strategize the way forward for systematic mapping of irrigated areas to obtain harmonized irrigated area maps and statistics in the country.

5.4　METHODS

5.4.1　STUDY AREA

India, which lies between 8°4' N and 37°6' N latitude and 68°7' E and 97°25' E longitude, has a total geographic area of 328.72 Mha. It is mainly a tropical country, but due to great altitudinal variations, almost all climatic conditions exist. The seasons in India are distinct, but the intensity of seasonal variations in the weather differs from region to region. There are four seasons: (1) winter (December–February), (2) summer (March–June), (3) southwest monsoonal season (June–September), and (4) postmonsoonal season (October–November). About 80% of the country's annual rainfall is mainly from the southwest monsoonal season from June to September, followed by the northwest monsoon in November–December. Also, the rainfall is highly variable, both spatially and in quantity, among the 35 meteorological subdivisions in the country. India is gifted with a river system comprising more than 20 major rivers. These river systems have a tremendous impact on irrigation systems in the country. Based on the physiography, soils, climate, length of growing period, land utilization, and forest types, 20 Agroecological regions (AERs) and 60 Agroecological subregions (AESRs) have been delineated in India [13]. As far as irrigated agriculture is concerned, each AER has its own significance in terms of cropping pattern, cropping intensity, and production and productivity levels.

5.4.2　DATA USED

5.4.2.1　Primary and Secondary Data on Irrigated Areas from GIAM

5.4.2.1.1　Primary Remote-Sensing Data Sets and Masks

The GIAM project mainly used freely available satellite data sets. AVHRR and MODIS data are of a relatively coarse scale, with resolutions from 10 km down to 250 m. The longest multitemporal series of remote-sensing data with global coverage is AVHRR 8-km (reprojected to 10 km). However, since this resolution is coarse, it has combined a 3-year monthly time series of AVHRR 10-km from 1997 to 1999 with a 1-km Système Pour l'Observation de la Terre (SPOT) Végétation mosaic of the world for 1999. The process starts with a number of publicly available data sets, which are processed into one large 159-layer time-series file, known as a megafile.

The time-series analysis is conducted on the megafile. The digital elevation model (DEM), temperature, and rainfall data are combined into the megafile to allow segmentation of a set of masks of different characteristic regions of the world, which are analyzed separately and then combined into the class-naming and area-calculation steps. A number of other data sets are used to provide contextual and detailed information, to assist in identifying, separating, and aggregating classes. The megafile used for the IWMI GIAM consisted of 159 data layers, which constitute 144 AVHRR 10-km layers for 3 years (12 layers from 1 band per year*4 bands including an NDVI band*3 years), 12 SPOT vegetation 1-km layers for 1 year, and single layers of DEM 1-km, mean rainfall for 40 years at 50 km, and AVHRR-derived forest cover at 1 km. The 159-band megafile data layers were all retained at a common resolution of 1 km by resampling the coarser resolution to 1 km. The second product was primarily dependent on 500-m MODIS 7-band, every 8-day time series for 2001–2003 and is commonly referred to as GIAM 500-m, which is used exclusively for irrigated area mapping in the Indian subcontinent.

5.4.2.1.2 Secondary Data Sets

Many secondary data sets used the 100-m resolution JERS-1 SAR tiles (http://southport.jpl.nasa.gov/GRFM/) for South America and Africa, to assist in mapping major rain-forest areas at higher resolution. The ESRI Landsat 150-m GeoCover was used to provide contextual information and pseudo-groundtruth, by geolinking to the class maps to identify and label classes. Google Earth (http://earth.google.com/) contains increasingly comprehensive image coverage of the globe at very high resolution of 0.61–4 m, allowing the user to zoom into specific areas in great detail, from a base of 30 m resolution data, based on GeoCover 2000. This assists in the identification and labeling of the GIAM classes, area calculations, and accuracy assessment of the classes. For every identified class, 20–50 sample locations were cross-checked using Google Earth data sets. There are two global archives of groundtruth data, one collected by IWMI and the other where public domain data from the degree confluence project (http://www.confluence .org/) were used.

5.4.2.1.3 Other Data Sets for Comparison Purposes

A number of existing global land use/land cover (LULC) products were used in the preliminary class identification and labeling process. These include the United States Geological Survey (USGS) LULC [14], the USGS seasonal LULC [14], Global Land Cover (GLC) 2000 [15], International Geosphere-Biosphere Programme (IGBP) [16], and Olson ecoregions of the world [17,18]. These data supplemented/complemented the groundtruth data during the preliminary class identification and labeling processes. The GLC 2000 (SPOT 1-km resolution) 10-day synthesis data from November 1, 1999, through December 31, 2000, were used for the classification (http://www.gvm.sai.jrc.it/glc2000/products/). The GLC characteristics database was developed on a continent-by-continent basis using 1 km, 10-day AVHRR data, spanning April 1992 through March 1993 [14]. The same primary data were used in the Global USGS LULC, seasonal USGS LULC, and IGBP LULC (http://edcdaac.usgs.gov/glcc/globe_int.html). Olson data provided 94 global unique ecosystem classes for the globe [17,18] (http://edcdaac.usgs.gov/glcc/globe_int.html).

5.4.2.2   Secondary Data on Irrigated Areas from National Statistics

In India, traditionally, village patwaris report irrigation statistics as a part of agricultural statistics that are, in turn, compiled at different levels, like village, tehsil, district, state, and national. Irrigation statistics at the national level are compiled and published by the DES, MoA. The material for compilation is supplied by the state governments and forms part of the land utilization statistics. The available secondary data on irrigated areas from the DES, minor irrigation census from the Ministry of Water Resources (MoWR), and available irrigated area maps from the Central Bureau of Irrigation and Power (CBIP) were used in the present study.

5.4.3   Concepts and Definitions of Irrigated Areas

5.4.3.1   Concepts and Definitions of Irrigated Areas in National Statistics

1. Major irrigation scheme are projects, which have a Cultivable Command Area (CCA) of more than 10,000 hectares (ha).
2. Medium irrigation schemes are projects, which have a CCA of 2,000–10,000 ha.
3. Minor irrigation scheme are all groundwater schemes and surface water schemes (both flow and lift), each having a CCA of up to 2000 ha.
4. Net area sown represents the total area sown with crops and orchards. An area sown more than once in the same year is counted only once.
5. Total cropped area represents the total area sown once and/or more than once in a particular year, i.e., the area is counted as many times as there are sowings in a year. This total area is known as gross cropped area.
6. Area sown more than once represents the area on which crops are cultivated more than once during the agricultural year. This is obtained by deducting net area sown from total cropped area.
7. NIA is the area irrigated through any source once a year for a particular crop.
8. GIA is the total area under crops, irrigated once and/or more than once a year. It is counted as many times as the number of times the areas are cropped and irrigated in a year.
9. Cropping intensity is the ratio of net area sown to the total cropped area.

5.4.3.2   Concepts and Definitions of Minor Irrigation Statistics by MoWR

1. Culturable command area is the area that can be physically irrigated from a scheme and is fit for cultivation.
2. GIA is the total irrigated area under various crops during a year, counting the area irrigated under more than one crop during the same year as many times as the number of crops grown and irrigated.
3. Gross irrigation potential created by a scheme is the total gross area proposed for irrigation under different crops during a year by a scheme. The area proposed for irrigation under more than one crop during the same year is counted as many times as the number of crops grown and irrigated.
4. Irrigation potential utilized is the gross area actually irrigated out of the gross proposed area to be irrigated by the scheme during the year.

5.4.3.3 Concepts and Definitions of Irrigated Areas of GIAM

1. Total Area Available for Irrigation (TAAI) is the area irrigated at any given point of time plus the area left fallow at the same point of time. The TAAI does not consider intensity. The NIA reported in national statistics and the "area equipped for irrigation," but not necessarily irrigated and reported in FAO's Aquastat are the nearest equivalent of TAAI. The GIAM TAAI is equivalent to the MoWR of India $NIA_{created}$ and nearly similar to $NIA_{utilized}$.

2. Annualized irrigated area (AIA) is the sum of the areas irrigated during different seasons. The AIA sums up areas irrigated during season 1, season 2, and continuously in a 12-month period. The AIA considers intensity. The GIA reported in national statistics is the nearest equivalent of AIA.

3. Major irrigation (major and medium surface water reservoirs) includes all areas irrigated from major and medium irrigation projects and almost exclusively from surface water reservoirs.

4. Minor irrigation (groundwater, small reservoirs, tanks) includes irrigated areas, predominantly from groundwater and also from small reservoirs and tanks having a CCA up to 2000 ha.

5. Data used and scales (or resolutions). Maps and statistics are produced for India by the IWMI GIAM project at 10-km resolution and 500-m resolution, using publicly available, high-quality, time-series, remote-sensing data, secondary data, groundtruth data, and Google Earth data.

5.4.3.4 Description of Methods

5.4.3.4.1 Methods of Data Collection on Irrigated Areas by the MoA

The responsibility for collecting and compiling statistics on irrigated area as part of an agricultural survey rests on the Director of Land Records, the Director of Agriculture, and the District Statistical Office under the MoA. These data are collected not only at the local but also, to some extent, at the district and state levels. In India, village patwaris report irrigation statistics as a part of agricultural statistics that are, in turn, compiled at different levels, like the village, tehsil, district, state, and national. Irrigation statistics at the national level are compiled and published by the DES, MoA. The material for compilation is supplied by the state governments and forms part of the land utilization statistics.

Agriculture census data on irrigation are collected with the help of a holding schedule. In this survey, information is collected in respect of 20% of the selected villages, and the data are presented at different administrative levels by adopting the statistical estimation procedure. In the agriculture census, the NIA is collected by different sources, like canals, wells, tube wells, tanks, and other sources. In this survey, the actual area irrigated by these sources is recorded numberwise for each operational holder. As per the definition of irrigated area, a particular parcel is classified as irrigated if it receives at least one irrigation turn during the reference period. For this purpose, the number of times a particular crop receives irrigation is of no consequence. By the above approach of collecting data on irrigation statistics, it is quite possible that the minor irrigation census and the agriculture census may be clubbed together.

There are some limitations in data collection on irrigated areas as well. The primary data for land utilization statistics are collected by village patwaris in prescribed forms by plot-to-plot enumeration in certain states, and are estimated on the basis of sample surveys in other states. The basic enumeration forms are not the same in all the states. In a few states, no separate columns for the sources of irrigation have been provided in the prescribed *Khasra* forms. The data collection system on irrigation is different in Kerala, Orissa, and West Bengal. In these states, land use and crop statistics are collected through sample surveys, which give estimates for the whole state or for some districts. Data gaps, incomplete sample frames and sample sizes, methods of selection, measurement of area, nonsampling errors, compilation errors, and gaps in geographical coverage are some of the limitations in data collection on irrigated areas.

5.4.3.4.2 Minor Irrigation Census by the MoWR

The first census on minor irrigation was started with the reference year as 1986–87. The criteria followed for the census of minor irrigation, which forms the basis for data collection, have varied from time to time. Under this, the groundwater scheme comprises dug wells, deep tube wells, shallow tube wells, and groundwater schemes up to 1 ha; surface flow irrigation and lift irrigation applied by major group of farmers were classified as minor irrigation schemes and were used for the purpose of the minor irrigation census. In addition to this, the data available from periodical progress reports and administrative reports of the state governments and ad hoc reports prepared by different agencies were used to cross-check and update the data on irrigation in different states. The census provided detailed information, giving statewise/districtwise number of schemes, CCA, potential created, potential utilized, etc., in respect of each type of minor irrigation scheme.

The second census of minor irrigation structures/schemes, with the reference year 1993–94, was conducted in all the states/Union Territories, and the All India Census report was published in March 2001. The third minor irrigation census has been conducted in 33 states (except Daman and Diu and Lakshdweep) with the reference year 2000–01 [19]. The main sources of minor irrigation statistical data are (i) land use statistics (LUS) of the MoA, (ii) periodical progress reports from state government departments, (iii) the annual administrative report compiled by the state government departments, and (iv) ad hoc reports prepared by various agencies from time to time on the basis of sample surveys to access the performance of minor irrigation works (MoWR; http://mowr.gov.in/micensus/mi3census/index.htm).

5.4.3.4.3 Methods of the GIAM Project in Deriving Irrigated Areas

The innovative methods of mapping irrigated areas at 10 km, 500 m, and 30 m resolutions were developed in the GIAM project [2,3,20–24]. The basic process involves segmenting the study area into characteristic regions that are easier to analyze, and then performing an unsupervised classification on each segment, containing all the items of 159-band information from the AVHRR time series and the single year of SPOT VGT data. The resulting classes are identified using a suite of spectral-matching techniques to interpret vegetation dynamics in multitemporal series. A number of classes could not be clearly identified, and so were subdivided and classified using simple decision trees and "groundtruth" data sourced from GeoCover 150-m and other secondary information [25]. This resulted in the generic class map of "unique" classes. As far as possible, class naming was harmonized with earlier GLC classifications. Irrigation classes were then derived by aggregating similar irrigated land use in the generic map, resulting in a 28-irrigation class map (GIAM 10-km-28 classes), and further aggregating this map into eight broad irrigated area classes of the world (GIAM 10-km-8 classes). The areas were estimated based on the cropping calendar, and then it was determined whether the crop was single, double, or continuous. Since pixel sizes are large at 1 km and dominated by AVHRR time series at 10 km, it is important to estimate the proportion of any one pixel that is irrigated in each season. The use of total pixel area would result in a massive overestimate. The Full Pixel Areas (FPAs) were converted to subpixel areas (SPAs) using irrigated area fractions for class (IAFs). In order to obtain reliable estimates of SPAs, three methods viz., Google Earth Estimates (GEE), high-resolution imagery (HRI), and Subpixel Decomposition Techniques (SPDT) were used. The SPDT and HRI approaches provide information on irrigated area intensities for different crop-growing seasons, whereas the GEE approach provides information on NIAs without intensity.

5.5 RESULTS AND DISCUSSION

5.5.1 OVERVIEW OF INDIAN IRRIGATED AGRICULTURE

5.5.1.1 General Land Use Pattern

Out of the total geographical area of 328.73 Mha, LUS are available for roughly 305 Mha, i.e., 93% of the total geographical area of the country. The analysis of land utilization pattern in India for the period from 1950–51 to 2000–01 reveals that the gross cropped area has increased significantly from 131.9 to 189.7 Mha (Figure 5.1). The net sown area has increased considerably from 118.8 to 141.1 Mha from 1950–51 to 2000–01. Significantly, during the same period, the gross area under irrigation has increased from 22.6 to 76.3 Mha, and the NIA in the country from 20.9 to 57.2 Mha. Interestingly, the area sown more than once has increased from 13.1 to 46.8 Mha, also during the same period. The increasing gap between gross cropped area and net sown area indicates the cropping intensity.

5.5.1.2 Trends in Cultivable Area Under Major Food Grain Crops

The analysis of areas under major food grain crops in India for the last 50 years (from 1950–51 to 2000–01) reveals that the area under rice has increased from 31.0 to 44.7 Mha (Figure 5.2). Significantly, for the same period, the area under wheat has increased from 10.0 to 25.8 Mha. However, the area under jowar, *bajra* (*Pennisetum typhoides*), and barley cultivation has declined in the last 50 years. The additional area added under rice and wheat has mainly come from irrigated agriculture.

5.5.1.3 Landholdings and Irrigated Areas

Based on the size, the landholdings in the country have been divided into five major categories. They are marginal (below 1 ha), small (1–2 ha), semimedium (2–4 ha), medium (4–10 ha), and large (10 ha and above). The analysis shows that the number of holdings increased from 71 million in 1970–71, to 115.6 million in 1995–96. During the same period, the size of the average holding in India decreased from 2.28 to 1.41 ha [26]. The analysis shows that the marginal landholdings (<1 ha) have

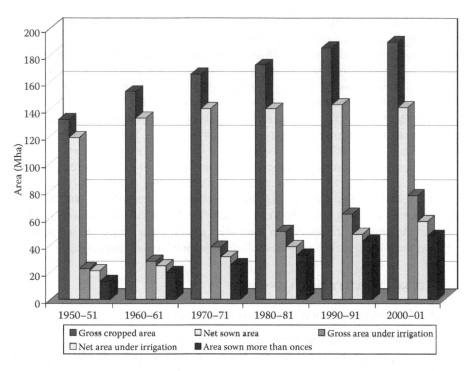

FIGURE 5.1 Land utilization pattern in India (1950–51 to 2000–01). (Source: Directorate of Economics and Statistics, *Agricultural Statistics at a Glance 2001*, Ministry of Agriculture, Government of India, New Delhi, 2001.)

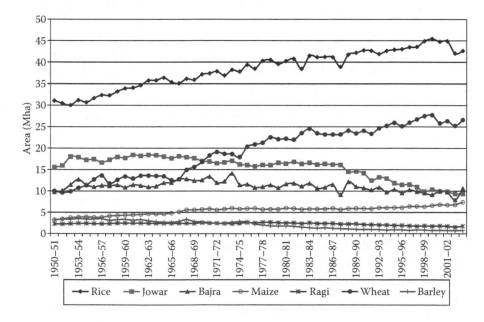

FIGURE 5.2 Trends in areas under major food grain crops in India (1950–51 to 2001–02).

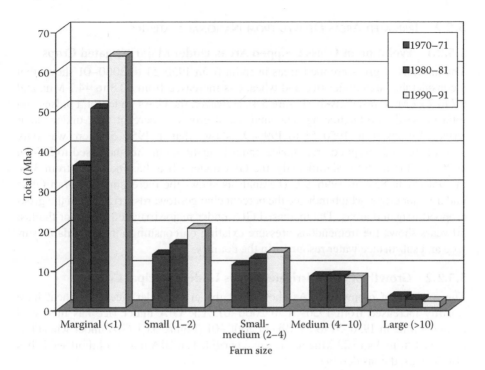

FIGURE 5.3 Total operational holdings in India (1970–71 to 1990–91). (Source: Directorate of Economics and Statistics, *Agricultural Statistics at a Glance 2001*, Ministry of Agriculture, Government of India, New Delhi, 2001.)

almost doubled within a span of 20 years (Figure 5.3). More than 50% of landholdings are less than 1 ha in size. Interestingly, the total number of large landholdings (>10 ha) has declined in the same period.

An analysis of irrigation status across the landholdings as per the data available for 1990–91 showed that nearly 28.9 Mha are wholly irrigated and 38.6 Mha partly irrigated. The size of nearly 70% of the wholly irrigated landholdings is less than 1 ha, each sown with a net area of 7.3 Mha. Most of the wholly irrigated areas are under small, semimedium, and medium landholdings. Interestingly, each landholding size of most of the partly irrigated areas is 4–10 ha (medium). The analysis shows that the area under wholly unirrigated conditions in the country is nearly 72.7 Mha.

5.5.1.4 Irrigated Areas of India in the Global Context

The FAO estimates show that China, India, and Pakistan alone account for about 45% of the world's irrigated areas. The estimates of the World Watch Institute show that the world-irrigated area rose by 3 Mha to reach 274 Mha, a gain of 1.1%. Asia, with an increase of 1.7%, is responsible for the worldwide irrigation expansion in 1999, and holds 70% of the total irrigated area. China and India claim 54 and 59 Mha NIA, respectively, and are equal to 41% of the total world irrigated area. However, the satellite sensor-based recent estimates of irrigated areas by IWMI show that China has the highest NIA with 108 Mha, followed by India with a NIA of 99.7 Mha [2].

5.5.2 IRRIGATED AREAS DERIVED FROM NATIONAL STATISTICS

5.5.2.1 Evolution of Gross Cropped Areas Under Major Irrigated Crops

The analysis of gross cropped areas in India from 1950–51 to 2000–01 shows that a considerable area under rice and wheat has increased from 30.8 to 44.7 Mha and 9.7 to 25.7 Mha, respectively (Figure 5.4). However, the GIAs under total pulses and total oil seeds are fluctuating. The analysis of cropwise percent GIA under major irrigated crops from 1950–51 to 1996–97 shows that in 1950–51 there was only 30% of the total cropped area of rice under irrigation, but this increased to 51% in 1996–97 (Figure 5.5). Significantly, the GIA under wheat has increased from 32% in 1950–51 to 85% in 1996–97. The analysis shows the increasing trends in GIAs under major crops, which indicate the percent changes towards irrigation in the gross cropped irrigated areas. The increased GIA under major irrigated crops for the last 50 years shows the tremendous pressure existing on consumption of available surface and subsurface water resources in the country.

5.5.2.2 Growth of Gross Irrigated Areas Under Principal Crops

The analysis of cropwise GIAs shows that the areas under rice and wheat have steadily increased from 1950–51 to 2000–01. The GIA under rice has increased from 9.9 Mha in 1950–51 to 23 Mha in 2000–01 (Figure 5.6). GIA under wheat has increased from 3.4 to 23 Mha in the same period. The GIA under total oil seeds has also increased considerably.

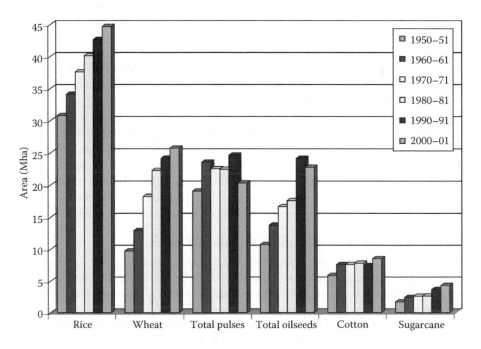

FIGURE 5.4 Gross irrigated area under major crops (in Mha). (Source: Directorate of Economics and Statistics, *Agricultural Statistics at a Glance 2001*, Ministry of Agriculture, Government of India, New Delhi, 2001.)

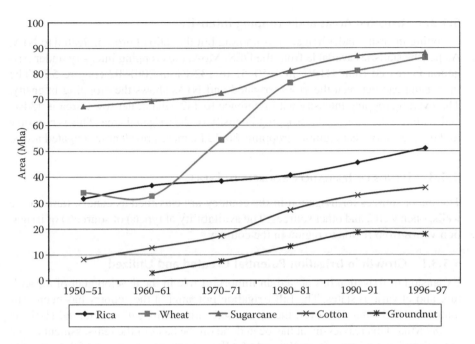

FIGURE 5.5 Percent gross cropped area under irrigation. (Source: Directorate of Economics and Statistics, *Agricultural Statistics at a Glance 2001*, Ministry of Agriculture, Government of India, New Delhi, 2001.)

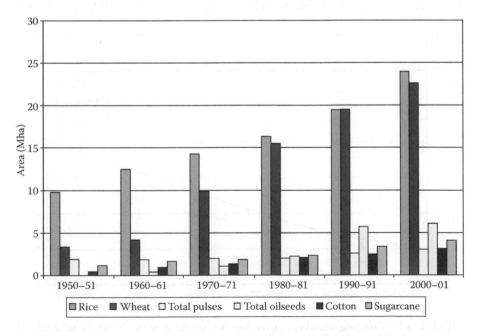

FIGURE 5.6 Area under irrigation by principal crops (1950–51 to 2000–01). (Source: Directorate of Economics and Statistics, *Agricultural Statistics at a Glance 2001*, Ministry of Agriculture, Government of India, New Delhi, 2001.)

5.5.2.3   Irrigated Areas and Cropping Intensity

Cropping intensity under irrigation is nothing but the ratio of gross irrigated to NIA. As per the statistics available from the DES, MoA, the cropping intensity under irrigation increased from 111.1% in 1950–51 to 132.1% in 2001–02 (Figure 5.7). The increasing gap between the gross irrigated and NIAs shows the cropping intensity. The overall cropping intensity with reference to 142 Mha of net sown area is 134%, and is almost the same as the cropping intensity of the irrigated area. This is contrary to the expectation that multiple cropping would be more prevalent in irrigated areas.

5.5.3   TRENDS OF IRRIGATION SOURCES IN INDIA

The major sources for irrigation in the country are classified as canals, tanks, tube wells, open wells, and other sources. The availability of type(s) of source(s) of irrigation varies from region to region in the country.

5.5.3.1   Growth in Irrigation Potential Created and Utilized

Expansion of irrigation has played an important role in the development of agriculture and of various states. The full irrigation potential of the country has been estimated at 113.5 Mha. However, the reassessed ultimate irrigation potential (UIP) is 139.89 Mha. This reassessment has been done on the basis of the reassessment of the potential of groundwater from 40 to 64.05 Mha, and the reassessment of the potential of minor irrigation with surface minor irrigation from 15 to 17.38 Mha. Thus, there has been an increase of 26.39 Mha in the UIP of the country, which was 113.5 Mha before reassessment. As per the MoWR, the total NIA created is 105 Mha and NIA

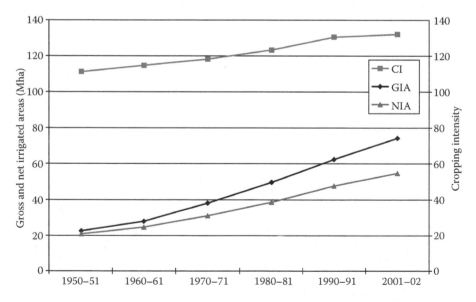

FIGURE 5.7   Cropping intensity, gross and net irrigated areas (1950–51 to 2001–02). (Source: Directorate of Economics and Statistics, *Agricultural Statistics at a Glance 2001*, Ministry of Agriculture, Government of India, New Delhi, 2001.)

utilized is 84 Mha in the country [19]. Even after achieving the full irrigation potential, nearly 50% of the total cultivated area is estimated to remain rain-fed.

5.5.3.2 Growth in Irrigation Areas Under Different Sources

The analysis of irrigated areas under different irrigation sources showed that the area under irrigation by government canals steadily increased from 7 Mha in 1950–51 to 17.18 Mha in 1993–94, and thereafter declined to 15.9 Mha in 2000–01 (Figure 5.8). Irrigated areas under tanks showed an increasing trend until the mid-1960s and a decreasing trend thereafter. Irrigated areas under tube wells, the main source under minor irrigation, greatly increased from the mid-1960s and reached 22.5 Mha by the end of 2000–01. Interestingly, the areas under all the irrigation sources except the tube wells showed a declining trend. The area irrigated by canals has steadily increased from 7 Mha in 1950–51 to 17 Mha in 1997–98. Irrigated area under tanks, tube wells, and other wells, which are the sources of minor irrigation, has increased from 13 to 37 Mha. Areas irrigated by tube wells and other wells have increased remarkably in the minor irrigation sector. As already indicated, the cropping intensity in irrigated and unirrigated areas remains almost the same. A possible reason is that water from irrigation sources is not available in all seasons for multiple cropping in most irrigated areas.

5.5.3.3 Dynamics of Different Irrigation Sources

Analysis of percent areas under different irrigation sources showed that the share of minor irrigation in the NIA is about 70%. The analysis of percent area under different irrigation sources reveals that areas irrigated by tube wells increased remarkably in the minor irrigation sector from 0.6% to 41.2%, from 1960–61 to 2001–02

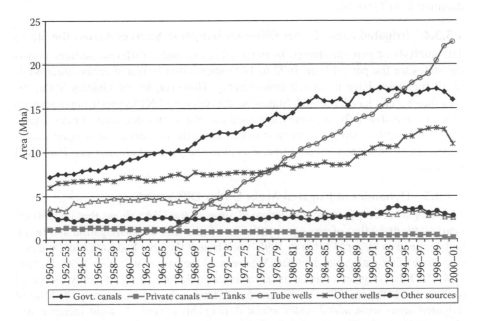

FIGURE 5.8 Irrigated areas under different irrigation sources (1950–51 to 2000–01).

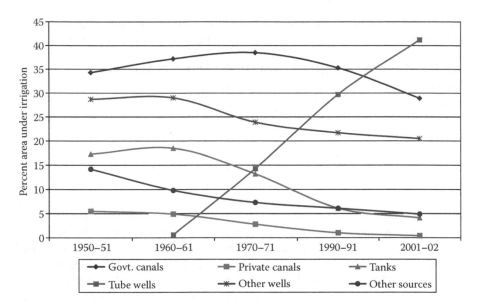

FIGURE 5.9 Percent area irrigated by different sources (1950–51 to 2001–02). (Source: Directorate of Economics and Statistics, *Agricultural Statistics at a Glance 2001*, Ministry of Agriculture, Government of India, New Delhi, 2001.)

(Figure 5.9). The percent share of government canals had increased from 34.3–38.5% from 1950–51 to 1970–71, and thereafter declined to 28.9% in 2000–01. The percent share of private canals, tanks, other wells, and other sources, has continuously declined from 1960–61.

5.5.3.4   Irrigated Areas Under Different Irrigation Sources Across the States

The analysis of percent change in irrigated areas under different sources across the states for the period from 1972 to 1993 shows that irrigated areas under wells in all the states have increased considerably. However, in the Gujarat State, the area under wells has decreased. Statewise percentage of NIAs under tanks showed that, in some states, the area had increased and thereafter declined in many states from 1972 to 1982. Similarly, in most of the states the percentage of irrigated areas under canal commands has declined, except in the states of Gujarat and Himachal Pradesh.

5.5.3.5   Dynamics of Irrigated Areas Under Different Crops

The analysis of cropwise ratio of irrigated areas to the GIAs shows that in 1950–51 nearly 44% (9.9 Mha) of gross cropped area of rice was under irrigation and that it had declined to 30% (23 Mha) in 1995–96. Though the area had increased, the percent share of irrigation had declined. Whereas, in the case of wheat, in 1950–51, only 15% (3.4 Mha) of gross cropped area was under irrigation, it significantly increased to 30% (23 Mha) in 1995–96 (Figure 5.10). This indicates that nearly 20 Mha of irrigated areas were added under wheat during this period. A slight increase was observed in the percentage of irrigated area under total oil seeds. The increased

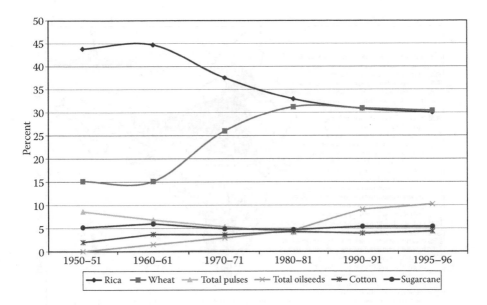

FIGURE 5.10   Cropwise percent irrigated area to gross irrigated area (1950–51 to 1995–96).

areas under irrigation, cropping pattern, and intensity have a significant impact on the water consumption pattern in the major irrigated crops.

5.5.4   IRRIGATED AREAS DERIVED FROM SATELLITE SENSORS-GIAM PROJECT

5.5.4.1   Irrigated Area Mapping Using NOAA AVHRR 10-km Data

The IWMI NOAA AVHRR 10-km irrigated area map with 28 classes shows watering method (in this case *irrigated*), irrigation type *(surface water, groundwater,* and *conjunctive use),* irrigation intensity *(single, double,* or *continuous crop),* and crop type (Figure 5.11). The GIAM statistics showed the TAAI (NIA) to be 100 Mha as per the 10-km resolution data. The GIAM AIA (GIA) is 132 Mha as per the 10-km resolution, which is about 27.53% of the global irrigated area [2].

5.5.4.2   Irrigated Area Mapping Using MODIS 500-m Data

The GIAM 500-m irrigated area map of India, produced exclusively for India, was based on the analysis of MODIS 500-m data of every 8-day composite from the period 2001–03 (Figure 5.12) [23]. The MODIS 500-m satellite sensor-based irrigation area statistics have been generated at the national, state, and district level for India. The same statistics of the GIAM project are globally accessible, with online data access at various administrative units (e.g., district, state) of India (http://www. iwmigiam.org). The GIAM 500-m statistics showed 113 Mha as TAAI (NIA) and 147 Mha as AIA (GIA) [23]. To analyze the spatiotemporal patterns, statewise irrigated areas could be generated from GIAM MODIS 500-m data. The irrigated area map of Uttar Pradesh is shown in Figure 5.13. The districtwise statistics on irrigated areas for the selected states could be generated from the state map to correlate with the national statistics.

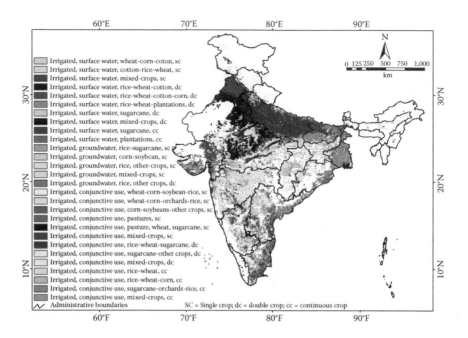

FIGURE 5.11   (See color insert following page 224.) Irrigated area map of India derived from NOAA AVHRR 10-km sensor data.

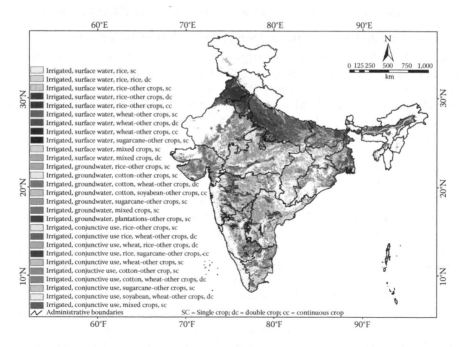

FIGURE 5.12   (See color insert following page 224.) Irrigated area map of India derived from MODIS 500-m sensor data.

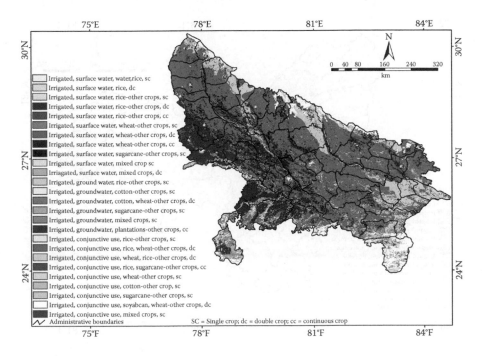

FIGURE 5.13 **(See color insert following page 224.)** Irrigated area map of Uttar Pradesh (GIAM MODIS 500-m).

5.5.4.3 Irrigated Area Mapping Using LANDSAT 30-m Data

As a part of the GIAM project, IWMI has carried out the irrigated area mapping in the Krishna River basin in India, using the time-series LANDSAT 30-m data (Figure 5.14). Extensive groundtruthing has been carried out in the basin for irrigated area class identification and labeling. Velpuri et al. [24] have discussed in detail the methods for irrigated area mapping based on Landsat 30-m.

5.5 COMPARATIVE ANALYSIS OF IRRIGATED AREAS OF GIAM WITH NATIONAL STATISTICS

An attempt has been made to compare the irrigated area statistics of AVHRR 10-km, MODIS 500-m, and LANDSAT 30-m generated in the GIAM project with the irrigated area maps of the CBIP, national statistics of the DES, and minor irrigation statistics of the MoWR of India.

5.5.1 COMPARATIVE ANALYSIS OF GIAM 10-KM AND CBIP MAJOR IRRIGATED AREAS

The NIA and the area equipped for irrigation (AEI) are equivalent to TAAI of GIAM. The NIA is almost exclusively derived from 162 major and 221 medium surface water reservoirs, as depicted in India's CBIP [27] irrigated area map for India for 1992 (Figure 5.15). Primarily, the CBIP map represents all the major and medium

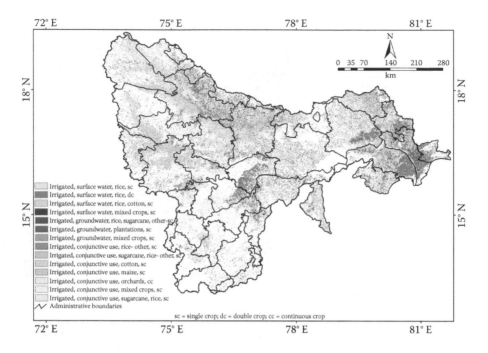

FIGURE 5.14   (See color insert following page 224.) Irrigated area map of Krishna basin (GIAM, LANDSAT 30-m).

irrigation command areas and misses out on massive areas outside these large and medium command areas. The total FPA of the CBIP map was 75 Mha. This itself exceeds the DES reported NIA of 57.2 Mha. In GIAM, 8-day time-series MODIS data have been used for the period 2001–03, to determine the exact irrigated areas within this CBIP area, and it has been found that 48% was surface water, 34% groundwater, and the remaining 18% was other LULC. This reduced the exact FPA irrigated within the CBIP map to 61.5 Mha (75 Mha*0.82 after deducting 18% for other LULC areas). The GIAM project developed a methodology for deriving SPA, which is the actual area irrigated, by multiplying the FPA with IAF [8]. The average IAF for 500 m was 0.68 [23]; so the area irrigated within the CBIP map (Figure 5.16) is 42 Mha (61.5 Mha*0.68). This probably does not include most of the minor irrigation.

5.5.2   COMPARATIVE ANALYSIS OF IRRIGATED AREAS OF GIAM 10-KM AND NATIONAL STATISTICS

The statewise AIA and TAAI of GIAM 10-km data set were compared with the gross and NIAs of the national statistics for the year 2000–01. The GIAM AIA is equal to the GIA in the national statistics. In the GIAM 10-km data, TAAI is equal to the NIA in the national statistics. The comparative analysis of GIAM AIA 10-km and GIA of the national statistics shows that AIAs of GIAM are higher than those in the national statistics, with a correlation coefficient (R^2) of .77 [28] (Figure 5.17). Similarly, the comparative analysis between the GIAM TAAI 10-km and NIA of the

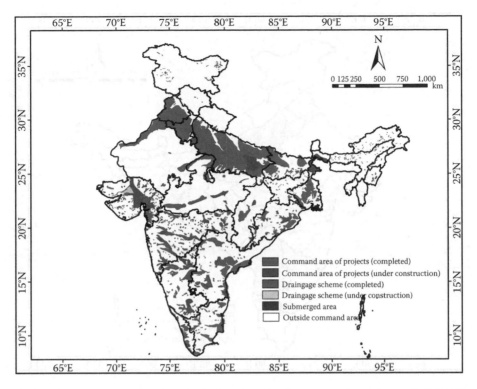

FIGURE 5.15 (**See color insert following page 224.**) Irrigated areas of India mapped by India's CBIP (1994).

national statistics shows that GIAM TAAI 10-km is higher than the national statistics, with an R^2 of .83 (Figure 5.18).

5.5.3 COMPARATIVE ANALYSIS OF IRRIGATED AREAS OF GIAM 500-M AND NATIONAL STATISTICS

The statewise statistics on irrigated areas from GIAM 500-m data sets and the gross and NIAs from the national statistics for 2000–01 were compared. The comparative analysis between the GIAM AIA 500-m and GIA of the national statistics shows that GIAM AIA is higher than the national statistics, with an R^2 of .84 [28] (Figure 5.19). Similarly, the comparative analysis between the GIAM TAAI 500-m and NIA of the national statistics shows that GIAM TAAI 500-m is higher than the national statistics, with an R^2 of .78 (Figure 5.20). The detailed statewise analysis of both data sets needs to be reanalyzed to ascertain the causes for similarity and differences.

5.5.4 COMPARATIVE ANALYSIS OF IRRIGATED AREAS OF GIAM 10-KM AND GIAM 500-M DATA SETS

An attempt has been made to compare the irrigated areas derived from GIAM 10-km and GIAM 500-m data sets of India. The analysis shows the TAAI (NIA) to be 100 Mha as per the 10-km resolution data and 113 Mha as per the 500-m resolution data.

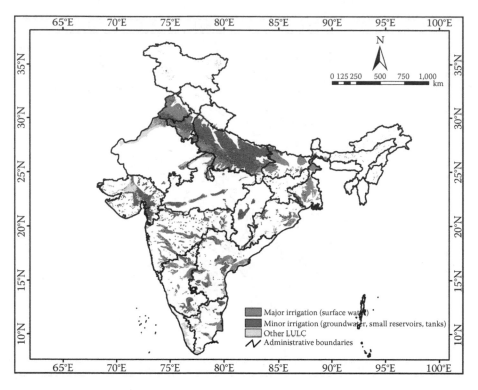

FIGURE 5.16 **(See color insert following page 224.)** Actual irrigated areas of India within India's CBIP (1994) map.

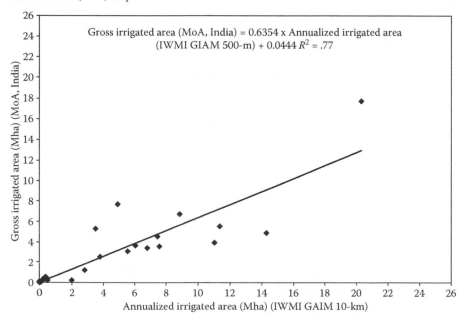

FIGURE 5.17 Analysis of annualized irrigated areas of GIAM and gross irrigated areas (MoA, India).

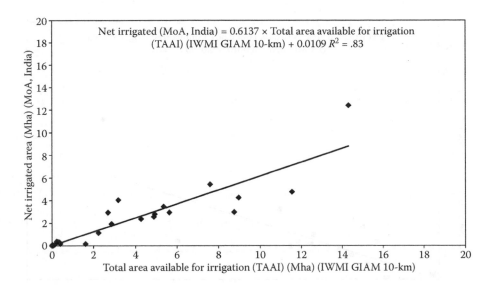

FIGURE 5.18 Analysis of total area available for irrigation of GIAM and net irrigated areas (MoA, India).

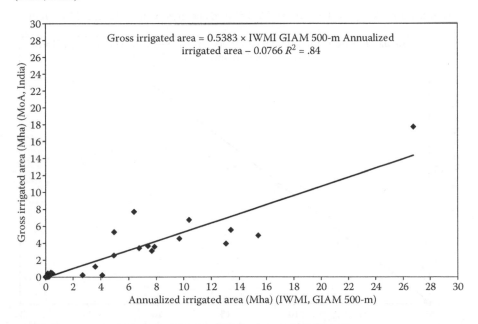

FIGURE 5.19 Analysis of annualized irrigated areas of GIAM 500-m and gross irrigated areas (MoA, India).

The GIAM AIA (GIA) is 132 Mha as per the 10-km resolution and 147 Mha as per the 500-m resolution [2]. The IWMI GIAM areas are consistently higher than the areas in the national statistics. The comparative analysis between the GIAM AIA 10-km and 500-m shows a significant correlation between the data sets, with an R^2 of .98 [28] (Figure 5.21). The comparative analysis of GIAM TAAI 10-km and

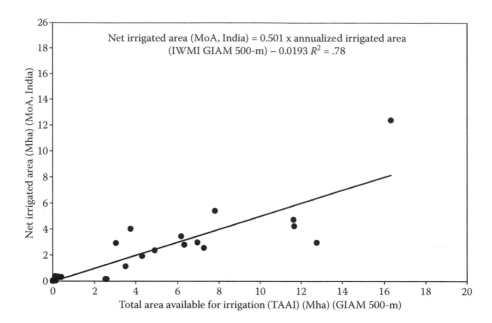

FIGURE 5.20 Analysis of TAAI of GIAM 500-m and gross irrigated areas (MoA, India).

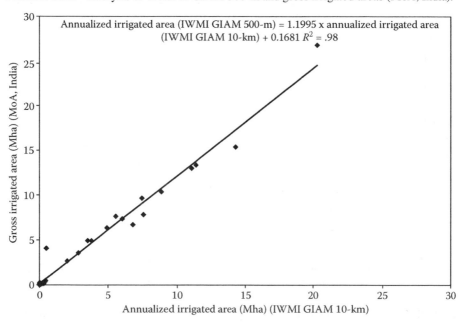

FIGURE 5.21 Analysis of annualized irrigated areas of GIAM AVHRR 10-km and MODIS 500-m data.

500-m data sets also show a significant correlation, with an R^2 of .97 (Figure 5.22). The high correlation between the GIAM 10-km and 500-m data sets indicates the strength of the methodology followed in deriving irrigated areas from two independent data sets under the GIAM project.

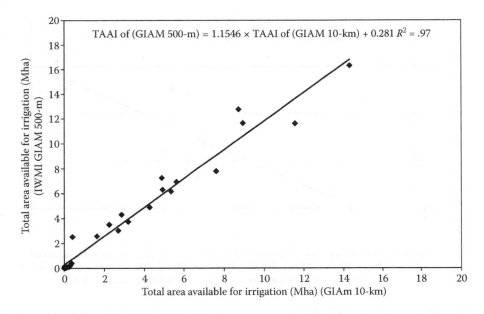

$$\text{TAAI of (GIAM 500-m)} = 1.1546 \times \text{TAAI of (GIAM 10-km)} + 0.281 \ R^2 = .97$$

FIGURE 5.22   Analysis of TAAI of GIAM AVHRR-10 km and MODIS 500-m data.

5.5.5   COMPARATIVE ANALYSIS OF IRRIGATED AREAS OF GIAM 30-M AND DIVISION-LEVEL IRRIGATED AREAS

To compare Landsat 30-m irrigated area statistics with a lower administrative level like the revenue division, the Bhongir and Nalgonda divisions in the Nalgonda District in the Krishna basin were selected, and irrigated area statistics from Landsat 30-m were generated at *mandal* (block) level to compare with the NIAs of the state government for the same mandals for 2000–01. The analysis shows a good correlation with the R^2 of .76 and .80 in Bhongir and Nalgonda divisions, respectively (Figures 5.23 and 5.24).

5.6   COMPARATIVE ANALYSIS OF IRRIGATION INTENSITY OF GIAM AND THE NATIONAL STATISTICS

5.6.1   COMPARATIVE ANALYSIS OF IRRIGATION INTENSITY OF GIAM 10-KM AND THE NATIONAL STATISTICS

Irrigation intensity is nothing but the ratio between the GIA and NIA. The comparative analysis of irrigation intensity of GIAM 10-km data and the national statistics (2001–02) shows the overall trend in both data sets. However, in some of the states like Haryana, Himachal Pradesh, and Punjab, the irrigation intensity was slightly lower in GIAM 10-km data than in the national statistics (Figure 5.25), whereas in the case of Kerala and Madhya Pradesh, the irrigation intensity of NOAA AVHRR 10-km data was slightly higher than that in the national statistics. As per the national statistics, it was considerably high in the states of Punjab (191%) and Haryana (180%), whereas it was very low in the states of Arunachal Pradesh (102%) and Madhya

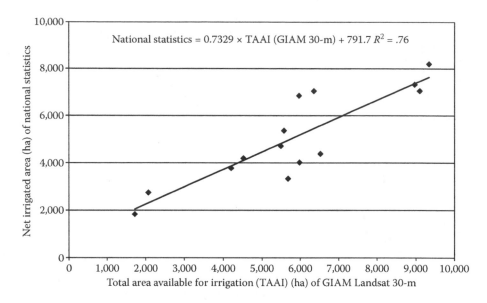

FIGURE 5.23 TAAI of GIAM 30-m and net irrigated area (DES, AP) of Bhongir Division in the Krishna basin.

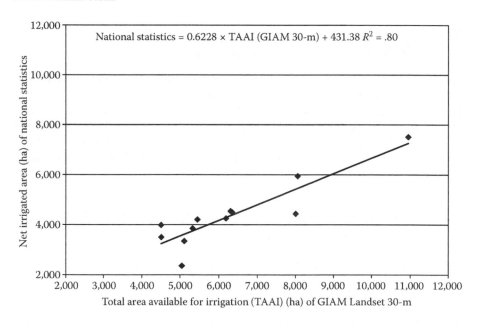

FIGURE 5.24 TAAI of GIAM 30-m and net irrigated area (DES, AP) of Nalgonda Division in the Krishna basin.

Pradesh (103%). However, in the majority of states, the irrigation intensity assessed through GIAM 10-km data was closely associated with that of the national statistics [28]. The reasons for deviations in irrigation intensity in some of the states need to be analyzed.

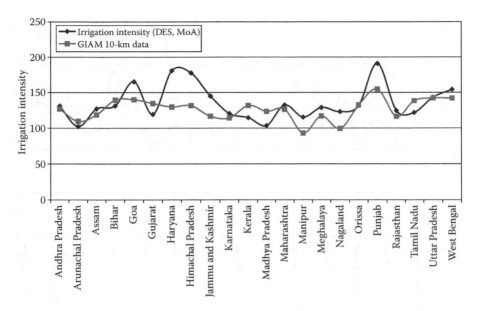

FIGURE 5.25 Irrigation intensity of GIAM 10-km and national statistics (2001–02).

5.6.2 COMPARATIVE ANALYSIS OF IRRIGATION INTENSITY OF GIAM 500-M DATA AND NATIONAL STATISTICS

The comparative analysis of irrigation intensity of GIAM 500-m data and the national statistics (2000–01) shows the overall trend in both data sets. In the states of Haryana and Punjab, the irrigation intensity was slightly lower in GIAM 500-m data than in the national statistics (Figure 5.26). In the case of northeastern states like Arunachal Pradesh, Assam, Meghalaya, Mezoram, and Nagaland, irrigation intensity of MODIS 500-m data was higher than that in the national statistics. As compared to the national statistics, the irrigation intensity in MODIS 500-m was also high in the states of Punjab (169%) and Haryana (163%), whereas it was very low in the states of Chhattisgarh (102%) and Jharkhand (103%). Irrigated areas of MODIS 500-m data also show that in the majority of states, the irrigation intensity assessed through GIAM 500-m data was closely associated with that of the national statistics.

5.7 COMPARATIVE ANALYSIS OF IRRIGATED AREAS OF GIAM AND IRRIGATION STATISTICS OF THE MoWR

The MoWR of India recently released the third minor irrigation census for India [19] (http://mowr.gov.in/micensus/mi3census/index.htm). As per these statistics, the minor irrigated areas (groundwater, small reservoirs, and tanks) in India for 2001–02 (a) utilized 53 Mha and (b) created 68 Mha. The major irrigated areas are primarily under major and medium surface water reservoirs. For the same period, the major irrigated areas utilized 31 Mha and created 37 Mha. In India, the total NIAs utilized ($NIA_{utilized-total}$) and the total NIAs created ($NIA_{created-total}$) from major and minor

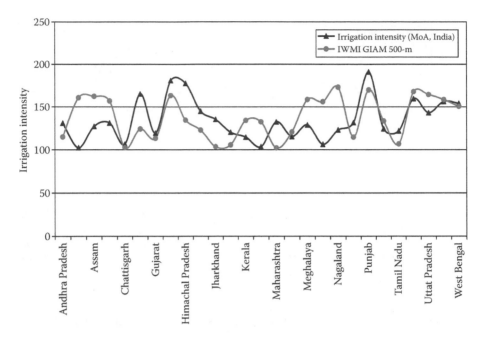

FIGURE 5.26 Analysis of irrigation intensity of GIAM 500-m and national statistics (2001–02).

irrigation in 2001–02 were 84 and 105 Mha, respectively (Figure 5.27), as reported by India's MoWR [19]. It is quite comparable to the GIAM 10-km TAAI of 101 Mha and GIAM 500-m TAAI of 113 Mha. The GIAM TAAI is equivalent to NIA$_{created}$ and nearly similar to NIA$_{utilized}$. The GIAM 10-km shows TAAI to be 101 Mha, of which 41 Mha are from major surface water reservoirs and 60 Mha from minor irrigation (groundwater, small reservoirs, tanks). The GIAM 500-m shows TAAI to be 113 Mha, of which 43 Mha are from major surface water sources and 70 Mha from minor irrigation sources (Figure 5.28). Interestingly, out of the UIP estimated at 140 Mha, only 59 Mha are through major and medium irrigation and 81 Mha from minor irrigation (64 Mha from groundwater and 17 Mha from surface water) [19].

A comparative analysis of irrigated areas of GIAM 10-km and GIAM 500-m was done with total irrigation potential created (MoWR). The GIAM 10-km TAAI statistics for statewise irrigated areas of India have been compared with India's MoWR statistics. The 10-km TAAI statistics for the Indian states are plotted against NIA$_{created}$ (Figure 5.29). The NIAcreated is the nearest equivalent for GIAM TAAI and had an R^2 value of .80 for 10 km. The analysis shows that, for many states, the irrigated areas of GIAM 10-km TAAI closely matched the total irrigation potential created (MoWR).

The GIAM 500-m TAAI statistics for the Indian states are plotted against NIA$_{created}$ (Figure 5.30). The analysis shows an R^2 value of .78 for 500 m. There are no systematic minor irrigated area statistics at the national level that account for irrigation intensity. Hence, comparisons of the MoWR national statistics were not made with GIAM AIA.

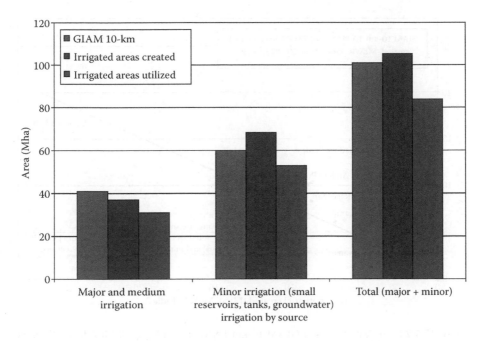

FIGURE 5.27   GIAM 10-km TAAI and minor irrigation statistics (MoWR) of India.

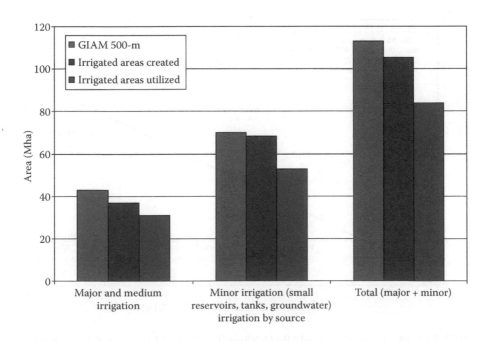

FIGURE 5.28   GIAM 500-m TAAI and minor irrigation statistics (MoWR) of India.

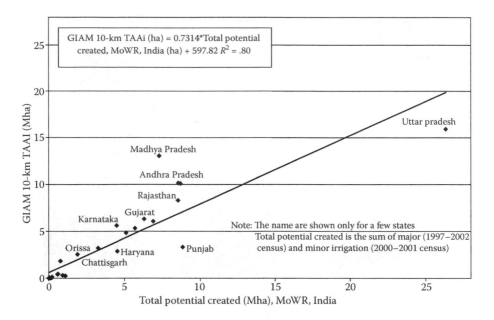

FIGURE 5.29 Irrigated areas of GIAM 10-km TAAI and NIA_{created} of MoWR for different states.

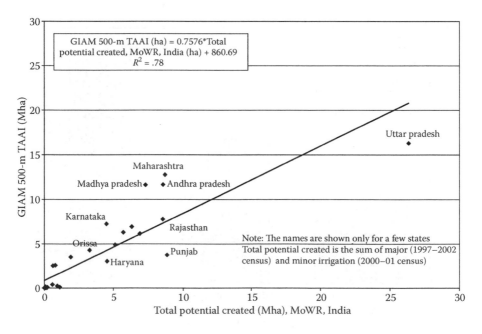

FIGURE 5.30 Irrigated areas of GIAM 500-m TAAI and NIA_{created} MoWR for different states.

5.8 CAUSES FOR DIFFERENCES IN IRRIGATED AREAS OF GIAM AND NATIONAL STATISTICS

The causes of differences in irrigated areas between GIAM data and the national statistics are many and include: (a) inadequate account of informal (e.g., groundwater, small tanks, small reservoirs) irrigation statistics in the national statistics; (b) need of localized IAFs in GIAM; (c) inconsistencies in how the national census data of irrigated areas are compiled; (d) definitions of how irrigated areas are mapped (e.g., IWMI GIAM considers supplemental irrigation under irrigated area classes); (e) misrepresentation of minimum mapping unit; and (f) accounting or nonaccounting of supplemental irrigated area classes, which have significant irrigation, mostly from informal sources like tube wells, but are often ignored and are accounted for or are mapped as rain-fed by many sources [2,8,22]. However, once the minor irrigation statistics are taken into consideration, the differences in irrigated areas between GIAM and India's national statistics become much smaller. The differences that still exist can be attributed to any one or a combination of causes mentioned above.

5.9 WAY FORWARD FOR MAPPING OF DETAILED IRRIGATED AREAS IN INDIA-GIAM EXPERIENCE

The need for the timely availability of irrigated area statistics at various administrative units of India cannot be overemphasized. Traditionally, irrigation statistics in India are collected as a part of agricultural statistics and compiled at different levels. In the generation of irrigation statistics, more emphasis has been given to the major and medium surface water irrigation schemes. This is probably the main reason for the underestimation of minor irrigated areas in the DES statistics. India's minor irrigation statistics by the MoWR also proved that, typically, irrigated area statistics do not adequately account for minor irrigation and are overly focused on major irrigation. Above all, in the existing statistical data it is difficult to precisely visualize the spatial pattern of irrigated areas of any given administrative unit. The size of irrigated landholdings also has significance in irrigated area mapping in terms of the minimum mapping unit and estimation of IAFs in the calculation of irrigated areas in the country. To overcome these difficulties, various satellite sensors with increases in spatial, spectral, and temporal resolutions provide valuable information on location, spatial distribution, and extent of irrigated areas for their accurate mapping and to analyze their spatiotemporal changes in the GIS environment. The analysis shows that irrigated area mapping through satellite-sensor data in fragmented landholdings, as in India, emphasizes the need for moderately high-resolution time-series satellite data sets, like Landsat 30-m, for their better estimation at the national scale. However, the available national statistics could be used as supplementary/complementary information to validate and cross-check the irrigated areas based on satellite sensors.

The study shows that IWMI GIAM irrigated areas are consistently higher than those in the national statistics. The reasons for this could be the absence of data on informal irrigated areas in the national irrigated area statistics, overestimation of IAFs for the GIAM classes, misrepresentation of the minimum mapping unit areas, and a host of others. So, understanding the real cause of such differences will make way for mapping of detailed irrigated areas and the development of an online Irrigated Area Mapping and Reporting System (IAMRS) for India, using moderately high-resolution satellite-sensor data to report irrigated areas at different administrative levels, like the district, state, and country, to effectively monitor the irrigated areas in the country.

5.10   IAMRS FOR INDIA

The IAMRS for India proposed to periodically map and report the irrigated area maps and statistics online to use in various real-time applications. This system could be an integral part of a Spatial Decision Support System (SDSS) through partnership between the Indian partners (ICAR/NBSS and LUP) and IWMI. This will be led, operated, and owned by national agricultural organization/institutes (ICAR/NBSS and LUP). The system will facilitate the provision of statistics and maps on irrigated areas at various administrative units (e.g., tehsil, district, state, and country) on a periodic basis. The IAMRS will provide online access for the user community, like the MoA, Department of Space, the MoWR, the Central Water Commission, and many other research institutes/universities. The IAMRS India will have two immediate goals: (a) to refine GIAM India irrigated areas, and (b) to develop an online IAMRS system using fusion of 500 m and 30 m data. IAMRS is also expected to lead to other synergetic portals, such as rain-fed croplands of India, wetlands of India, drought reports of India, and so on.

The specific objectives of IAMRS would be the following:

- To determine how the irrigated area statistics are derived for India in the national statistics. This should involve how irrigated areas are calculated, whether informal irrigation statistics are involved, and definitions of irrigated areas.
- To investigate the main causes of the differences in irrigated areas between the GIAM-10 km, GIAM 500-m, and the Indian national statistics, to refine GIAM irrigated areas of India based on adequate groundtruth.
- To work out the modalities to set up an online operational IAMRS for India, to map and monitor the spatiotemporal changes of irrigated areas.

The presently available 10-km and 500-m resolution irrigated area maps generated by IWMI as a part of GIAM could be verified in the selected locations with the available national statistics to find out reasons for differences in irrigated area statistics, to frame a mature IAMRS for India. The understanding of the real cause for such differences will be invaluable in formalizing the online IAMRS India using 500 m or even higher-resolution satellite-sensor data, to provide information on irrigated areas at different administrative levels, like district, state, and country.

As a collaborative effort, ICAR and IWMI could work out the possibilities of setting up an operational IAMRS for India based on GIAM 10-km and 500-m data sets.

Under this system, periodical (preferably every 2 years) changes in irrigated areas could be mapped and monitored for their effective management. The freely available satellite-sensor data sets could be used in the system to assess the irrigated areas. The goal of IAMRS would be to update the irrigated area statistics in the country, depicting the nature and spatial distribution of irrigated areas on a periodical basis.

5.11   NEED FOR HARMONIZING OF IRRIGATED AREA STATISTICS

The need for timely availability of harmonized irrigated area statistics at various administrative units of India cannot be overemphasized. The IWMI GIAM study showed significant differences in the irrigated area statistics when compared with the national statistics. India's DES reported 57.2 Mha as NIA for 2001–02. The equivalent of NIA in GIAM is TAAI. The GIAM 10-km V2.0 TAAI of 101 Mha and GIAM 500-m TAAI of 113 Mha show a staggering difference relative to NIA of DES. Similarly, the GIA of DES for the baseline year 2001–02 was 76.3 Mha, which is dramatically lower than the GIAM 10-km AIA of 132 Mha and GIAM 500-m AIA of 146 Mha.

The need to understand the causes of such differences has become critical in order to harmonize and synthesize the irrigated area mapping and the reporting system for India. Since the GIAM project has completed the mapping of irrigated areas at 10 km, and 500 m resolutions for India and extensive secondary data on irrigated area is available for India, the harmonized irrigated area statistics could be generated.

5.12   GIS-BASED SDSS

The harmonized irrigated area maps and statistics derived from satellite sensors at a moderate scale will be of immense use to feed into an SDSS. It could be possible to feed the other relevant thematic databases to SDSS, to arrive at logical conclusions to address different issues related to crop production. A GIS-based SDSS has many more advantages over the conventional methods in data handling and management. First, digital data are rapidly replacing traditional methods of data collection through a tedious administrative hierarchy. Second, it is hard to maintain data consistency and accuracy when a few thousand reporting officials are involved. Third, satellite-sensor data offer an opportunity to consistently update data every few years. Fourth, the digital data, once verified and refined, will be a reliable and accurate resource that can be easily updated. Fifth, the entire work can be done using freely available, guaranteed, and high-quality satellite-sensor data from sensors such as MODIS, Landsat, and IRS. Sixth, the system can be maintained and updated by a few specialists. Seventh, the data can be made available as maps and/ or statistics for online access. Eighth, the approach, methods, and data compiled and processed through the project are available for a multitude of other applications. Next, the protocols for online data sharing are an attractive mechanism for sharing many other data, such as rain-fed croplands, wetlands, and droughts. Last, they enable different users to interact online to update the irrigated area statistics with their local knowledge and data.

5.13 CONCLUSIONS

As per the national statistics for 2000–01, the gross and NIAs are 76.3 and 57.2 Mha, respectively. In contrast, the GIAM statistics show the TAAI (equivalent to NIA) to be 100 Mha as per the 10-km resolution data and 113 Mha as per the 500-m resolution data. As per the Indian national statistics, the GIA was 76.3 Mha in 2000–01. However, GIAM AIA (equivalent to GIA) is 132 Mha as per the 10-km resolution data and 147 Mha as per the 500-m resolution data. The irrigated areas in the GIAM data sets are consistently higher than those in the national statistics. The reason for this could be ascertained with detailed studies in selected locations.

As per the MoWR minor irrigation statistics for the period ending 2000–01, the $NIA_{utilized-minor irrigation}$ was 53 Mha and $NIA_{created-minor irrigation}$ was 68 Mha; besides the irrigated areas from major and medium irrigation, $NIA_{utilized-major irrigation}$ was 31 Mha and $NIA_{created-major irrigation}$ was 37 Mha. The total (major and medium plus minor) $NIA_{utilized-total}$ was 84 Mha and $NIA_{created-total}$ was 105 Mha. As per definitions, the equivalent of GIAM TAAI is $NIA_{created}$. The GIAM 10-km TAAI of 101 Mha and GIAM 500-m TAAI of 113 Mha match very well with their equivalent $NIA_{created}$, 105 Mha. These results have highlighted the strength of remote-sensing approaches used in irrigated area mapping in the GIAM project. The remote-sensing approaches have many advantages in irrigated area mapping, such as in providing temporal characteristics, mapping crop types and/or dominance, and locating the precise areas of irrigated areas. The differences that still exist between the GIAM and the national statistics could be evaluated to refine the irrigated area maps and statistics leading to their precise estimation.

The study demonstrates the strengths of the GIAM work in establishing irrigated areas by considering intensity, mapping informal irrigation, determining crop calendars of irrigated areas, and simulating trends in biomass dynamics of irrigated areas over time (e.g., trends every month from 1981 to 2001). The study also demonstrates the potential and cost-effectiveness of satellite remote-sensing techniques to map irrigated areas at various scales or pixel resolutions, to update irrigated areas of India in a consistent way over time and space. Effective integration of various temporal irrigated thematic layers in GIS enhances the evaluation of performance for diagnostic analyses of various parameters. The study indicates that enhanced operational use of such techniques, using the irrigated area maps and statistics, enables the development of simulation models, like optimum crop classification algorithms, yield prediction models, more appropriate vegetation indices, satellite data normalization procedures, models for refinement of existing AERs, crop productivity regions, and food surplus and deficit regions in the country.

ACKNOWLEDGMENTS

The authors are grateful to Dr. Mangla Rai, Director General (ICAR), Dr. J.S. Samra, Deputy Director General (NRM), Dr. P.D. Sharma, Assistant Director General (Soils), and Dr. P.S. Minhas, Assistant Director General (Integrated Water Management) of ICAR for giving the opportunity and encouragement to carry out India-focused irrigated area mapping in the GIAM project. The authors are very grateful to Professor Frank Rijsberman, Director General of IWMI, for his vision and great support for GIAM. The enthusiastic support of Dr. A.K. Maji, Director (Acting), NBSS and LUP is highly recognized. Dr. Bharat Sharma, senior researcher of IWMI (Delhi office) has always been a pillar of support.

REFERENCES

1. Directorate of Economics and Statistics, *Agricultural Statistics at a Glance 2001*, Ministry of Agriculture, Government of India, New Delhi, 2001.
2. Thenkabail, P.S. et al., *An Irrigated Area Map of the World (1999) Derived from Remote Sensing*, Research Report, International Water Management Institute, Colombo, Sri Lanka, 2006.
3. Thenkabail, P.S. et al., Global irrigated area map (GIAM) for the end of the last millennium derived from remote sensing. International *Journal of Remote Sensing*. 2009 (accepted, in press).
4. Endersbee, L., *A Voyage of Discovery: A History of Ideas about the Earth, with a New Understanding of the Global Resources of Water and Petroleum, and the Problems of Climate Change*, Monash University Bookshop, Frankston, Victoria, Australia, 2005.
5. Sinha, S., Rs 24,500 cr plan for harvesting rain water, *Times of India*, January 6, 2003.
6. Shah, T. et al., Sustaining Asia's groundwater boom: An overview of issues and evidence, *Natural Resources Forum*, 27, 130–40, 2003.
7. Shah, T., Singh, O.P., and Mukherjee, A., Groundwater irrigation and South Asian agriculture: Empirical analyses from a large-scale survey of India, Pakistan, Nepal Terai, and Bangladesh, Presented at IWMI-Tata Annual Partners' Meeting, February 17–19, Anand, India, 2004.
8. Thenkabail, P.S. et al., Spectral matching techniques to determine historical land use/ land cover (LULC) and irrigated areas using time series AVHRR Pathfinder Datasets in the Krishna River Basin, India, *Photogrammetric Engineering and Remote Sensing*, 73, 9, 1029–40, 2007.
9. Siebert, S. et al., Development and validation of the global map of irrigation areas, *Hydrology and Earth System Sciences*, 9, 535–47, 2005.
10. Siebert, S., Hoogeveen, J., and Frenken, K., *Irrigation in Africa, Europe and Latin America – Update of the Digital Global Map of Irrigation Areas to Version 4*, Frankfurt Hydrology Paper 05, Institute of Physical Geography, University of Frankfurt, Frankfurt am Main, Germany and Food and Agriculture Organization of the United Nations, Rome, Italy, 2006.
11. Droogers, P., *Global Irrigated Area Mapping: Overview and Recommendations*, Working Paper 36, International Water Management Institute, Colombo, Sri Lanka, 2002.
12. Turral, H., *Global Irrigated Area – Working Document*, International Water Management Institute, Colombo, Sri Lanka, 2002.
13. Velayutham, M. et al., Agro-ecological subregions of India for planning and development, *NBSS, Publ.* 35, 372, NBSS & LUP, Nagpur, India, 1999.
14. Loveland, T.R. et al., Development of a global land cover characteristics database and IGBP DISCover from 1 km AVHRR data, *International Journal of Remote Sensing*, 21, 6/7, 1303–30, 2000.
15. Bartholomé, E. and Belward, A.S., GLC 2000: A new approach to global land cover mapping from earth observation data, *International Journal of Remote Sensing*, 26, 9, 1959–77, 2005.
16. IGBP (International Geosphere-Biosphere Programme), *Global Change*, Report No. 12, Stockholm, Sweden, 1990.
17. Olson, J.S., *Global Ecosystems Framework: Translation Strategy*, Internal Report, USGS EROS Data Centre, Sioux Falls, SD, 1994.

18. Olson, J.S., *Global Ecosystems Framework: Definitions*, Internal Report, USGS EROS Data Centre, Sioux Falls, SD, 1994.
19. MoWR (Ministry of Water Resources), *3rd Census of Minor Irrigation Schemes (2000–01)*, Ministry of Water Resources, New Delhi, Government of India, 2005, http://mowr.gov.in/micensus/mi3census/index.htm.
20. Thenkabail, P.S., Schull, M., and Turral, H., Ganges and Indus river basin land use/land cover (LULC) and irrigated area mapping using continuous streams of MODIS data, *Remote Sensing of Environment*, 95, 3, 317–41, 2005.
21. Biggs, T. et al., Vegetation phenology and irrigated area mapping using combined MODIS time series, ground surveys, and agricultural census data in Krishna River Basin, India, *International Journal of Remote Sensing*, 27, 19, 4245–66, 2006.
22. Biradar C.M. et al., A global map of rainfed cropland areas (GMRCA) at the end of last millennium using remote sensing. *International Journal of Applied Earth Observation and Geoinformation*. 11, 2, 114–129. doi:10.1016/j.jag.2008.11.002. January, 2009.
23. Velpuri, M., et al,. Influence of Resolution or Scale in Irrigated Area Mapping and Area Estimations. *Photogrammetric Engineering and Remote Sensing (PE&RS)* (accepted, in press).
24. Dheeravath, V., et al., Irrigated areas of India derived using MODIS 500m data for years 2001-2003. *ISPRS Journal of Photogrammetry and Remote Sensing* (in review).
25. Tucker, C.J., Grant, D.M., and Dykstra, J.D., NASA's Global Orthorectified Landsat data set, *Photogrammetric Engineering & Remote Sensing*, 70, 3, 313–22, 2005.
26. Directorate of Economics and Statistics, *Agricultural Statistics (Provisional)*, Ministry of Agriculture, New Delhi, 2001–2002.
27. CBIP (Central Board of Irrigation and Power), *Central Board of Irrigation and Power Irrigated Area Map of India for 1994*, New Delhi, India, 1994.
28. Reddy, G.P.O. et al., Irrigated areas of India based on satellite sensors and national statistics: Issues and way forward from global irrigated area mapping (GIAM)-2006, Presented in First International Workshop on Global Irrigated Area Mapping (GIAM), 25–27 September, IWMI, Colombo, Sri Lanka, 2006.

6 Mapping Irrigated Lands across the United States Using MODIS Satellite Imagery

Jesslyn F. Brown
U.S. Geological Survey

Shariar Pervez and S. Maxwell
Stinger Ghaffarian Technologies Inc.*

CONTENTS

* Contractor to the USGS EROS, work performed under USGS contract 08HQ CN0005.

6.1   INTRODUCTION

Although there are over 22 million hectares (Mha) of irrigated land in the United States [1], consistent, current, and detailed geospatial information on the distribution and location of all irrigated fields is not readily available. The United States Department of Agriculture (USDA) National Agricultural Statistics Service (NASS) publishes the amount of irrigated area by county (administrative unit) and by crop every year (in acres) in its annual crop statistics and every five years in the US Census of Agriculture [1]. The statistics published by the USDA are collected via questionnaires and summarized by county. The current wall-to-wall land cover database of the conterminous United States [2], mapped with 2001-era Landsat satellite imagery at 30-m^2 resolution, does not include an irrigated lands class. All agricultural land cover is categorized as either cultivated crops or pasture/hay. Also, within the last few years, complementary research has resulted in geospatially continuous estimates of land surface evapotranspiration based on remote-sensing inputs [3,4]. These efforts model evapotranspiration across rain-fed and irrigated landscapes and have been successful in modeling surface energy fluxes. However, up to now, these and other efforts have not separated land use types, are subnational in scope (Afghanistan is the study area in the case of Senay et al. [4]), and are not yet operational. To our knowledge, few efforts have resulted in a cohesive and contemporary (post-2000) map of irrigation status across the entire United States with a spatial resolution that provides subcounty mapping details.

What is the value of information on irrigation status for water management or research uses? Irrigation in the United States accounts for about 65% of total water use (excluding thermoelectric power generation) and remains the largest user of freshwater [5]. In 2000, irrigation water use was estimated to be 137 billion gallons (518.6 billion liters) per day [5].

In areas with arid climates, irrigation is required to grow crops. In semiarid and humid regions, irrigation increases crop yields and reduces risk associated with climatic variability. For example, in 2002, the average crop yields for soybean, corn, barley, and wheat on irrigated farms in the United States surpassed the yields for rain-fed farms by 15, 30, 99, and 8%, respectively [6].

Due to the rising population and the potential impacts of climatic change, freshwater may well be a heavily contested resource in the future, thus making timely, detailed, objective, and accurate information on the location and amount of irrigated areas in managed systems even more valuable to water managers and policy makers [7]. Accurate, current, and detailed information on irrigation status could improve crop yield models as well as the estimation of crop water use or groundwater withdrawals for irrigation.

Satellite earth observations provide synoptic and objective geospatial information related to water use and management, including irrigation status. The first attempt to map global irrigation incorporating satellite observations was recently released (the satellite sensor-based Global Irrigated Area Mapping; GIAM) by the International Water Management Institute [8]. This data set represents global irrigated lands for 1999 and is based mainly on 1-km^2 satellite imagery from the Advanced Very High Resolution Radiometer (AVHRR) and Système Pour l'Observation de la Terre (SPOT) vegetation sensors.

Here, we present a consistent, objective, and repeatable methodology, initiated at the USGS Earth Resources Observation and Science Center for mapping contemporary (post-2000) irrigation status in the United States, incorporating satellite imagery from NASA's Moderate Resolution Imaging Spectroradiometer (MODIS) instrument. The resulting geospatial irrigation data are referred to as the MODIS Irrigated Agriculture Dataset for the United States (MIrAD-US; Figure 6.1). Validation of this type of national-level data set is challenging because few, if any, data sets on contemporary irrigated lands have parallel details and coverage for comparison. The collection of comprehensive groundtruth data is currently beyond the scope of this project. To illuminate the strengths and weaknesses of the MIrAD-US data set, we present comparisons between MIrAD-US and the GIAM database and provide a plan for future validation of MIrAD-US.

6.2 MIrAD-US METHODS AND DATA

Our methodological approach for mapping irrigation in the United States incorporated three data sources: time-series satellite-derived vegetation index (Normalized Difference Vegetation Index; NDVI) observations, USDA county-level irrigation

FIGURE 6.1   (See color insert following page 224.) The MIrAD-US map. Irrigated pixels are shown in red on a background showing peak vegetation growth.

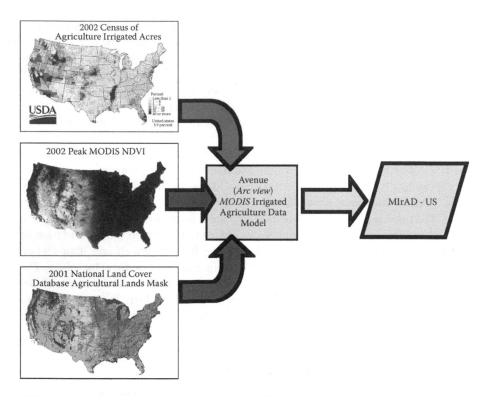

FIGURE 6.2   **(See color insert following page 224.)** MIrAD-US model processing flow.

summary statistics for 2002, and general land cover information. Figure 6.2 is an illustration of the methodological flow.

The NDVI is well known in the literature on remote sensing. The NDVI represents a dimensionless, radiometric measure that capitalizes on the differential response of incident visible red (absorbed by leaf chlorophyll) and near-infrared (reflected by spongy mesophyll) radiation in the vegetation canopy, and has been found to be correlated with the relative condition and amount of green vegetation [9,10]. As a result, time-series NDVI observations have proven useful for the quantification of seasonal events [11–13] and biophysical vegetation characteristics (e.g., leaf area index and biomass) [14–17].

The overall concepts guiding our remote-sensing mapping approach are as follows:

1. Irrigated crops should have higher peak NDVI values than nonirrigated crops [18].
2. The spectral values (and consequently, the NDVI) of irrigated and nonirrigated crops ought to be similar in seasons with optimum precipitation. Severe drought conditions will maximize the difference between irrigated and nonirrigated crops. The annual period with the most severe and widespread drought conditions should be selected for the model.
3. The growing season peak NDVI, whenever it occurs, is preferred as opposed to choosing a specific time period, because peak greenness time varies for each crop and for each geographic region of the United States [18].

6.2.1 MODIS Annual Peak Vegetation Index

The NDVI calculated from surface reflectance red and near-infrared measurements collected by the MODIS instrument aboard the Earth Observing System (EOS) Terra platform [19] provided a data source for estimating the annual peak (or maximum) of growing season productivity at a 250-m spatial resolution. MODIS (MOD13Q1, Collection 4) 16-day composite NDVI data were obtained from the Land Processes Distributed Active Archive Center (LPDAAC) and accessed from the EOS Data Gateway (http://edcimswww.cr.usgs.gov/pub/imswelcome/). We calculated the annual peak (or maximum) NDVI from a 2002 annual MODIS time series to incorporate into the MIrAD-US irrigation model.

6.2.2 Land Cover

The methodology incorporated a land cover mask to correct potential misclassification by screening out nonagricultural lands from the model. The mask was calculated from land cover information accessed from the 2001 National Land Cover Database (NLCD) [2]. The dominant land cover in each 250-m cell was calculated using geospatial techniques. Tests determined that forestlands and agricultural lands both exhibit high annual peak NDVI. Table 6.1 shows the average statistics for the 2002 MODIS annual peak NDVI within each cover class in the NLCD.

To avoid misidentification of irrigation in forested lands that also exhibit high peak NDVI, areas that were not classified in the NLCD land cover categories of pasture/hay or cultivated crops were eliminated or masked from consideration by the model. However, the modest agreement between NASS county farmlands and NLCD county agricultural lands (pasture/hay and cultivated crops) in Figure 6.3 suggest a possible source of error in spatial distribution of irrigated lands in the model result.

TABLE 6.1
Peak Annual NDVI Statistics from 2002 MODIS
for Specific NLCD Classes

NLCD Land Cover Class	2002 Peak Annual MODIS NDVI (class mean)
Perennial ice/snow	0.33
Barren land	0.24
Deciduous forest	0.85
Evergreen forest	0.75
Mixed forest	0.85
Shrub/scrub	0.38
Grassland/herbaceous	0.51
Pasture/hay	0.79
Cultivated crops	0.75
Woody wetlands	0.83
Emergent herbaceous wetlands	0.68

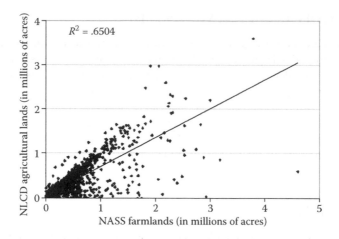

FIGURE 6.3 Comparison of NASS county farmlands [1] and NLCD county agricultural lands. [2]

TABLE 6.2
Recent Trends in Irrigated Area in the United States

Year	2002	1997	1992	1987	1982
Irrigated lands (1000 acres/Mha)	55,311/22.38	56,289/22.78	49,404/19.99	46,386/18.77	49,002/19.83
Difference from 2002 (1000 acres/Mha)	–	+978/400 (0.02%)	–5,907/–2,386 (–0.1%)	–8,925/–3,608 (–0.16%)	–6,309/–2,549 (0.12%)

6.2.3 COUNTY IRRIGATION STATISTICS

The irrigated area for each of the 3114 counties in the United States in 2002 was provided by the 2002 Census of Agriculture. The Census is a full accounting of agriculture for the United States, performed by the USDA every 5 years (1992, 1997, 2002, and so on). The total irrigated area in the United States appears to be quite stable over time, from 1982 to 2002 (Table 6.2). The largest difference from 2002 occurred in 1987 when 3.608 Mha less (0.16%) were irrigated [6].

6.2.4 MIRAD-US AUTOMATED MODEL

In an automated classification environment (ArcView Avenue), the 2002 county statistics (that is, number of irrigated hectares) provided the criteria (or target threshold) for dynamically identifying and selecting the appropriate area (in a number of 250-m² cells) with the highest annual peak MODIS NDVI within the appropriate land cover categories. This step required an iterative approach implemented county by county. In the first iteration, the cells with the highest peak NDVI within the land cover masked cells in a county were identified as irrigated lands, and the accumulated area covered by those pixels was calculated. Then, this accumulated area was compared with the target area provided by the 2002 Census of Agriculture. If the total area of the selected cells was less than the target area, then cells with the next highest annual peak NDVI were selected in descending order and the accumulated area of the cells identified by the two highest annual peak NDVI values, which were then compared with the target number of irrigated hectares for the county. These steps were repeated until the target area for the county was exceeded. In a final step, all lone 250-m^2 pixels were filtered from the irrigated cells, based on the assumption that, in the United States, irrigated fields are generally larger than 62,500 m^2 (15.4 acres/6.3 ha). The size of a typical center pivot irrigation is about 125 acres/50 ha [20]. This methodology may, therefore, not resolve very small, isolated, irrigated fields. The model was run for each county in the conterminous United States and a two-class map was produced, delineating irrigated and nonirrigated lands at 250-m^2 resolution.

6.3 EVALUATION OF MIrAD-US AND COMPARISON WITH GIAM

Because there are few, if any, detailed, current, published maps of irrigation status (independent of the USDA Census of Agriculture statistics) across the conterminous United States, we performed a comparison of the MIrAD-US data set with GIAM. GIAM has eight irrigation classes downscaled to a 1-km resolution with fractional irrigation expressed in each cell for each class. GIAM incorporated several methods to calculate class-specific irrigation fractions. In the United States, most irrigation is generally delivered in a single season [5], so the fraction coefficients computed using a combination of Google Earth, high-resolution imagery, and fallow were the preferred method. In this study, these GIAM fraction coefficients are incorporated.

The purpose of the comparison was to illuminate the similarities and differences in the two data sets and to identify potential regional strengths and weaknesses. We present geospatial analyses of the spatial distribution patterns of irrigated areas from both maps, the cluster sizes of the irrigated areas, areas of agreement and disagreement, and areas of potential overestimation and underestimation of irrigated lands. Two additional comparisons were performed. In one, the GIAM and the MIrAD-US irrigated areas were compared with the county irrigated area statistics published by the USDA in the 2002 Census of Agriculture [1]. In the other, an analysis of the relationship between irrigated area and average annual precipitation is presented.

6.3.1 EVALUATION METHODOLOGY

A simple image-differencing technique can clarify the differences and similarities between two raster image data sets. However, the significant differences in the spatial resolutions (GIAM is 1-km and MIrAD-US has 250-m resolution), classification methods, and different timing of the source satellite imagery of these irrigation data sets meant that a cell-by-cell analysis would be challenging to interpret. The resulting areas of disagreement related to resolution differences would be difficult to separate from differences in the data sets related to methodologies, source data, or classification errors. We chose geospatial analytical techniques to investigate the spatial distribution patterns of variables associated with spatial domain for this comparison. We have used linear regression [21], Moran's I statistics [22,23], and simultaneous autoregression methods [22] in a geospatial analysis and comparison of irrigated areas in MIrAD-US and GIAM.

6.3.2 REGIONALIZED APPROACH

We used a regional framework for our analysis and comparison of MIrAD-US and GIAM. This regional approach was supported by the geographic array of irrigation patterns and irrigation water sources. Most of the irrigated lands in the United States are located in the western and central regions. Surface water is the primary source for irrigation in the intermountain west, whereas groundwater is the main source for irrigation in the central region [5]. The eastern United States is comparatively humid, and irrigation is less prevalent in this part of the country, except in some areas in the Mississippi Flood Plain (MFP) and in Florida. Groundwater is also the main source of irrigation for the MFP region [5].

Four regions (shown in Figure 6.4) were identified (western, central, MFP, and eastern) based on the primary water source for irrigation, climate, geographic homogeneity, and extent of the groundwater aquifer (where groundwater was the main source). Three of these regions (western, central, and MFP) include 18 of the 20 states with the largest amount of irrigated lands, according to the 2002 Census of Agriculture [6]. The eastern region was not considered in this analysis, but will be analyzed at a future date.

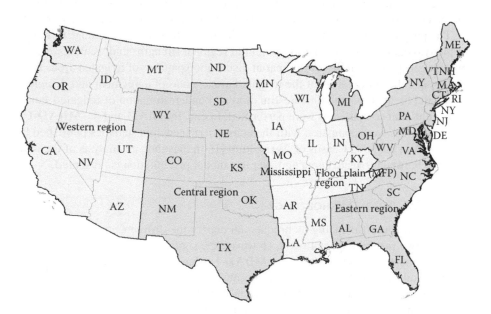

FIGURE 6.4 Analysis regions based on main irrigation water source, climate, geographic heterogeneity, and extent of the groundwater aquifer.

6.3.3 DATA PREPARATION

Data preparation steps included reprojection and grouping of the GIAM data categories, data normalization, and calculation of the fraction of irrigated area by county, or County Irrigated Area Fraction (CIAF). We chose the Lambert Azimuthal Equal Area projection for this comparison to match the coordinate system of the MIrAD-US data set.

The US counties (administrative units) were used as the spatial unit for this comparison. The CIAF, the fractional area of each county designated as irrigated, was calculated from both maps. In this calculation, all eight categories of the GIAM data were considered irrigated. The MIrAD-US data contain two categories, irrigated and nonirrigated lands, so the CIAF calculation was relatively straightforward.

Of the 3,114 counties in the United States, there are 346 counties in the western region, 715 in the central region, and 1,005 in the MFP region (Table 6.3). There are some limitations in the analytical strategy based on counties, which we will discuss later.

An *arcsin square root* transformation was applied to the CIAF summaries from GIAM and MIrAD-US to ensure that the data were normally distributed [21]. This transformation provided a remedy for outliers and skewness in each distribution, because the geospatial statistical techniques are based on an assumption of normally distributed data.

An ArcInfo shapefile format was used to manage the CIAF calculation for GIAM and MIrAD-US for the three regions of analysis. We used the R statistical software package to investigate the spatial autocorrelation of each regional irrigation data set and to compare the GIAM and MIrAD-US. Region-specific neighborhood objects were created for the counties based on Queen's continuity rules with row standardized weight matrices. The Queen's continuity rules identify the neighbors of each county that share the border or vertices with that specific county [21,24]. A weighted matrix describes the relationship between an element (in this case, county) and its neighboring elements. Here, in the weighted matrix, a value of 1 represents pairs of elements that are adjacent, and a value of 0 represents pairs that are not adjacent. For the computation of Moran's I and spatial regression, the weighted matrix was standardized by row to yield a meaningful interpretation of the results [21], by dividing each element in a row by the corresponding row sum.

6.3.4 DISTRIBUTION PATTERNS OF IRRIGATED AREAS IN GIAM AND MIrAD-US

We analyzed the distribution patterns of the irrigated areas, as summarized by county (i.e., CIAFs) from both maps, using Moran's scatterplots. Moran's scatterplot

TABLE 6.3
County Area Statistics by Region [1]

Regions	Total Area (km²)	No. of Counties	Average County Area (km²)
Western	2,413,971	346	6,977
Central	2,318,369	715	3,242
MFP	1,527,000	1,005	1,519

describes an observation in respect to its surrounding neighbor's observations. It is a plot of x_i (the value at each county vs. $\Sigma w_{ij}x_i$), which is the weighted average of the x_i value based on neighbors [22]. Not surprisingly, the scatterplots for all three regions show evidence of spatial autocorrelation for CIAF in both maps, indicating that irrigated areas were mainly mapped in cluster patterns in both maps. Figure 6.5a and b present very similar spatial distribution characteristics for irrigated areas in the western region. For both maps, the numbers of counties with low CIAF surrounded by the low weighted average of CIAF for neighboring counties (in other words, the low-low distribution) were high, and there were relatively few counties with high CIAF surrounded by the high weighted average of CIAF for neighboring counties (that is, high-high distribution). This suggests that both maps effectively mapped irrigation patterns with similar types of spatial distributions. In the western region, average farm sizes were larger and over 50% of the farms were irrigated (Table 6.4), so irrigation occurred at relatively larger scales/clusters in this region, according to the USDA Census of Agriculture. This appears to support the distribution pattern result (Figure 6.5a and b). In other words, the spatial autocorrelation characteristics, or pattern of distribution of both data sets are similar.

The central region shows markedly different distribution patterns from the western region in the two maps. In Figure 6.6a and b, the low-low distribution was high for MIrAD-US, but comparatively lower for GIAM. On the other hand, the high-high distribution was comparatively low for MIrAD-US, but high for GIAM. This difference in the two CIAF distributions indicates that the MIrAD-US was effective at mapping sparsely irrigated areas across the central region, whereas GIAM appeared to be less successful at mapping sparse irrigation and predominantly captured larger, spatially aggregated, irrigated areas in the region.

In the MFP region, the low-low distribution was again comparatively high for the MIrAD-US, while the high-high distribution was low (Figure 6.7a and b). Results in the central region are similar in that the GIAM low-low distribution was comparatively low, but the high-high distribution was high. This analysis indicated that the total number of counties with high CIAF was higher in the GIAM data set than that for the MIrAD-US data.

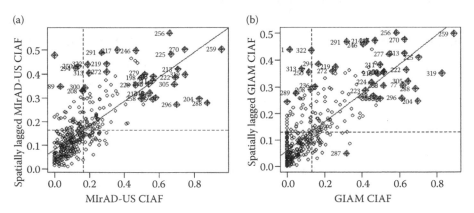

FIGURE 6.5 Distribution pattern of irrigated areas by county in the western region, (a) CIAF extracted from MIrAD-US irrigated area map, (b) CIAF extracted from GIAM.

TABLE 6.4
Farm Size and Number of Irrigated Farms from the USDA 2002 Census of Agriculture [1]

Region	Average Farm Size (acres/ha)	Farms with Irrigation (%)	Total No. of Farms	No. of Irrigated Farms
Western	813.5/329.2	50.4	264,674	133,356
Central	799.6/323.6	15	513,692	76,842
MFP	264.4/107	4	779,963	31,189

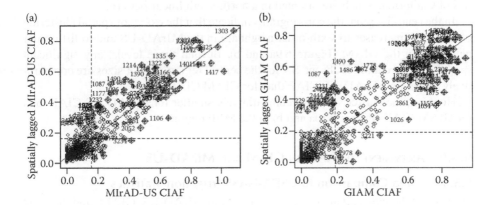

FIGURE 6.6 Distribution pattern of irrigated areas by county in the central region, (a) CIAF extracted from MIrAD-US, (b) CIAF extracted from GIAM.

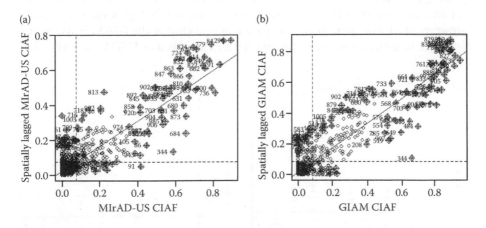

FIGURE 6.7 Distribution pattern of irrigated areas by county in the MFP region, (a) CIAF extracted from MIrAD-US irrigated area map, (b) CIAF extracted from GIAM.

A feature found in the distribution of all three regional GIAM plots was the large number of zero CIAF data points with wide variability in the spatially lagged GIAM CIAF. This was very obvious for the western and central regions.

6.3.5 GIAM AND MIrAD-US SPATIAL STRUCTURE

Another geospatial tool, Moran's I correlogram, provided information on the spatial patterns of irrigation in the CIAF distributions for both maps. Moran's I provides a description of spatial structure (i.e., cluster structure or distribution) and an indication of the extent of significant spatial clustering of similar values around any set of observations [23]. The correlograms for the western region for CIAF in Figure 6.8a and b showed very similar structures of spatial clustering up to the third order of neighbors from both maps. The similar Moran's I values in the correlograms between lag 4 and 7 indicate significant autocorrelation of similar values in a linear pattern, which corresponds well with the high density of irrigation in the central valley counties of California, which are oriented in a north-south linear pattern.

In the central region, the correlograms indicate that the extent of spatial clustering of the CIAF increases to sixth-order neighbors in the MIrAD-US and to fifth-order neighbors in the GIAM (Figure 6.9a and b). Beyond these levels, any significant spatial autocorrelation is absent. So, for the central region, cluster sizes are comparatively bigger for MIrAD-US CIAF than for GIAM CIAF.

However, the cluster sizes are comparatively smaller for MIrAD-US CIAF than for GIAM CIAF (Figure 6.10a and b) in the MFP region.

6.3.6 ASSESSMENT OF CIAFs IN GIAM AND MIrAD-US

6.3.6.1 Area Comparison to USDA-NASS Irrigation Statistics

Statistics of the irrigated area in each county from MIrAD-US, GIAM, and the 2002 USDA Census of Agriculture [1], summarized by region, are shown in Table 6.5. The MIrAD-US model is derived from the 2002 USDA Census of Agriculture county irrigated area statistics, and because the USDA statistics are not an independent measure for analyzing the MIrAD-US, we expect and see a relatively high level of agreement. However, the USDA irrigation statistics are independent of the GIAM irrigated area statistics. In the western region, the irrigated area identified by USDA was 83,600 km²

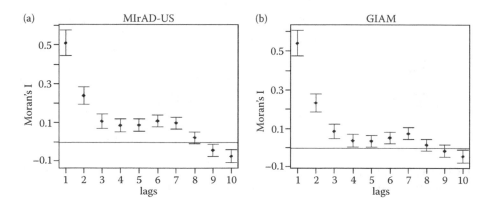

FIGURE 6.8 Moran's I correlograms showing pattern of spatial clustering in the western region, from (a) MIrAD-US, and (b) GIAM.

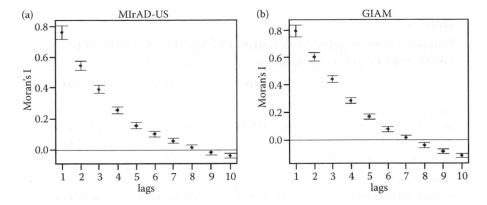

FIGURE 6.9 Moran's I correlograms showing pattern of spatial clustering in the central region, from (a) MIrAD-US, and (b) GIAM.

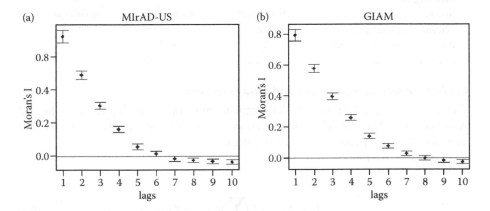

FIGURE 6.10 Moran's I correlograms showing pattern of spatial clustering in the MFP region, from (a) MIrAD-US irrigated area map, and (b) GIAM.

(about 3.47% of the total area in the region). In comparison, the GIAM irrigated area was 64,900 km^2 (about 2.7% of the total area in the region). For the western region, the GIAM irrigated area was 22.5% less than that given by the 2002 USDA Census of Agriculture [1]. The central region GIAM irrigated area estimate was 141,400 km^2, or 164.5% of the USDA areal estimate (86,000 km^2). GIAM estimated 6.1% of the entire region as irrigated, while MIrAD-US and the USDA estimated 3.7% of the area as irrigated. The USDA statistics reported 36,600 km^2 of irrigated area in the MFP region, but the GIAM estimate was 56,400 km^2, or 154% of the USDA areal estimate. Therefore, the most significant differences in irrigated area statistics are in the central and MFP regions between GIAM and the MIrAD-US and between GIAM and the 2002 USDA Census of Agriculture irrigation statistics. We further analyzed the spatial distribution and magnitude of these differences, focusing on the GIAM and MIrAD-US. Local Moran's I statistic [22,24] Maps and ordinary least squarest (OLS) regression models [22] were applied for all three regions.

TABLE 6.5

Comparison of Irrigated Area Statistics of MIrAD-US, GIAM, and USDA 2002 Census of Agriculture

	MODIS	GIAM	USDA
Western Region			
Irrigated area (km²/Mha)	83,849/8.39	64,972/6.49	83,625/8.36
Percent of the total area	3.5	2.7	3.4
Percent difference from USDA	0.3	−22.3	−
Central Region			
Irrigated area (km²/Mha)	87,319/8.73	141,469/14.14	86,013/8.60
Percent of the total area	3.7	6.1	3.7
Percent difference from USDA	1.1	64.5	−
MFP Region			
Irrigated area (km²/Mha)	39,679/3.97	56,466/5.64	36,609/3.66
Percent of the total area	2.6	3.7	2.4
Percent difference from USDA	8.5	54.2	−

6.3.6.2 Areas of Agreement and Disagreement

The local Moran's I (Equation 6.1) statistic map of Local Indicator of Spatial Association (LISA) provided an indication of locations of so-called hot spots, or spatial high-density clusters of similar values.

Local Moran's I:

$$I_{\text{local}} = (x_i - \bar{x}) \sum_{j=1}^{n} w_{ij}(x_j - \bar{x}), \qquad (6.1)$$

where x_i is the observation (CIAF), $\bar{x}$ is the mean of the observations, w_{ij} is the weight, and x_j is the neighbor's observation. The values of Moran's I generally vary between 1 and −1. Positive autocorrelation in the data translates into positive values of I; negative autocorrelation produces negative values. Values close to zero indicate no autocorrelation [24]. The maps of local Moran's I for GIAM and MIrAD-US can be subsequently compared for the geographic patterns of similarity and dissimilarity. The western region maps of local Moran's I (Figure 6.11a and b) generally show strong agreement in the higher density irrigated counties in California, Washington, and Idaho. In the central region, the agreement is good in the irrigated counties in Nebraska and Texas, but it is poor for western Kansas (Figure 6.11c and d), where the GIAM data set indicates a much higher density of irrigated lands. The central region maps also show evidence of magnitude differences in irrigated areas within the clusters of high-density irrigation.

In the MFP region (Figure 6.11e and f), the LISA maps show close agreement for the large cluster of irrigated lands covering parts of Missouri, Arkansas, Mississippi, and Louisiana. However, GIAM identified a few clusters of irrigation in southern

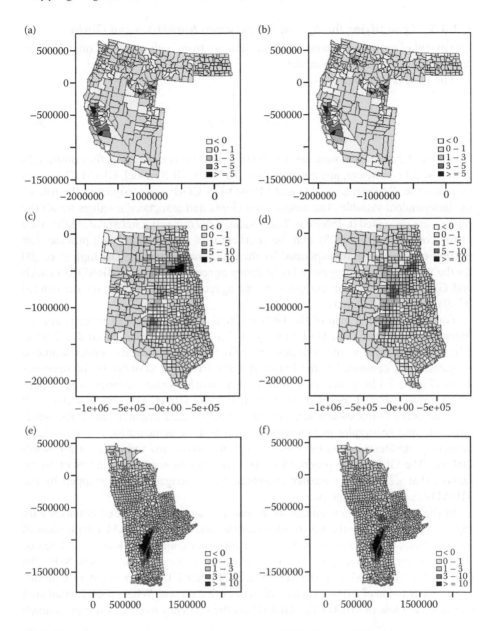

FIGURE 6.11 **(See color insert following page 224.)** The local Moran's statistic maps of LISA for the western region (a) MIrAD-US CIAF, (b) GIAM CIAF; the central region (c) MIrAD-US CIAF, (d) GIAM CIAF; and the MFP region (e) MIrAD-US CIAF, (f) GIAM CIAF.

Illinois and central Iowa where MIrAD-US did not, and the MIrAD-US irrigated area map identified a cluster of irrigation in northern Indiana. The LISA maps provided an effective and heuristic tool for qualitative comparison and visual assessment of spatial clustering of irrigated areas from both the GIAM and the MIrAD-US, but the maps do not support a quantitative evaluation of spatial agreement.

6.3.6.3 Quantifying the Agreement Between MIrAD-US and GIAM

The extent of spatial agreement in irrigated areas from both maps was quantified by applying an OLS regression model.

The OLS regression:

$$y = \beta_0 + x\beta_1 + \varepsilon, \tag{6.2}$$

where y is the dependent variable, x is the independent variable, β_0 is the constant, β_1 is the model coefficient, and ε is the noise. To fit the OLS model, GIAM CIAF was assigned the dependent variable and MIrAD-US CIAF was designated the role of the independent variable. The model coefficients and associated p-values for all the regions are presented in Table 6.6. The p-value determines the statistical significance of the coefficients (∞=0.05), and the coefficient of determination (r^2) provides the portion of total variation explained by the regression equation. The high r^2 of .80 for the western region suggests a fairly strong agreement between MIrAD-US CIAF and GIAM CIAF. But, in comparison, the agreements are weaker for the central (r^2=0.63) and MFP regions (r^2=0.58).

For each region, a map of the OLS residuals and a scatterplot of residuals vs. independent variable (MIrAD-US) supported an analysis of the spatial distribution of residuals (in other words, counties where GIAM overestimated or underestimated irrigated areas compared to the irrigated areas from MIrAD-US). In the residuals maps (Figure 6.12a, c, and e), counties with positive values identify areas where GIAM overestimated irrigated lands compared to MIrAD-US, and counties with lower negative values show where GIAM underestimated irrigated lands. The residuals map and scatterplot in Figure 6.12a and b show comparatively low residuals occurring randomly across the western region without any trend except for North Dakota. The GIAM map produced CIAF estimates in northern counties of North Dakota that were 10% or greater compared to the irrigated areas mapped by the MIrAD-US in those counties.

In the central region, residuals were spatially autocorrelated, and counties with high residuals were mostly in Nebraska, Kansas, and Texas. In GIAM, results showed overestimates in irrigated areas by 20% or more for counties in Kansas and Texas, and underestimates in irrigated lands by 20% or more for a few counties in Nebraska (Figure 6.12c). GIAM and MIrAD-US have higher CIAF agreement in counties where the fraction of irrigated area is high in both data sets. Where the irrigated area estimates are low in MIrAD-US, GIAM had the tendency to either estimate a much

TABLE 6.6

The OLS Model Coefficients and Associated p-Values

Parameters	Western Region Value	Western Region P-Value	Central Region Value	Central Region p-Value	MFP Region Value	MFP Region p-Value
Slope (β_1)	0.9	~0	1.09	~0	1	~0
R^2	0.8		0.63		0.58	

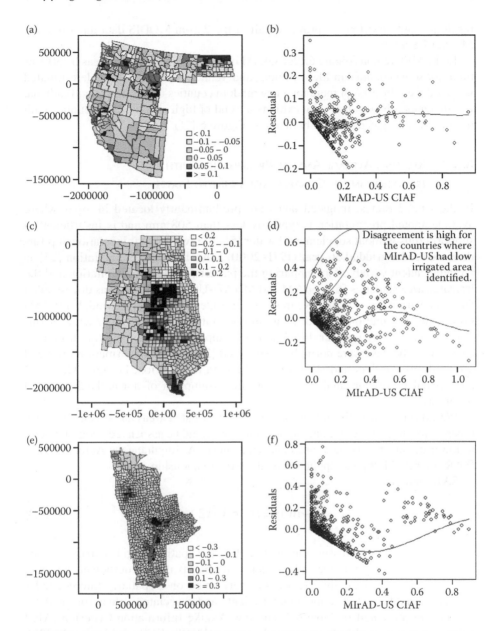

FIGURE 6.12 (See color insert following page 224.) (a) The OLS regression residuals map, (b) scatterplot of MIrAD-US CIAF and OLS residuals for the western region; (c) OLS regression residuals map, (d) scatterplot of MIrAD-US CIAF and OLS residuals for the central region; and (e) OLS regression residuals map, (f) scatterplot of MIrAD-US CIAF and OLS residuals for the MFP region.

higher irrigated fraction or to estimate none (zero) (Figure 6.12d). These results indicate that the MIrAD-US model has a better capability of mapping geographically sparse irrigation, whereas the GIAM strategy mapped the larger clusters of irrigated areas fairly well in the central region. This is not surprising, considering the finer resolution and, therefore, improved detail of the 250-m MODIS data input into the MIrAD-US model.

In the MFP region (Figure 6.12e), GIAM overestimated irrigated areas by 30% or more for some counties in Iowa, Illinois, and Tennessee, whereas it underestimated irrigated areas by 30% or more for a few random counties in Minnesota and Indiana. In this region, the residuals plot exhibits a trend of high residuals for counties with high-density irrigated areas in both maps (Figure 6.12f).

6.3.7 RELATING AVERAGE ANNUAL PRECIPITATION WITH IRRIGATION IN THE CONTERMINOUS UNITED STATES

In the United States, irrigated lands are predominantly located in areas where average annual precipitation is typically less than 508 mm and is insufficient to support crops without supplemental water [5]. To investigate this relationship further, we used a gridded 30-year (1971–2000) annual average precipitation [25] as an independent variable for comparing the patterns between precipitation and the irrigated area statistics from GIAM and MIrAD-US. The data set was downloaded from the Parameter-elevation Regressions on Independent Slopes Model (PRISM) group data distribution web portal at Oregon State University (http://www.prism-climate.org), and a *log* transformation was applied to the precipitation data to ensure that the data were normally distributed [21]. This transformation provided a remedy for outliers and skewness in the data distribution, because Moran's I and spatial regression analyses rely on the assumption of normally distributed observations.

We expected a negative relationship between irrigated areas and precipitation (in other words, higher average annual precipitation would be associated geographically to lower density in irrigated lands and vice versa). A Simultaneous Autoregressive (SAR) model [24,26] was applied to analyze the relationship.

SAR regression:

$$(1-\rho W)y = (1-\rho W)X\beta + \varepsilon, \tag{6.3}$$

where ρ is the autoregressive parameter, W is the spatial weight matrix, y is the dependent variable, X is the independent variable, β is the coefficient, and ε is the noise. This is basically a regression performed on a spatially filtered, independent variable [26]. The model coefficients and associated values for all three regions are presented in Table 6.7. The low Akaike Information Criterion (AIC) values suggest a good fit for the model compared to the linear model results [26]. The β coefficients show that average annual precipitation was indeed negatively autocorrelated with irrigated areas in both maps for all three regions. However, not all the coefficients were statistically significant (p-value <.05). Statistically significant autocorrelation was not found between GIAM or MIrAD-US CIAF and precipitation in the MFP region; however, these results support a conclusion that the MIrAD-US had a higher spatial correlation with precipitation in the western and central regions.

TABLE 6.7
Simultaneous Autoregressive (SAR) Model Coefficients and Associated Values for CIAF as a Function of Historical Annual Average Precipitation

Regions	Source Map	β	p-Value	ρ	AIC	Linear Model
Western	MIrAD-US	−0.156	2.39E-13	0.85	−506.1	−291.7
	GIAM	−0.132	3.93E-10	0.839	−508.8	−278.7
Central	MIrAD-US	−0.128	0.0002	0.924	−1436.2	−541.4
	GIAM	−0.087	0.06	0.925	−1058.4	−155.8
MFP	MIrAD-US	−0.002	0.344	0.907	−2292.2	−1206.1
	GIAM	na	na	na	−2132.1	−564.3

6.3.8 LIMITATIONS AND CAVEATS

Counties were the spatial units for this comparison and they vary substantially in size across the United States (Table 6.3). Across the eastern United States, counties are relatively small, and they are generally larger in the west. Thus, the associated data sets of these counties were subjects of a Modifiable Area Unit Problem (MAUP). Further investigation is required to examine the effects of MAUP on the spatial autocorrelation. It may especially be a factor in the higher agreement between the CIAFs in the western region. Moran's I and regression models assume normally distributed data sets. As CIAFs were not normally distributed data sets, there were violations of this assumption for these two functions.

The GIAM data set was created mainly from remote-sensing data collected in 1999. This means that there is a potential additional source of disagreement because the MIrAD-US was created from 2002 MODIS imagery. We did not account for this temporal mismatch in this analysis. However, we assumed that this mismatch was comparatively insignificant, since recent trends in irrigated area (from 1997 to 2002) appear to be holding steady (see Table 6.2).

6.4 DISCUSSION OF THE GEOSPATIAL ASSESSMENT OF MIrAD-US AND COMPARISON WITH GIAM

Comparison of remote-sensing-based maps with differences in spatial resolution and methods is a challenge. In this evaluation, we used geospatial-analysis techniques to compare geospatial information derived from two national-level irrigation data sets (note: GIAM is global, so the United States portion is a subset of GIAM) as opposed to conventional methods. The county-summarized, irrigated-area information (expressed as a fraction) extracted from MIrAD-US and GIAM was compared regionally across much of the conterminous United States. A regional analysis revealed that the two maps showed the highest level of agreement in the western region, where irrigation patterns are relatively homogeneous across comparatively larger farms. However, GIAM and MIrAD-US differed across the central and MFP regions, where average farm sizes are smaller and irrigation is more scattered and sparse. The GIAM showed a tendency to estimate more irrigated lands in the central and MFP regions when compared to the MIrAD-US. There may actually be less clustering in irrigation than indicated in this analysis because the eastern region was not compared and postprocessing eliminated single MODIS pixels from the MIrAD-US model results.

We compared the MIrAD-US and GIAM to two additional data sets, irrigated area reported by the USDA 2002 Census of Agriculture [1] and PRISM historical average annual precipitation [25]. The GIAM data were calculated independently from the USDA statistics and showed a total irrigated area that was 22.3% lower in the western region, 64.5% higher in the central region, and 54.2% higher in the MFP region. Our conclusion was that the resolution of the source data for GIAM (that is, 1-km AVHRR and SPOT Vegetation) may be at the root of the large differences in these irrigation amounts. In the comparison to historical average annual precipitation for the United States, the MIrAD-US exhibited a stronger relationship than GIAM over all three regions, but the relationship was strongest in the western region where the climate is the most arid.

In the analysis, Moran's scatterplots were helpful in explaining the distribution of associated geospatial information, and local Moran's I statistic maps showed the locations of spatial agreement and disagreement. An OLS regression model quantified the agreement in geospatial information and an autoregression model identified the strength of the relationship between geospatial variables. These geospatial-analysis techniques provided effective heuristic tools for comparing geospatial information derived from raster maps with differences in spatial resolutions and classification methods.

Where the two data sets showed the best agreement, we have gained higher confidence in both maps, based on this convergence of evidence. This is true for much of the western region, with the exception of the northern border of North Dakota and the eastern edge of Idaho. In the central region, the areas with the best agreement include much of Wyoming, Colorado, and New Mexico, along with eastern Oklahoma and Kansas. The northern, eastern, and southwestern edges of the MFP region generally show good agreement as well. The residuals plots (Figure 6.10b, d, and e) reveal different issues in all three regions.

6.5   FUTURE EVALUATION EFFORTS AND CONCLUSIONS

Plans are underway to perform further evaluation of the MIrAD-US. In the absence of other national-level irrigation maps with similar mapping detail, we plan to perform state-level comparisons with "best available" published and contemporary land cover maps, primarily focusing these efforts on the states with the largest amount of irrigated farmland (e.g., California, Nebraska, Texas, Arkansas, Idaho).

A recent evaluation that focused on Nebraska used a detailed 2005 Landsat-derived land use map containing an irrigated land use category [27]. The overall classification accuracy of the irrigation layer in the 2005 Nebraska land use map was 93.6% based on comparisons to 3,375 field reference points. Our comparisons between the MIrAD-US data for 2002 and the irrigated class in the Nebraska land use map [27] showed the proportion of agreement at 90.2%.

Irrigation water use is important to the productivity of agriculture in the United States, and it is linked to future land and water uses and to the cycling of water-borne chemicals. Monitoring irrigated lands and related agricultural water use will continue to be an important part of the national water budget of the United States and contribute to making projections about water supply and demand in the future.

ACKNOWLEDGMENTS

The MODIS time-series data set used in this research was created at the Kansas Applied Remote Sensing Program at the University of Kansas through funding from the USGS AmericaView Program (Grant No. AV04-KS01) and the Department of the Army (Grant No. 9SR-04-C-0006). The authors thank peer reviewers Dr. Eugene Fosnight and Dr. Darrell Napton for their thoughtful comments and suggestions. The authors also thank Ronald Smith, USGS, for the calculation of the land cover mask. Any use of trade, product, or firm names is for descriptive purposes only and does not imply endorsement by the US Government.

REFERENCES

1. USDA-NASS (U.S. Department of Agriculture, National Agricultural Statistics Service), *2002 Census of Agriculture, Summary and State Data, Volume 1, Geographic Area Series, Part 51, AC-02-A-51,* 2004, http://www.nass.usda.gov/census/census02/volume1/us/USVolume104.pdf.
2. Homer, C. et al., Development of a 2001 National Landcover Database for the United States, *Photogrammetric Engineering and Remote Sensing,* 70, 829–40, 2004.
3. Anderson, M.C. et al., A climatological study of evapotranspiration and moisture stress across the continental United States based on thermal remote sensing: 1. Model formulation, *Journal of Geophysical Research D: Atmospheres,* 2, D107, 2007.
4. Senay, G.B. et al., A coupled remote sensing and simplified surface energy balance approach to estimate actual evapotranspiration from irrigated fields, *Sensors,* 7, 979–1000, 2007.
5. Hutson, S.S. et al., *Estimated use of water in the United States in 2000,* U.S. Geological Survey, Reston, VA, Circular 1268, 2004.
6. Veneman, A.M., Jen, J.J., and Bosecker, R.R., *2002 Census of Agriculture—Farm and Ranch Irrigation Survey (2003), Volume 3, Special Studies, Part 1,* U.S. Department of Agriculture, National Agricultural Statistics Survey (NASS), 2004, http://www.agcensus.usda.gov/Publications/2002/FRIS/fris03pdf.
7. Bastiaanssen, W.G.M., Molder, D.J., and Makin, I.W., Remote sensing for irrigated agriculture: Examples from research and possible applications, *Agricultural Water Management,* 46, 137–55, 2000.
8. Thenkabail, P.S. et al., Global irrigated area map (GIAM) for the end of the last millennium derived from remote sensing, *International Journal of Remote Sensing,* 2009 (accepted, in press).
9. Asrar, G., Myneni, R.B., and Kanemasu, E.T., Estimation of plant canopy attributes from spectral reflectance measurements, in *Theory and Applications of Optical Remote Sensing,* G. Asrar, Ed., Wiley, New York, 1989, 252–96.

10. Baret, F. and Guyot, G., Potentials and limits to vegetation indices for LAI and APAR assessments, *Remote Sensing of Environment*, 35, 161–73, 1991.
11. Reed, B.C. et al., Measuring phenological variability from satellite imagery, *Journal of Vegetation Science*, 5, 703–14, 1994.
12. Reed, B.C., Loveland, T.R., and Tieszen, L.L., An approach for using AVHRR data to monitor US Great Plains grasslands, *Geocarto International*, 13–22, 1996.
13. Yang, L. et al., An analysis of relationships among climate forcing and time-integrated NDVI of grasslands over the US northern and central Great Plains, *Remote Sensing of the Environment*, 65, 25–37, 1998.
14. Sannier, C.A.D., Taylor, J.C., and Du Plessis, W., Real-time monitoring of vegetation biomass with NOAA-AVHRR in Etosha National Park, Namibia, for fire risk assessment, *International Journal of Remote Sensing*, 23, 71–89, 2002.
15. Wang, Q. et al., On the relationship of NDVI with Leaf Area Index in a deciduous forest site, *Remote Sensing of Environment*, 94, 244–55, 2005.
16. Gonzalez-Alonso, F. et al., Forest biomass estimation through NDVI composites: The role of remotely sensed data to assess Spanish forests as carbon sinks, *International Journal of Remote Sensing*, 27, 5409–15, 2006.
17. Wessels, K.J. et al., Relationship between herbaceous biomass and 1-km^2 Advanced Very High Resolution Radiometer (AVHRR) NDVI in Kruger National Park, South Africa, *International Journal of Remote Sensing*, 27, 951–73, 2006.
18. Wardlow, B.D. and Egbert, S.L., Large-area crop mapping using time-series MODIS 250 m NDVI data: An assessment for the US Central Great Plains, *Remote Sensing of Environment*, 2, 1096–1116, 2008.
19. Justice, C.O. and Townshend, J.R.G., Special issue on the Moderate Resolution Imaging Spectroradiometer (MODIS): A new generation of land surface monitoring, *Remote Sensing of Environment*, 83, 1–2, 2002.
20. Evans, R.G., *Center Pivot Irrigation*, Research Report, USDA-Agricultural Research Service, Sidney, MT, 2001.
21. Anselin, L., *Spacestat Tutorial. A Workbook for Using Spacestat in the Analysis of Spatial Data*, University of Illinois at Urban-Champaign, IL, 1992.
22. O'Sullivan, D. and Unwin, D.J., *Geographic Information Analysis*, John Wiley, Hoboken, NJ, 2002.
23. Barbaro, L. et al., The spatial distribution of birds and carabid beetles in pine plantation forests: The role of landscape composition and structure, *Journal of Biogeography*, 34, 652–64, 2007.
24. Overmars, K.P., de Koning, G.H.J., and Veldkamp, A., Spatial autocorrelation in multiscale land use models, *Ecological Modelling*, 164, 257–70, 2003.
25. Daly, C. et al., A knowledge-based approach to the statistical mapping of climate, *Climate Research*, 22, 99–113, 2002.
26. Anselin, L., Under the hood issues in the specification and interpretation of spatial regression models, *Agricultural Economics*, 27, 247–67, 2002.
27. Dappen, P. et al., *Delineation of 2005 Land Use Patterns for the State of Nebraska*, Final Report, Nebraska Department of Natural Resources, Lincoln, 2007, http://www.calmit.unl.edu/2005landuse/data/2005_NE_landuse_finalreport.pdf.

7 Use of Remote Sensing to Map Irrigated Agriculture in Areas Overlying the Ogallala Aquifer, United States

Bethany Kurz and Santhosh Seelan
University of North Dakota

CONTENTS

7.1   INTRODUCTION

The global population has more than doubled in the past five decades, from 3 billion in 1959 to 6.6 billion in 2007 [1]. It is estimated that by 2025, one-third of the world's population will face severe water shortages (see Ref. [2] and references therein), and to meet increased demands, the world's water supply will need to increase by an additional 22% from 1995 use levels [3]. Considering that irrigation consumes about 60% of the world's diverted freshwater resources [4], competition for water between irrigation, and municipal and industrial uses will only increase.

Examples of the growing competition for water can be seen around the world. For example, 70% of the grain produced in China is from irrigated land, yet burgeoning population growth, increased industrialization, and rapid urban development have led to increased demand for water that would typically be used for irrigation [5].

In the Shandong Province of China, which provides one-fifth of China's corn and one-seventh of its wheat, a major source of irrigation water, the Yellow River, has been dry for long periods of time due to increased upstream municipal use [5]. A similar situation is found in Quetta, a city with a population of over one million that is the capital of Pakistan's southwestern province of Balochistan. Overexploitation of groundwater resources for municipal use coupled with drought and dramatic population increase, have resulted in groundwater declines of over 200 ft (60.96 m) [6]. Increased demand for water has led to the diversion of water from irrigation to municipal use, and thousands of farmers have been forced to abandon irrigation of their lands [6].

The importance of irrigated agriculture for the world's food supply cannot be understated. It is widely believed that irrigation accounts for 40% of the world's food supply, yet accounts for only about 17% of cultivated area [7]. As the global population increases, there will be additional demand not only for water, but also for food; however, the rate of increase in irrigated agriculture is declining compared to population growth in the latter half of the twentieth century [7]. Irrigated agriculture rapidly increased from the mid-1900s to the late 1900s, primarily as a result of extensive national investment in surface water storage and diversion projects following World War II; however, since the 1980s, the level of investment in water supplies for irrigation has dramatically decreased [7]. Reduced national investment in surface water storage, coupled with improved pumping technologies and more affordable and accessible energy, resulted in a shift of irrigation water use from primarily surface water to groundwater. This creates problems since many of the groundwater resources now used for irrigated agriculture are finite and the rates of depletion for irrigated use are higher than those of recharge by natural systems.

Thus, irrigated agriculture faces multiple challenges, including competition from municipal and industrial sectors, less national and international investment in water supply projects for irrigation, and increased dependence on groundwater resources that are often limited. It is imperative that water use for irrigation is managed as efficiently as possible to ensure food production from these areas is maintained at a consistent and sustainable rate.

7.2   GLOBAL IRRIGATED AREA MAPPING (GIAM)

One of the key organizations focusing on the sustainable use of water and land resources in agriculture is the International Water Management Institute (IWMI). IWMI is a nonprofit scientific research organization that is developing tools and methods necessary to help developing countries eradicate poverty through effective management of water and land resources [8].

One of the key research themes of IWMI is *Integrated Water Management for Agriculture*. The goal of this program is to understand how to increase food production for a growing population while simultaneously meeting the water-quality and -quantity requirements of other economic and environmental sectors [9]. As part of this initiative, IWMI started a project in 2002 to map global irrigated agricultural areas in order to gain a better understanding of the following key questions [10]:

- How much irrigation do we have now?
- How much do we need in the future?
- How much do we want in the future to achieve a sustainable balance with the environment?
- How much water does it require, and will this be available?

In order to address these questions, key goals of IWMI's GIAM effort were to more precisely define the actual area and spatial distribution of irrigation and to elaborate the extent of multiple cropping over a year, particularly in Asia, where two or three crops may be planted in one year, but where cropping intensities are not accurately known [10].

The GIAM effort focused primarily on the use of satellite imagery to better define and map irrigated areas. Satellite imagery is ideal for conducting global-scale mapping efforts because it is consistent, continuously updated, timely, widely available, and is becoming increasingly free of charge [10]. The overall approach used in GIAM was to utilize time-series data sets to evaluate changes in spectral reflectance on the earth's surface over time. This approach is particularly effective in identifying seasonal vegetation, such as crops, since the spectral reflectance will change quite noticeably over the life cycle of the crop. This is also an effective method for differentiating rain-fed from irrigated crops since, during the period for which irrigation is necessary, the irrigated crops will typically be more vigorous than the same type of crop that is rain-fed.

As a collaborative partner with IWMI in the GIAM effort, the Space Studies Department at the University of North Dakota (UND) is conducting research to better map and understand irrigated agriculture in the Midwest. The overarching goal of this collaboration is to investigate the trends in groundwater-dominant irrigated areas of the United States over the last three decades using satellite-sensor data. A key task of this research is to establish the relationship between groundwater use, irrigated area expansion/reduction, and the implications on global food production.

In order to investigate the trends in irrigated agriculture in the United States, a region of intense irrigation and groundwater use was selected as a focal area. This area is located in the Midwest and overlies the extensive Ogallala aquifer, also referred to as the High Plains aquifer. The Ogallala aquifer is perhaps one of the best areas to study the impacts of declining groundwater levels on irrigated agriculture because it is an excellent example in the United States, of poor and unsustainable water-management practices. Overutilization of groundwater resources in the aquifer, primarily due to intensive irrigation practices, has resulted in a significant decline in the groundwater level since the predevelopment of the region.

There are two principal goals of the Ogallala aquifer research. The first goal is to better refine the GIAM irrigated area class maps in the Midwest region by utilizing the GIAM methodology with higher resolution data sets (Moderate Resolution Imaging Spectroradiometer [MODIS] 500 m or 250 m), and to verify these classes through extensive groundtruthing. The second goal is to conduct a temporal analysis of irrigated agriculture from the 1970s through 2005, utilizing higher resolution imagery, namely Landsat, and advanced land classification techniques. If possible, the correlation between water level decline and changes in irrigated agriculture will be better defined.

7.3 OGALLALA AQUIFER BACKGROUND

The Ogallala aquifer underlies 450,658 km^2 of land located in portions of eight states within the High Plains of the United States, including Wyoming, South Dakota, Nebraska, Kansas, Colorado, New Mexico, Texas, and Oklahoma (Figure 7.1). It is the largest freshwater aquifer in the world and one of the most intensively developed groundwater resources in the country [11]. More than 90% of the water pumped from the Ogallala is for irrigated agriculture [12], which accounts for 30% of all irrigation water use in the United States [13].

The Ogallala is an unconfined aquifer comprised of clay, silt, sand, and gravel of Miocene to Pliocene ages that eroded from the eastern face of the Rocky Mountains approximately 17 to 3 million years ago [15,16] (Figure 7.2). The elevation of the High Plains region ranges from 7,800 ft (2377.44 m) at the western boundary to 1,100 ft (335.28 m) at the eastern edge [15]. As of 1996–97, the saturated thickness of the aquifer ranged from 0–1,200 ft (0–365.76 m) [16], with an average saturated thickness of 200 ft (60.96 m) [13] (Figure 7.3). The average depth to the aquifer ranges from just beneath the surface to 300 ft (91.44 m) [15]. The total estimated volume of water within the Ogallala as of 2003 was approximately 2,940 million acre-feet (approximately 363 million hectare-meters [ha-m]) [17]. This water resource is considered "fossil" water, since most of it had accumulated approximately 10,000 years ago during the wetter climatic regime of the Pleistocene. Once this water resource is depleted, the only means of recharge (aside from localized recharge in Nebraska by the Platte River system) is precipitation, which is drastically lower now compared to the colder and wetter conditions experienced 10,000 years ago.

The High Plains region was originally viewed as unfit for agriculture due to a lack of surface water resources and its semiarid climate, with an average annual precipitation of less than 20 in. (510 mm) [13]. In fact, exploration of the region in 1820 by Major Stephen H. Long and others, led one member of that expedition to comment that the region "is almost wholly unfit for cultivation, and of course uninhabitable by a people depending on agriculture for their subsistence" [15]. Despite this assessment and many others like it, farmers began trickling into the region as a result of cheap land and fallacies like "the rain follows the plow" [15]. It was not until the mid-1900s that irrigated agriculture increased dramatically as a result of several key factors, including: (1) increased access to more advanced deep-well turbine pumps; (2) increased availability of low-cost energy (gasoline and natural gas) to run pumps of irrigation wells; (3) access to affordable irrigation equipment following World War II; (4) the introduction of sprinkler irrigation systems in the 1950s; and (5) increased availability of inexpensive aluminum piping, as opposed to more-expensive steel [13,19,20]. As a result of these technological advances, irrigated agriculture in the High Plains increased from an estimated 2.1 million acres (approximately 850,000 ha) in 1949 to 13.7 million acres (5,544,250 ha) in 1980 (Figure 7.4).

The latter decades of the twentieth century followed a different trend. Irrigation increased by only 0.2 million acres (80,938 ha) from 1980–97, and declined by 1.2 million acres (485,628 ha) from 1997 to 2002 [17]. Whereas the region was

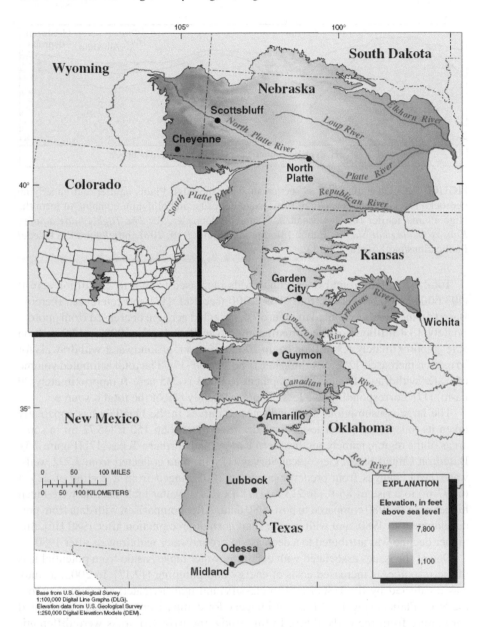

FIGURE 7.1   Aerial extent of the Ogallala aquifer (From Qi, S.L. et al., *Classification of Irrigated Land Using Satellite Imagery, The High Plains Aquifer, Nominal Date 1992*, USGS Water-Resources Investigation Report 02-4236, USGS, Denver, CO, 2002.).

primarily unfit for agriculture (except for marginal wheat production) prior to development of irrigation, current crops include corn, wheat, sorghum, cotton, peanuts, dry beans, and alfalfa [15].

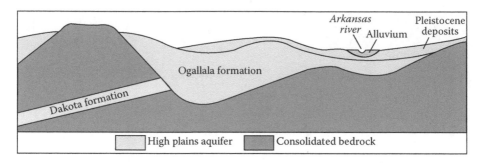

FIGURE 7.2 Generalized cross section showing the High Plains aquifer and underlying bedrock. The Ogallala Formation, Pleistocene deposits, and alluvium combine to form the High Plains aquifer (From Buchanan, R. and Buddemeier, R., *The High Plains aquifer, Kansas Geological Survey*, Public Information Circular 18, 2001, http://www.kgs.ku.edu/Publications/pic18/pic18_1.html.).

Total estimated groundwater withdrawals increased from 4 to 23 million acre-ft (493,600–2,838,200 ha-m) from 1949–2000 (see Ref. [17] and references therein). Based on the trends exhibited in Figure 7.4, irrigated acreage decreased from approximately 13.8 million acres (approximately 5.6 million ha) in 1995 to 13.2 million acres (approximately 5.3 million ha) in 2000; however, groundwater withdrawals for irrigation increased 12% over this same time-period [17]. The total estimated volume of water withdrawn from predevelopment to 2003 is 235 acre-ft (approximately 29 ha-m) [17], corresponding to a loss of approximately 8% of the total resource.

The large consumption of groundwater resources in the Ogallala has apparently taken its toll. Groundwater levels have declined more than 150 ft (45.72 m) in some areas of the region, namely southwestern Kansas and northern Texas [17] (Figure 7.5). Based on United States Geological Survey (USGS) data collected from 3,792 wells, water level changes from predevelopment to 2003 ranged from a decline of 223 ft (67.97 m) to a rise of 86 ft (26.213 m) [17]. Levels of water table decline were much higher prior to 1980 compared to post-1980 data. After comparison with data from predevelopment to 1980, and with greater than normal precipitation after 1980 [16], this higher decline was attributed to a decrease in groundwater withdrawals after 1980.

One of the issues associated with declining water tables, aside from potential loss of the resource, is increased costs of energy and pumping [12,17]. In 2002, a study was conducted by the USGS [15] to classify and map irrigated agriculture across the High Plains using 1992 Landsat imagery for comparison with data on irrigated agriculture from the early 1980s. In this study, the irrigated areas were differentiated from nonirrigated areas using threshold values based on vegetation indices. The results of the study estimated the total irrigated acreage of the High Plains at 13.1 million (5.3 million ha) in 1992, compared to 1982 estimates of 13.7 million acres (5.54 million ha) based on a similar data set. Although these numbers are slightly different from those reported by McGuire [17] and shown in Figure 7.4, a key finding of the research was a shift in irrigated agriculture away from areas of high groundwater decline from 1980–92. Figure 7.6 shows the comparison in percentage of irrigated

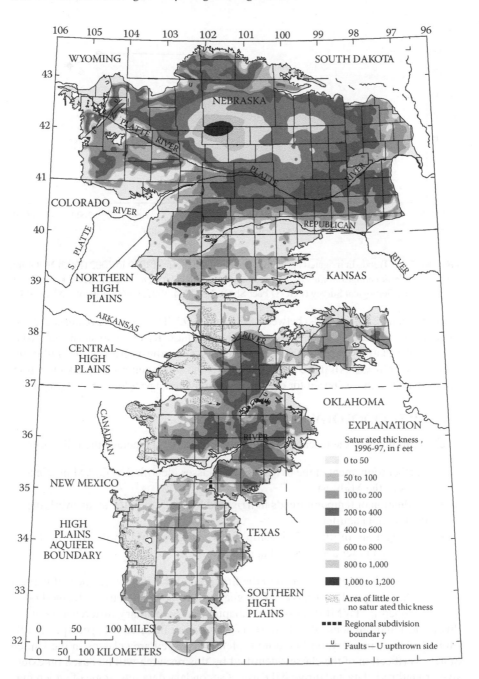

FIGURE 7.3 **(See color insert following page 224.)** Saturated thickness of the Ogallala aquifer. (From McGuire, V.L. and Fischer, B.C., *Water-level changes, 1980 to 1997, and saturated thickness, 1996–97, in the High Plains aquifer*, USGS Fact Sheet 124-99, USGS, Lincoln, NE, 1999.).

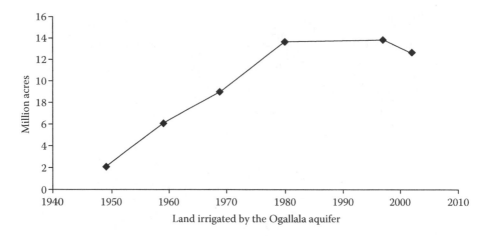

FIGURE 7.4 Trends in irrigated agriculture (million acres) from 1949 to 2002 (From McGuire, V.L., *Water-level changes in the High Plains aquifer, predevelopment to 2003 and 2002 to 2003*, United States Geological Survey (USGS) Fact Sheet, FS-2004-3097, USGS, Lincoln, NE, 2004.).

land (based on 2×2 km cells) between 1980 and 1992. The areas of greatest decrease in irrigated agriculture occur along the North Platte River in western Nebraska, south of the Platte River in south-central Nebraska, southwestern Kansas, and in the Panhandle of Texas. All these areas, except for the south-central Nebraska location, are the same as those chosen for evaluation of irrigated agriculture in this project.

7.4   METHODOLOGY

The overall goals of the research being conducted at UND include the following:

- Further refinement of the irrigated area classes identified by GIAM in select areas of the Midwest.
- Evaluation of temporal trends in irrigated agriculture in the areas overlying the Ogallala aquifer.
- Investigation of the relationship between groundwater decline in the Ogallala aquifer and irrigated agriculture.

The focus of this chapter is related to the first goal of the project: further refinement of the GIAM irrigated area classes. This goal will be completed using the methodology developed by IWMI discussed in Chapter 3. The primary difference between this project and previous GIAM efforts is that a more detailed data set, specifically MODIS 500-m imagery, will be used to define irrigated area classes in the United States. Verification of the classes identified by this research will be based on extensive groundtruth data and through the use of secondary data sets, some of which may be more detailed than what is available for other areas of the globe. For example, in some states, the National Agricultural Statistics Service provides maps showing what crops were grown on individual farms during a given year. Data sets such as this can help greatly in the identification and naming of irrigated crop types.

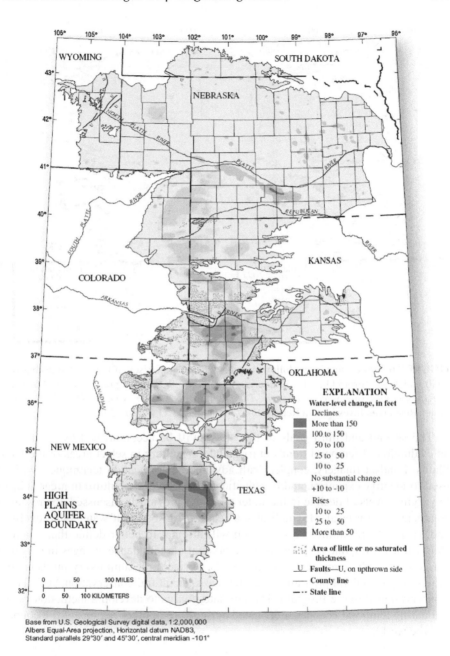

FIGURE 7.5 (See color insert following page 224.) Changes in the water levels of the Ogallala aquifer from predevelopment to 2003. (From McGuire, V.L., *Water-level changes in the High Plains aquifer, predevelopment to 2003 and 2002 to 2003*, United States Geological Survey (USGS) Fact Sheet, FS-2004-3097, USGS, Lincoln, NE, 2004.).

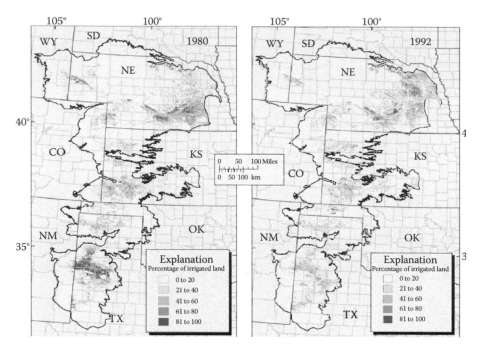

FIGURE 7.6   (See color insert following page 224.) Comparison of irrigated areas between 1980 and 1992. (From Qi, S.L., Konduris, A., Litke, D.W., and Dupree, L. *Classification of Irrigated Land Using Satellite Imagery, The High Plains Aquifer, Nominal Date 1992*, USGS Water-Resources Investigation Report 02-4236, Denver, CO, 31 pp., 2002.)

The second and third goals of this study will be accomplished by conducting an evaluation of temporal trends in irrigated agriculture in select areas of the High Plains. Landsat imagery, coupled with advanced classification techniques, will be used to identify expansion and/or reduction in irrigated agriculture in areas subject to varying levels of decline in the water table. As previously discussed, the research conducted by Qi et al. [15] indicates a relationship between water table declines and irrigated agriculture. This research will attempt to better define that relationship, while taking into account other factors that may influence changes in irrigated agriculture, such as regional climatic trends, agricultural commodity markets, and availability of federal programs to encourage agricultural conservation, such as the Conservation Reserve Program. The ultimate goal is to develop a methodology for evaluating temporal changes in irrigated agriculture in other regions of the country and the world.

7.5   RESULTS

The research is ongoing; therefore, the following sections will focus on the progress made to date and preliminary results as they relate to the refinement of the GIAM classes. Efforts to date have focused on site selection, imagery collection, groundtruth data collection, and preliminary comparison of groundtruth data with the GIAM classes.

7.5.1 Site Selection and Imagery Collection

To achieve the overall goals of this research, including the evaluation of temporal trends in irrigated agriculture overlying the Ogallala aquifer, an appropriate study area was determined. Because a key goal of this research was to evaluate the relationship between declining water levels and changes in irrigated agriculture, three focus areas of intense irrigation and varying levels of water decline were selected for detailed analyses. These locations comprise a 17,430 km^2 area in south-central Nebraska, a 19,690 km^2 area in northern Texas, and a 12,430 km^2 area in southwest Kansas. Levels of water table decline from predevelopment (prior to the 1940s) until 2003 range from 10–50 ft (3.048–15.24 m) at the Nebraska site, from 25–100 ft (7.62–30.48 m) at the Texas site, and from 50–150 ft (15.24–45.72 m) at the Kansas site [17].

Based on 2003 irrigation census data, the key types of irrigation in these three focus areas include sprinkler, gravity flow, and low-flow or drip irrigation systems. Sprinkler systems account for the majority of irrigated agriculture in the region [21].

For each location, 30-m Landsat imagery was obtained for July or August in the mid-1970s, mid-1980s, mid-1990s, and for every year from 2001 through 2005. Most of the imagery was purchased from the USGS Earth Resources Observation and Science (EROS) data center. Additional imagery was obtained through the Global Land Cover Facility at the University of Maryland (http://glcfapp.umiacs. umd.edu:8080/esdi/index.jsp).

7.5.2 Groundtruth Data Collection

A groundtruth campaign was conducted within the three focus areas during the summer of 2006. In preparation, the 2005 Landsat scenes for each of the areas were visually examined and enhanced using the ER Mapper by Leica Geosystems. A band combination of 7-4-1 was used.

At least 100 groundtruth points were selected for each area, based on visual interpretation of irrigated vs. nonirrigated areas. Most of the irrigation at the three sites is center pivot; thus, the crop circles created by this practice were very easy to identify (Figure 7.7). Based on visual interpretation of the imagery, the groundtruth points primarily included irrigated and nonirrigated agricultural land (harvested and non-harvested) and grasslands/rangelands. Each groundtruth point was given a number and plotted on the corresponding Landsat image as a point in an annotation layer. A map of the groundtruth points, including the most efficient driving routes, was compiled using the Microsoft Streets and Trips software program. This allowed for labeling of each groundtruth point with the appropriate number and also helped in estimating total driving time and distance.

The groundtruth methodology employed in this effort was a modified version of one developed by IWMI and included the following data for each groundtruth point: location information (latitude/longitude); multiple photographs; crop type, growth stage, health and vigor; irrigation type (if any) and water source; land use cover (defined by percent) and classification. This information was recorded on a modified version of the groundtruth form developed by IWMI; however, the form had to be modified to account for land access restrictions within the United States

FIGURE 7.7   Landsat image of irrigated areas in Kansas. The white squares are locations at which groundtruth data were collected.

that were not issues for IWMI. Because it is not possible to walk onto private land without landowner permission, and obtaining permission would have been overly time consuming, some of the data typically collected by IWMI during groundtruth campaigns were not collected for this effort. The data omitted include: random plant counts per square meter, handheld spectrometer readings, and information on irrigation intensity.

The groundtruth campaign was conducted from mid-July to mid-August 2006 and included a total of 364 points—100 from Texas, 108 from Kansas, and 156 from Nebraska. At least three photographs were collected from each site. In addition, to aid in verification of IWMI's irrigated area mapping effort, data from additional 45 points were collected from areas classified by GIAM as predominantly irrigated. At some of the groundtruth points, where landowners were readily available, additional information pertaining to crop rotations and irrigation practices was collected. This information was recorded on the groundtruth forms completed for each site. The groundtruth information, data, and photographs are being compiled in digital format and will be made available to IWMI and other interested researchers.

7.5.3 PRELIMINARY COMPARISON BETWEEN GIAM AND GROUNDTRUTH POINTS

Although an in-depth evaluation of the groundtruth data has yet to be completed, several key conclusions can be made. The irrigation schemes in the Kansas and Texas focus areas were primarily groundwater-irrigated, center-pivot systems that are quite easily identified using visual interpretation of the imagery. The situation

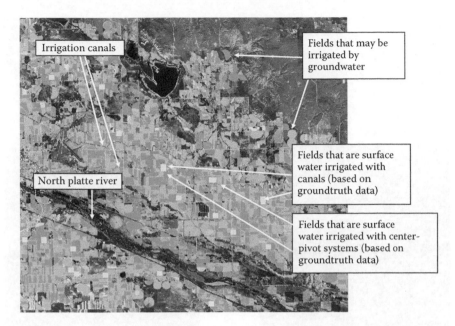

FIGURE 7.8 Surface water and groundwater irrigated fields along the North Platte River in Nebraska.

is not so straightforward in the Nebraska focus area. This area also has a very large percentage of irrigated agriculture; however, the irrigation source water is a combination of surface water and groundwater. Various types of irrigated fields, groundtruth data points (yellow), and large, public canal systems can be seen in an image of this area (Figure 7.8). The surface water irrigation is typically canal-fed and occurs primarily along the North Platte River. Most of the canal-fed fields are square in shape and thus differ from the round fields exhibited by center-pivot irrigation schemes; however, there are instances where surface water is used in center-pivot irrigation systems. In addition, there are also groundwater-fed center-pivot irrigation systems adjacent to the river; thus, a differentiation between groundwater-fed and surface water-fed irrigation schemes cannot be made based solely on the shape of the field. Although a time-series analysis can be applied to this area to differentiate irrigated from nonirrigated crops, additional methodologies or data will be needed to differentiate between groundwater and surface water irrigation.

Another potential source of error in using satellite imagery will likely be the identification of alternative irrigation practices, such as drip or low-flow irrigation (Figure 7.9). Although the percentage of these practices is still low throughout the High Plains region, they will probably increase as water supplies become increasingly scarce and more conservative irrigation practices are warranted. These areas are likely to be classified as irrigated, based on time series; however, they would be very difficult to differentiate from conventional surface water irrigation practices, such as canal irrigation. This could prove problematic in trying to calculate a water use budget, since low-flow irrigation uses less water than typical canal irrigation.

FIGURE 7.9 An edible bean field in Nebraska that employs low-flow irrigation. Because the water is conveyed to the end of each row by pipe rather than canal, less water is lost to evaporation.

The groundtruth data were also used to begin assessing the accuracy of the classes identified in GIAM (V2.0). The 108 groundtruth points collected in Kansas were compared to the GIAM data, both visually and quantitatively. Based on a comparison with the groundtruth data, the errors of omission in the GIAM dataset were 20.4% and those of commission were 18.5%. Considering that these errors were similar in magnitude, ideally they would cancel each other out in calculating total irrigated acreage to give a reasonably accurate estimate. A visual assessment of the data was conducted by overlying the 2005 Landsat image for the Kansas area with the irrigated area class map (Figure 7.10). In Figure 7.10, the gray areas were classified by GIAM as "other," meaning nonirrigated, while the colored areas were classified as irrigated. Based on the visual identification of center-pivot irrigation systems found within the gray areas, it is quite apparent that in this localized area, not all of the irrigated areas were included in the GIAM assessment. It is important to remember that the GIAM assessment was based on 10-km/1-km data resolution, so there are bound to be errors in comparison to the 30-m Landsat data.

7.6   NEXT STEPS

The next step in this study will be to apply the GIAM methodology to select regions within the High Plains area using 500-m MODIS data. The MODIS data and other

FIGURE 7.10 **(See color insert following page 224.)** The irrigated area classes identified in GIAM overlying a 2005 30-m Landsat image for an area in Kansas.

applicable data sets will be combined into a megafile and an unsupervised classification will be conducted. The classes will be grouped and further refined, based on spectral-matching techniques and secondary data sets. An accuracy assessment of these classes will be conducted based on the groundtruth data collected during the summer of 2006. An accuracy assessment will also be conducted between the classes identified by this work and those already identified through GIAM efforts.

The groundtruth data will also be used to identify the percentage of area within a pixel that is irrigated agriculture. This is an important step in attempting to quantify the total irrigated area found within each of the classes. It is based on the premise that not all of the land contained within a 1-km or 500-m pixel area is irrigated. Groundtruth data and high-resolution imagery help establish what percentage of land is actually irrigated within a given pixel, as compared to surrounding land which, in areas of the Midwest, may be tree rows, wetlands, ponds, or other types of nonirrigated land use.

In order to conduct a temporal analysis of irrigation patterns in the High Plains region, the Landsat imagery collected for the three focus areas in Texas, Kansas, and Nebraska will be classified using advanced techniques. One of the techniques being investigated is the use of object-oriented image classification using the eCognition software program. The basis for object-oriented classification is that, rather than using a pixel-based classification technique that relies solely on the spectral characteristics of individual pixels, classes are identified based on groupings of pixels, called objects. Objects are identified within an image prior to classification based primarily on size, shape, and spectral characteristics. Once objects are identified and look acceptable to the user, a classification can be conducted based on multiple parameters, including spectral signature, tone, texture, shape, area, context, and even information from other data layers. It allows for the incorporation of data from a variety of other sources, such as other satellite sensors, vector files, digital elevation models (DEMs), and theme maps. Certain factors can be weighted more heavily than others based on prior knowledge and interpretation by the person processing the image; thus, it allows for a "smarter" classification while still utilizing high-powered quantification and statistical methods found within the program. This type of classification is particularly appealing to areas like the High Plains region, which has a large percentage of center-pivot irrigation systems. A person can look at an image and easily identify center-pivot irrigated areas based on shape; however, a classification program based solely on the spectral characteristics of individual pixels would classify an irrigated wheat field in the same manner as a nonirrigated wheat field (assuming the spectral characteristics were the same).

REFERENCES

1. United States Census Bureau, *World population information*, 2007, http://www.census. gov/ipc/www/idb/worldpopinfo.html.
2. Keller, A., Sakthivadivel, R., and Seckler, D., *Water Scarcity and the Role of Storage in Development*, IWMI Research Report 39, IWMI, Colombo, Sri Lanka, 2000.
3. IWMI, *World Water Supply and Demand*, Colombo, Sri Lanka, 2000.
4. Thenkabail, P.S., Schull, M., and Turral, H., Ganges and Indus river basin land use/land cover (LULC) and irrigated area mapping using continuous streams of MODIS data, *Remote Sensing of Environment*, 95, 317–41, 2005.
5. Brown, L.R. and Halweil, B., China's water shortage could shake world food security, *World Watch*, July/August, 10–21, 1998.
6. UNOCHA (UN Office for the Coordination of Humanitarian Affairs), *Pakistan: Focus on water crisis*, 2006, http://www.irinnews.org/print.asp?ReportID=27814 (accessed September 2006).
7. Thenkabail, P.S. et al., *An Irrigated Map of the World (1999) Derived from Remote Sensing*, IWMI Research Report 105, IWMI, Colombo, Sri Lanka, 2006.
8. IWMI (International Water Management Institute), *Overview of IWMI*, 2006, http:// www.iwmi.cgiar.org/about/intro.htm.
9. IWMI, *Annual Report 2004–2005*, IWMI, Colombo, Sri Lanka, 2005.
10. Thenkabail, P.S. et al., *A Methodology to Develop Global Map of Irrigated Area (1999) Using Remote Sensing*, IWMI, Colombo, Sri Lanka, 2005.
11. Albrecht, D.E., The adaptations of farmers in an era of declining groundwater supplies, *Southern Rural Sociology*, 7, 47–62, 1990.
12. Guru, M.V. and Horne, J.E., *The Ogallala Aquifer*, produced by the Kerr Center for Sustainable Agriculture, Inc., Poteau, OK, 2000, http://www.Kerrcenter.com/publica tions/ogallala_aquifer.pdf#search=%22Ogallala%20aquifer%22.
13. Rosenberg, N.J. et al., Possible impacts of global warming on the hydrology of the Ogallala aquifer region, *Climatic Change*, 42, 677–92, 1999.
14. Qi, S.L. et al., *Classification of Irrigated Land Using Satellite Imagery, The High Plains Aquifer, Nominal Date 1992*, USGS Water-Resources Investigation Report 02-4236, USGS, Denver, CO, 2002.
15. Zwingle, E. and Richardson, J., Ogallala aquifer: Wellspring of the high plains, *National Geographic Magazine*, 183, 80–109, 1993.

16. McGuire, V.L. and Fischer, B.C., *Water-level changes, 1980 to 1997, and saturated thickness, 1996–97, in the High Plains aquifer*, USGS Fact Sheet 124-99, USGS, Lincoln, NE, 1999.

17. McGuire, V.L., *Water-level changes in the High Plains aquifer, predevelopment to 2003 and 2002 to 2003*, United States Geological Survey (USGS) Fact Sheet, FS-2004-3097, USGS, Lincoln, NE, 2004.

18. Buchanan, R. and Buddemeier, R., *The High Plains aquifer, Kansas Geological Survey*, Public Information Circular 18, 2001, http://www.kgs.ku.edu/Publications/pic18/pic18_1.html.

19. Beaumont, P., Irrigated agriculture and ground-water mining on the High Plains of Texas, USA, *Environmental Conservation*, 12, 119–30, 1985.

20. Bruce, C. and Triplett, L. (Eds.), *The Great Plains Symposium: The Ogallala Aquifer, Sharing Knowledge of the Future*, The Great Plains Foundation, Kansas, MO, 1996.

21. United States Census of Agriculture, *Land Irrigated by Method of Water Distribution: 2003 and 1998*, Farm and Ranch Irrigation Survey, US National Agricultural Statistics Service, 2003, http://www.nass.usda.gov/Data_and_statistics/SVG/index.asp.

8 Assessing the Extent of Urban Irrigated Areas in the United States

Cristina Milesi
California State University

Christopher D. Elvidge
NOAA National Geophysics Data Center

Ramakrishna R. Nemani
NASA Ames Research Center

CONTENTS

8.1 INTRODUCTION

Monitoring of irrigated areas is of critical importance for ensuring global food security and managing the sustainable use of freshwater resources. As such, mapping efforts in irrigated areas have traditionally focused on agricultural irrigated areas

217

[1,2]. However, in the United States and in other developed countries that have undergone more recent low-density urbanization, such as Canada and Australia, a growing sector of yet mostly undocumented irrigated areas is to be found within urban environments [3–5]. While not contributing to significant food and fiber production, but rather serving a functional, recreational, and aesthetic purpose [6], the irrigation of urban landscape is responsible for a large component of urban water use in these countries. Estimates of North American cities indicate that, depending on climate, the fraction of residential water use for outdoor purposes ranges from 22–38% in cooler climates to 59–67% in cities with hot and dry climates [7]. During the past several decades, factors such as the low-density character of urbanization accompanied by a cultural predilection towards living in single-family housing surrounded by lush landscaped vegetation have contributed to the large expansion of irrigation in urban areas across all climatic regions of the United States. This expansion is likely to have reached considerable proportions in the United States, where turfgrass, the most common type of urban landscaping vegetation, is reported to cover an area ranging from 11.1–20.2 million hectares (Mha) [8–11]. As a preferred type of urban landscaping vegetation in the United States, turfgrass can be found in all urban settings in a country where the majority of the population live in single-family houses, each surrounded by a lawn. Beyond residential housing, turfgrass is also found along many sidewalks, business parks, institutional buildings, recreational parks, cemeteries, golf courses, etc. Being characterized by non-native grass species, turfgrass needs to be irrigated at some point during the year to grow in optimal conditions in most parts of the country.

Assessing the extent of urban irrigated areas is, therefore, becoming increasingly important for planning future urban water demand by local water resource planners, to understand the opportunities of water conservation and to help devise regional water policies. However, given the heterogeneous and highly fragmented nature of urban landscaping, mapping and monitoring of urban irrigated areas require the use of high-resolution remote sensing images (of the order of 1-m [12]). The cost of such images and the computation requirements of their analysis still make direct mapping of irrigated areas elusive for regional assessments of urban irrigated areas. Even at a local scale, estimates of urban irrigated areas, let alone maps, are hard to find. In this study, we explore remote sensing-based and census-based methodologies to provide a first attempt at estimating the likely urban irrigated area within the conterminous United States, to understand how relevant they are in comparison to irrigated cropland areas.

8.2   HISTORICAL BACKGROUND ON THE ORIGINS OF URBAN IRRIGATED AREAS IN THE UNITED STATES

The diffusion of urban irrigated areas in the United States can be traced back to the planning of the first suburban developments in the late 1800s, when the first landscaped low-density residential areas started to be built for the new middle class that desired escaping the increasingly unhealthy conditions of industrial cities [10,13,14]. The style of English gardens with a lawn was soon adopted as the preferred type of landscape. Campaigns of neighborhood beautification promoted by the various gardening clubs and the individual homeowners' need to affirm their social status through the exterior appearance of their houses, resulted in increasing yard maintenance. After the first lawnmowers and rubber garden hoses became available, the standards of appearance of a lawn were also soon defined as "a plot with a single type of grass with no intruding weeds, kept mown at a height of an inch and a half, uniformly green, and neatly edged" [13]. Suburban areas accelerated their expansion after World War II, when the availability of loans from the Federal Housing Authority stimulated a housing boom that made relocation to the suburbs and owning a yard affordable to a large proportion of the population. The diffusion of the use of fertilizers and pesticides further contributed to making lawns an urban landscape type of choice for their versatility of surrounding the houses with a formal and neat look, while serving as a private playground. Lawns can be found in urban and suburban areas across the country, from New England to the Southwest. Manicured lawns and lush landscapes remain a form of status symbol, contributing considerably to real estate appreciation [15]. However, because of the significant impact on urban water consumption, water conservation outreach programs and watering restriction ordinances are increasingly targeting lawn-watering practices.

8.3   METHODS

8.3.1   Approaches to Estimating the Extent of Irrigated Urban Vegetation

As a first approximation of the extent of urban irrigation in the conterminous United States, we start from estimates of turfgrass area and then derive irrigated turf area by making adjustments to the total area, based on watering practices.

Recent estimates of turfgrass area are available from turfgrass and green industry surveys for 10 states:

- Illinois [16]
- Iowa [17]
- Kansas [18]
- Maryland [19]
- Michigan [20]
- New Jersey [21]
- New York [22]
- North Carolina [23]
- Virginia [24]
- Wisconsin [25]

While providing detailed information on the extent of turfgrass area in different residential and commercial sectors of these states, these reports are not sufficient to produce a continental estimate. To derive continental estimates, we have two options: (1) use an indirect remote sensing-based approach, as suggested by Milesi et al. [9]; or (2) use data on the characteristics of new housing provided by the US Construction Census, similar to the method proposed by Vinlove and Torla [26] to estimate the area of residential lawns, and then adjust these estimates to include the total turfgrass area associated with urban and suburban settlements in the United States.

8.3.2   REMOTE SENSING APPROACH

Given the ubiquity of turfgrasses in the American urban areas, it can be assumed that a proportion of turfgrass area can be associated with each portion of a built-up area (also called impervious surface area; ISA). Measures of lawn and built-up area, measured from 1-km tiles of high-resolution aerial photography collected along transects from sparsely developed outskirts to dense downtowns of 13 major US centers, indicated that the two variables had a moderately strong relationship (Figure 8.1). The fraction of turfgrass area is inversely related to the fractional ISA, with turfgrass areas increasing as the ISA decreases down to 10%, where the samples of aerial photography suggested that nonsignificant portions of lawn were present. Based on these data, we developed a predictive model that allowed rapid mapping of the turfgrass area for the continental United States, starting from a map of impervious areas [9].

Two national datasets of ISA are currently available for the continental United States. One was produced by the National Oceanic and Atmospheric Administration (NOAA) and provides subpixel fraction of ISA at 1 km^2 spatial resolution. The ISA estimates reported in this map refer to the year 2000/2001 and were obtained by relating satellite-derived radiance-calibrated nighttime city lights, road density, and three Landsat-derived urban land cover classes, to observations of ISA from the above-mentioned high-resolution aerial photography [27]. This ISA data set was used to map the area in the United States potentially under turfgrass [9]. The other national ISA data set was produced by the United States Geological Survey (USGS) by relating Landsat data to measures of imperviousness derived from digital orthophoto quadrangles through regression-tree algorithms [28]. The USGS ISA data set provides a subpixel fraction of ISA at 30-m resolution.

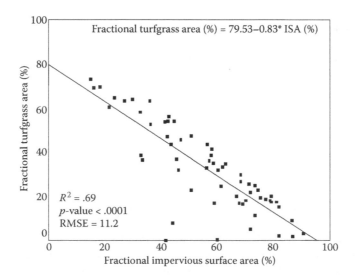

FIGURE 8.1   Scatterplot of the fractional ISA and turfgrass area measurements derived from high-resolution aerial photographs of urban areas across the United States.

Here we apply the turfgrass predictive model developed in Milesi et al. [9] to the USGS fractional ISA. We resample the original USGS ISA 30-m data set to a 1-km grid in Albers Conical Equal Area projection and apply the fractional turfgrass area vs. impervious area predictive equation reported in the plot in Figure 8.1.

8.3.3 HOUSING DATA APPROACH

The data on the characteristics of new housing [29] offer the possibility to perform a back of the envelope calculation and an attempt at validating of the remote sensing-based estimates of turfgrass area for the conterminous United States. These data also allow us to gain a perspective on the annual expansion of urban landscaped area.

We use time series of median lot sizes, median house-floor area, number of stories, and type of parking of sold, new, single-family houses, collected by the US Census Bureau [29]. Nationally aggregated data are available from 1978–2006 (Tables 8.1 and 8.2).

For each year from 1978 to 2001, we produce two estimates of newly added turfgrass area, one based on median lot and house size (Lawn$_{med,t}$; Equation 8.1) and the other based on average lot and house size (Lawn$_{avg,t}$; Equation 8.2):

$$\text{Lawn}_{med,t} = L_{med,t} - H_{med,t} - P_t - D_t, \tag{8.1}$$

$$\text{Lawn}_{avg,t} = L_{avg,t} - H_{avg,t} - P_t - D_t, \tag{8.2}$$

where, $L_{med,t}$ is the total new lot area in year t based on median lot area, and is obtained by multiplying the total number of new single-family units built in year t (SF$_t$) by the median lot size of year t; $L_{avg,t}$ is the total new lot area in year t based on average lot area and is obtained as done for $L_{med,t}$ except that average rather than median lot size is used; $H_{med,t}$ (Equation 8.3) and $H_{avg,t}$ (Equation 8.4) refer to total house footprint area; P_t (Equation 8.5) is the total area of parking in year t; D_t (Equation 8.6) is the total area of paved driveways.

The total house footprint area, $H_{med,t}$, is based on median house size ($h_{med,t}$) and number of stories of the newly constructed units, and is calculated as:

$$H_{med,t} = 1\text{-story SF}_t * h_{med,t} + 2\text{-story SF}_t * (h_{med,t}/2) + \text{split level SF}_t(h_{med,t}/1.5). \tag{8.3}$$

The calculation of $H_{avg,t}$ (Equation 8.4) is similar to that of $H_{med,t}$ except that median house size is replaced with average house size ($h_{avg,t}$):

$$H_{avg,t} = 1\text{-story SF}_t * h_{avg,t} + 2\text{-story SF}_t * (h_{avg,t}/2) + \text{split level SF}_t(h_{avg,t}/1.5). \tag{8.4}$$

P_t (Equation 8.5) is calculated assuming the following garage sizes: 23 m^2 for a one-car garage, 37 m^2 for a two-car garage, and 56 m^2 for a three-car garage. Half

TABLE 8.1

Characteristics of New Housing and Estimated New Residential Lawn, Based on Median Lot and House Size

Year	SF×1,000	Median Lot Size (m²)	Median House Aize (m²)	1-Story SF (×1,000)	2-Story SF (×1,000)	Split Level SF (×1,000)	1-Car SF (×1,000)	2-Car SF (×1,000)	3-Car or more SF (×1,000)	Carport SF (×1,000)	$A_{Res,med}$ (ha)	Average Median Lawn Size (m²)
1978	816	910	153	477	241	98	196[a]	359[a]	0[a]	57[a]	59,195	725
1979	708	890	153	402	231	75	163[a]	333[a]	0[a]	42[a]	49,991	706
1980	531	853	146	316	163	52	117[a]	260[a]	0[a]	32[a]	35,671	672
1981	436	804	145	246	151	39	92[a]	227[a]	0[a]	22[a]	27,211	624
1982	411	768	142	234	147	30	82[a]	222[a]	0[a]	21[a]	24,223	589
1983	623	778	147	327	255	41	118[a]	355[a]	0[a]	25[a]	37,252	598
1984	639	791	150	314	286	39	115[a]	377[a]	0[a]	26[a]	38,981	610
1985	688	825	148	330	316	42	117[a]	427[a]	0[a]	21[a]	44,322	644
1986	750	832	153	347	363	40	116	487	0[a]	21	48,586	648
1987	670	864	164	295	340	35	100	473	0[a]	14	44,809	669
1988	676	857	167	289	357	30	83	497	0[a]	12	44,607	660
1989	650	883	173	278	345	27	72	492	13[a]	10	44,037	677
1990	534	929	176	224	283	27	59	410	33[a]	7	38,219	716
1991	509	906	177	210	269	30	53	393	50[a]	6	34,975	687
1992	610	906	177	266	311	33	57	425	64	7	42,206	692
1993	666	899	177	292	344	30	57	463	81	5	45,554	684
1994	670	892	177	300	343	26	60	463	86	5	45,243	675
1995	667	883	175	308	334	25	52	467	90	5	44,399	666

1996	757	855	180	347	383	28	59	528	106	6	47,890	633
1997	804	836	182	366	417	22	64	555	121	6	49,238	612
1998	886	835	186	399	472	15	65	620	135	8	53,951	609
1999	880	843	189	381	489	11	68	615	141	5	54,096	615
2000	877	830	193	387	475	16	62	601	156	5	52,344	597
2001	908	829	195	397	499	12	58	634	163	4	54,009	595
Average		854	168									650

[a] Information on garage type was not available. Values were obtained by polynomial interpolation of the available data.

TABLE 8.2

Characteristics of New Housing and Estimated New Residential Lawn, Based on Average Lot and House Size

Year	SF ×1,000	Average Lot Size (m²)	Average House Size (m²)	1-Story SF (×1,000)	2-Story SF (×1,000)	Split Level SF (×1,000)	1-Car SF (×1,000)	2-Car SF (×1,000)	3-Car or more SF (×1,000)	Carport SF (×1,000)	$A_{Res,avg}$ (ha)	Average Lawn Size (m²)
1978	816	1,743	163	477	241	98	196[a]	359[a]	0[a]	57[a]	126,580	1,551
1979	708	1,623	164	402	231	75	163[a]	333[a]	0[a]	42[a]	101,341	1,431
1980	531	NA	158	316	163	52	117[a]	260[a]	0[a]	32[a]	NA	NA
1981	436	NA	159	246	151	39	92[a]	227[a]	0[a]	22[a]	NA	NA
1982	411	1,287	157	234	147	30	82[a]	222[a]	0[a]	21[a]	45,061	1,096
1983	623	1,524	162	327	255	41	118[a]	355[a]	0[a]	25[a]	83,012	1,332
1984	639	1,343	166	314	286	39	115[a]	377[a]	0[a]	26[a]	73,466	1,150
1985	688	1,636	164	330	316	42	117[a]	427[a]	0[a]	21[a]	99,339	1,444
1986	750	1,472	168	347	363	40	116	487	0[a]	21	95,698	1,276
1987	670	1,635	177	295	340	35	100	473	0[a]	14	95,868	1,431
1988	676	1,321	182	289	357	30	83	497	0[a]	12	75,252	1,113
1989	650	1,286	186	278	345	27	72	492	13[a]	10	69,665	1,072
1990	534	1,381	190	224	283	27	59	410	33[a]	7	61,759	1,157
1991	509	1,326	190	210	269	30	53	393	50[a]	6	55,865	1,098
1992	610	1,660	191	266	311	33	57	425	64	7	87,535	1,435
1993	666	1,625	191	292	344	30	57	463	81	5	93,133	1,398
1994	670	1,625	190	300	343	26	60	463	86	5	93,686	1,398

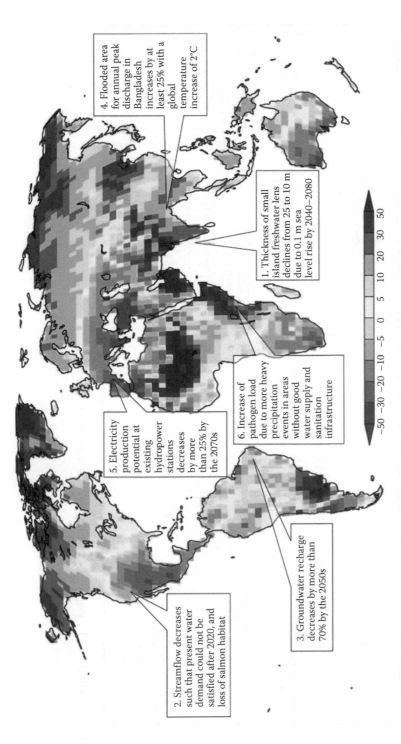

COLOR FIGURE 1.2 Illustrative map of future climatic change impacts on freshwater, which are a threat to the sustainable development of the affected regions. The background shows an ensemble mean change of annual runoff, in percent, between the present (1981–2000) and 2081–2100 for the SRES A1B emissions scenario; blue denotes increased runoff, and red denotes decreased runoff. (From IPCC, *Climate Change Impacts, Adaptation and Vulnerability. Contribution of Working Group II to the Fourth Assessment Report of the Intergovernmental Panel on Climate Change*, Cambridge University Press, Cambridge, 2007.)

The following labels appear on the map:

1. Thickness of small island freshwater lens declines from 25 to 10 m due to 0.1 m sea level rise by 2040–2080

2. Streamflow decreases such that present water demand could not be satisfied after 2020, and loss of salmon habitat

3. Groundwater recharge decreases by more than 70% by the 2050s

4. Flooded area for annual peak discharge in Bangladesh increases by at least 25% with a global temperature increase of 2°C

5. Electricity production potential at existing hydropower stations decreases by more than 25% by the 2070s

6. Increase of pathogen load due to more heavy precipitation events in areas without good water supply and sanitation infrastructure

Color scale: −50 −30 −20 −10 −5 0 5 10 20 30 50

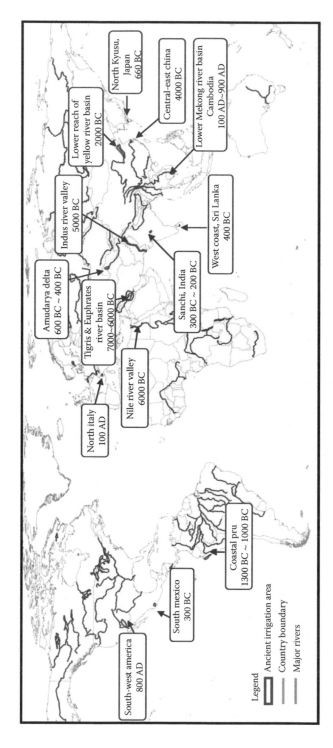

COLOR FIGURE 2.1 Regions of ancient irrigation systems (since 7000 BC).

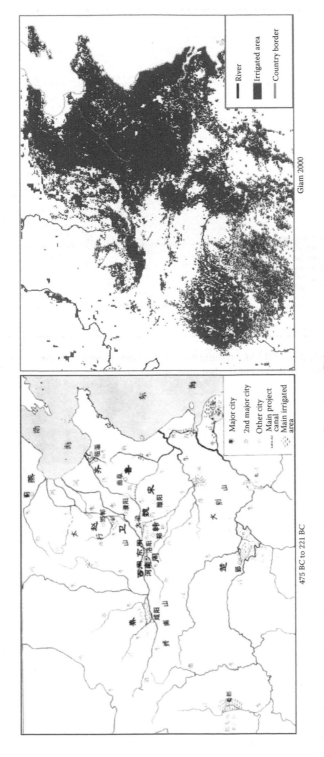

COLOR FIGURE 2.3 The comparison of irrigated areas in Central-east China.

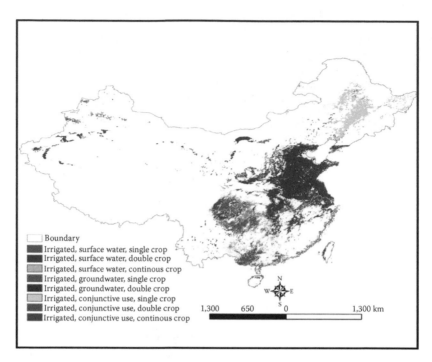

COLOR FIGURE 2.5 GIAM irrigated areas for China. Spatial distribution of GIAM irrigated areas for China at the end of the last millennium, 2000.

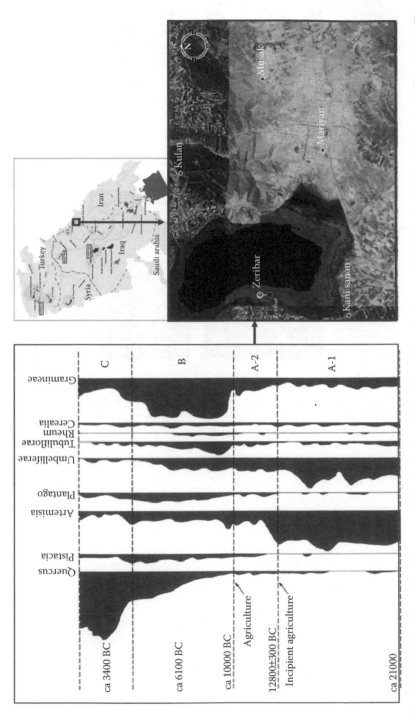

COLOR FIGURE 2.11 Pollen diagram of the Lake Zeribar, indicating early agriculture in the highlands of Tigris and Euphrates River basin. (From Van Zeist, W. and Wright, H.E., Preliminary pollen studies at Lake Zeribar, Zagros mountains, southwestern Iran, Washington, D.C., *Science*, 140, 3562, 65–69, 1963.)

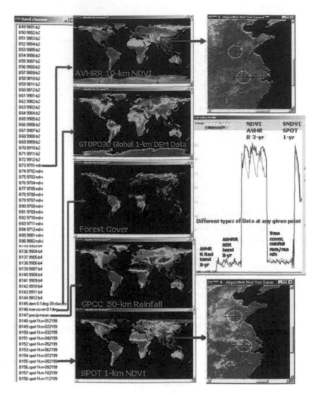

COLOR FIGURE 3.1 Megafile data cube using time-series satellite-sensor data and secondary data. The time-series satellite-sensor data and secondary data are resampled into a single resolution MFDC of hundreds or even thousands of layers.

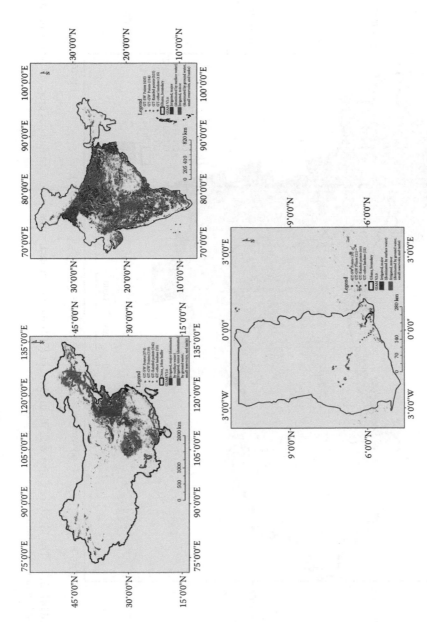

COLOR FIGURE 3.4 Groundtruth data points of China, India, Ogallala (the United States), and Ghana. Extensive groundtruth data points were collected in a standard format during this project with the goal of using them for (a) ISDB generation, (b) class identification, and (c) accuracy assessment.

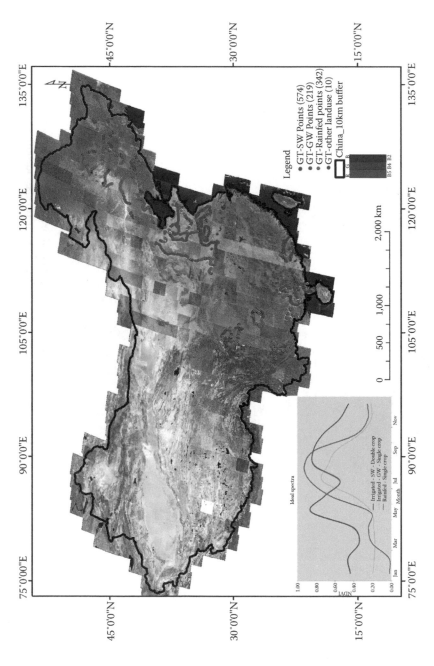

COLOR FIGURE 3.5 The ISDB for China with illustrations of ISDB for (a) irrigated, surface water, continuous crop; (b) irrigated, groundwater, single crop; and (c) rain-fed, single crop.

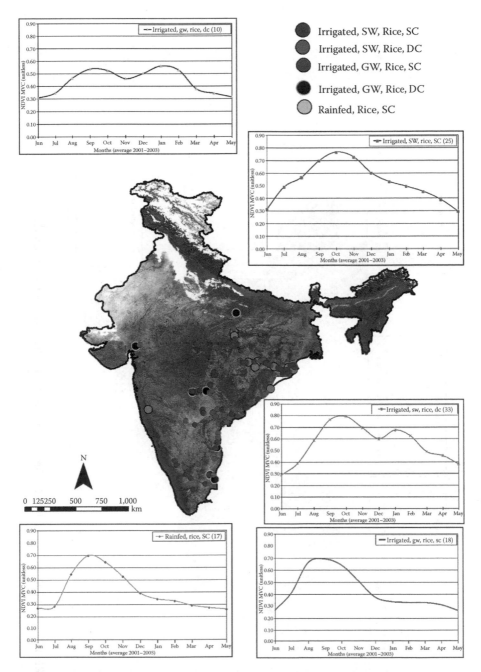

COLOR FIGURE 3.6 The ISDB for India with illustrations of ISDB for (a) irrigated, surface water, rice, double crop; (b) irrigated, surface water, rice, single crop; (c) irrigated, groundwater, rice, double crop; (d) irrigated, groundwater, rice, single crop; and (e) rain-fed, single crop.

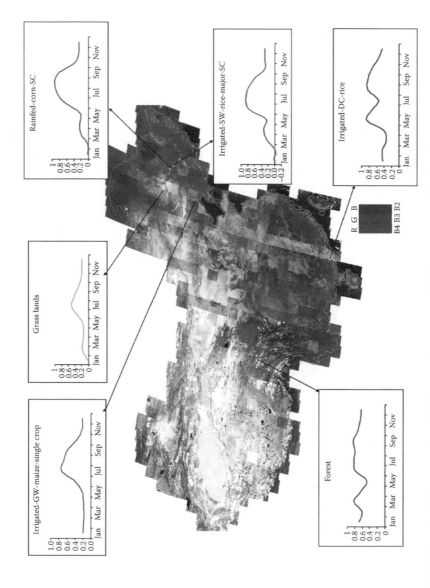

COLOR FIGURE 3.7 The ISDB for various LULC with illustrations of ISDB for various LULC in China.

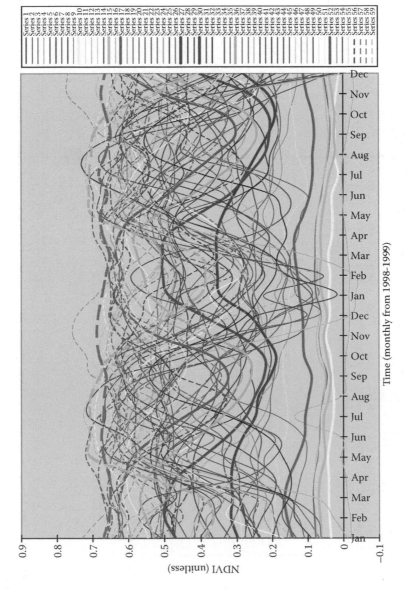

COLOR FIGURE 3.9 Class spectra. The NDVI class spectra were generated for all the 250 classes for one of the segments.

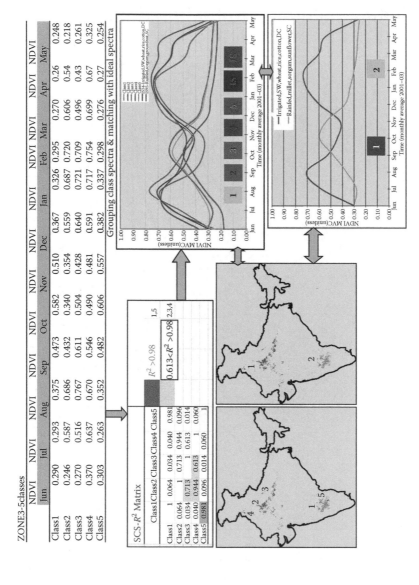

COLOR FIGURE 3.11 Matching class spectra with ideal spectra. All the classes in the class spectra that match ideal spectra are given the same label as the ideal spectra.

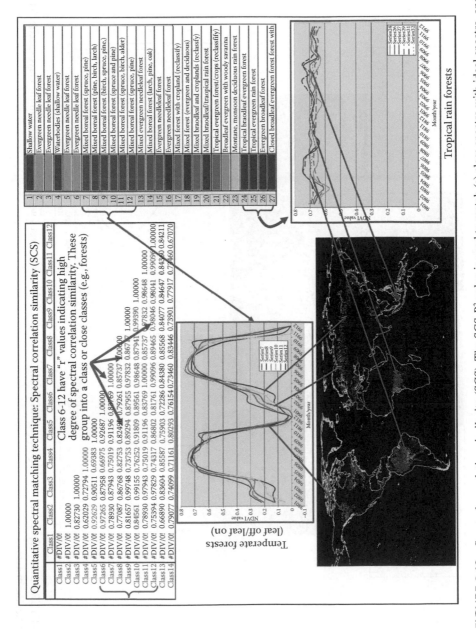

COLOR FIGURE 3.12 Spectral correlation similarity (SCS). The SCS R^2-value is used to match (a) class spectra with ideal spectra, and (b) classes with similar spectral characteristics in the class spectra. The higher the SCS R^2-value, the greater the match.

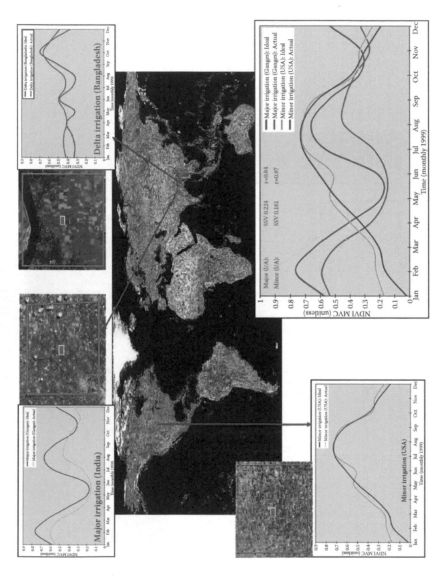

COLOR FIGURE 3.13 Spectral similarity value (SSV). The SSV is used to group classes with similar spectra in the class spectra. The lower the SSV, the better the match.

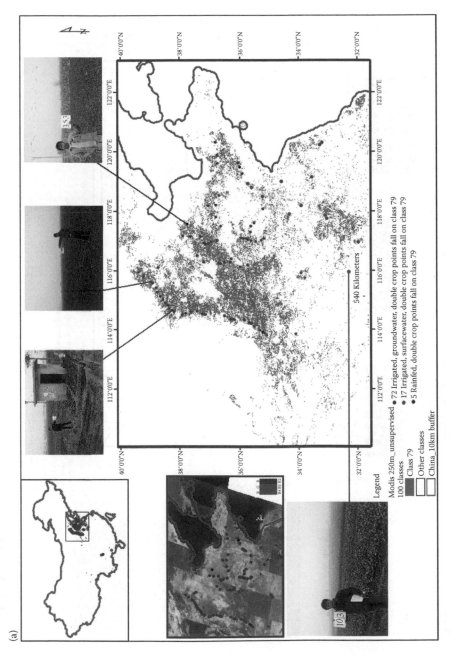

COLOR FIGURE 3.14 Groundtruth data in the class identification process.

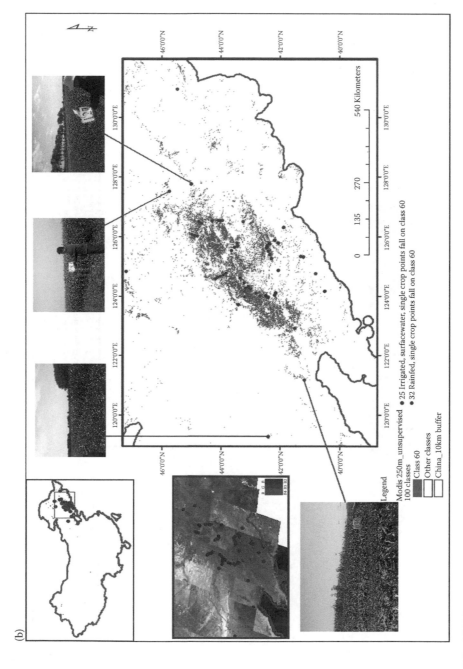

COLOR FIGURE 3.14 (Continued)

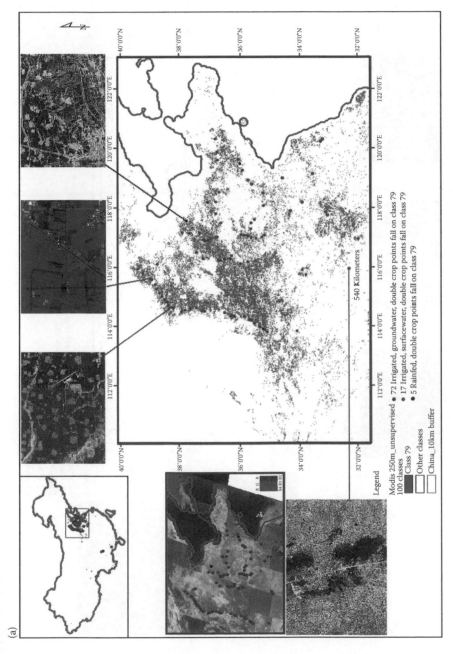

COLOR FIGURE 3.15 Google Earth data in the class identification process. The Google Earth very high-resolution imagery (GE VHRI) data points falling on class 79 (a) and class 60 (b) are illustrated.

Legend

Modis 250m_unsupervised
100 classes
■ Class 79
□ Other classes
□ China_10km buffer

● 72 Irrigated, groundwater, double crop points fall on class 79
● 17 Irrigated, surfacewater, double crop points fall on class 79
● 5 Rainfed, double crop points fall on class 79

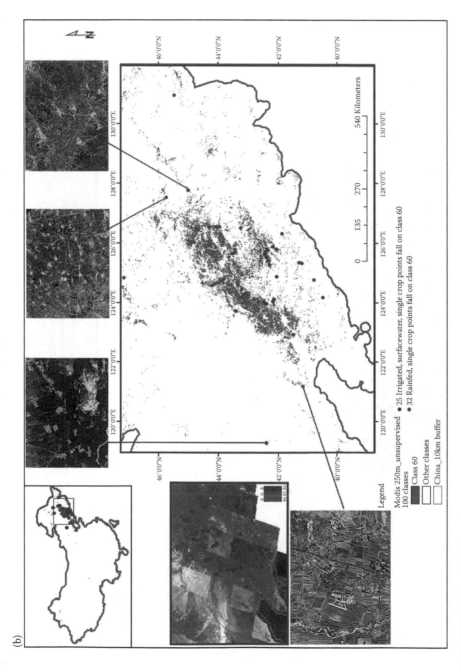

COLOR FIGURE 3.15 (Continued)

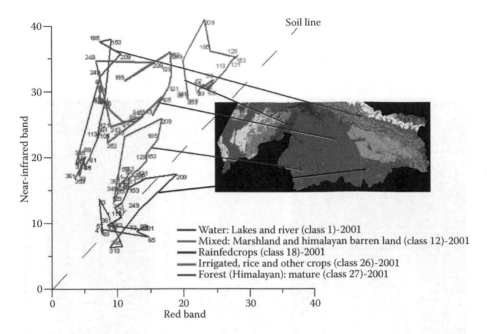

COLOR FIGURE 3.18 The space time spiral curve (STSC) plots. The STSCs depict time series in 2-D feature space. The STSCs track the movement of the classes over time and space.

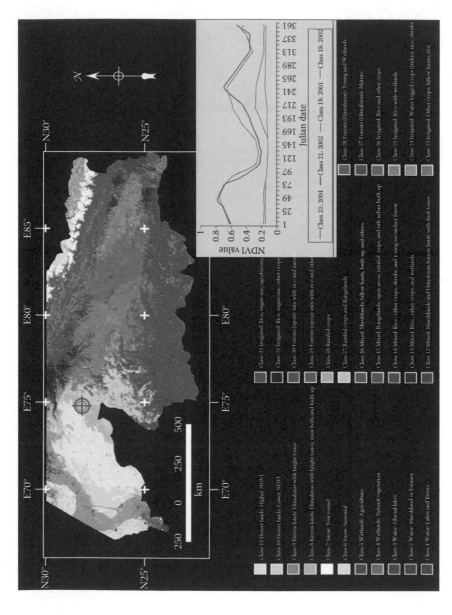

COLOR FIGURE 3.19   The NDVI time-series plots. The time-series NDVI plots for understanding and labeling classes.

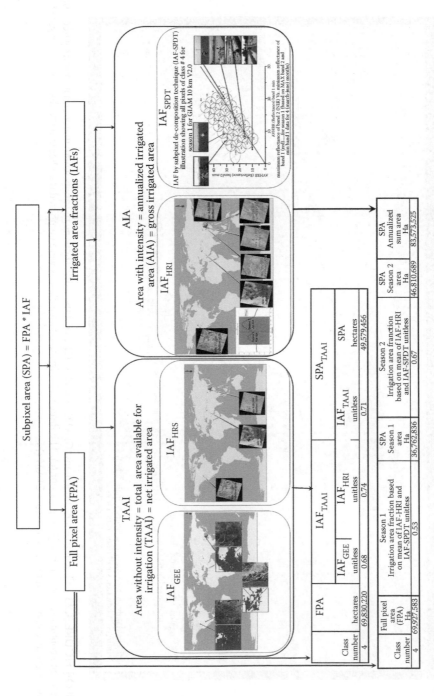

COLOR FIGURE 3.22 Subpixel area (SPA) or actual area calculation process for GIAM. The SPA calculation without considering the intensity (TAAI) and considering the intensity (AIA) is illustrated.

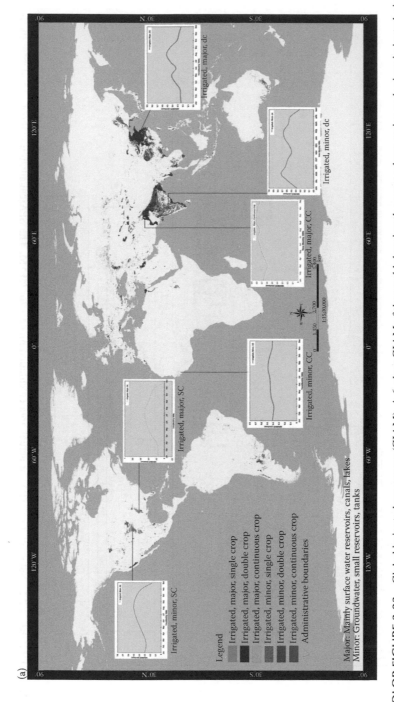

COLOR FIGURE 3.23 Global irrigated area maps (GIAM). A 6-class GIAM of the world showing the areas where major and minor irrigated, single, double, and continuous (or perennial) crops occur (a). The three major irrigated area classes were combined together and the three minor irrigated area classes were also combined to show a 2-class GIAM (b).

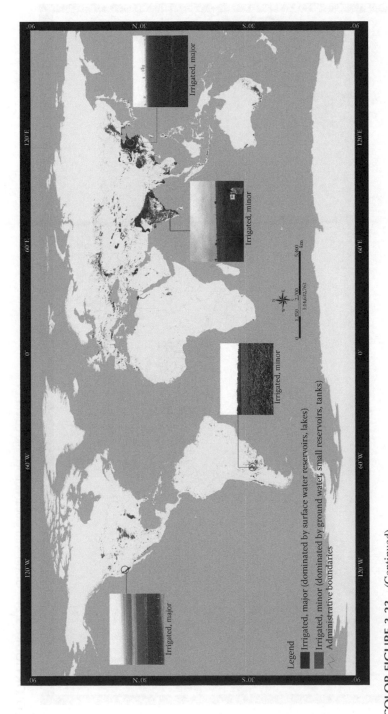

COLOR FIGURE 3.23 (Continued)

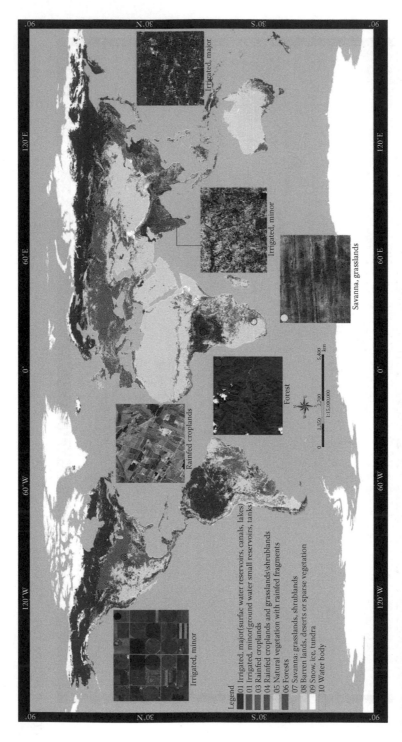

Legend
01 Irrigated, major(surfac water reservoirs, canals, lakes)
01 Irrigated, minor(ground water small reservoirs, tanks)
03 Rainfed croplands
04 Rainfed croplands and grasslands/shrublands
05 Natural vegetation with rainfed fragments
06 Forests
07 Savanna: grasslands, shrublands
08 Barren lands, deserts or sparse vegetation
09 Snow, ice, tundra
10 Water body

COLOR FIGURE 3.24 GIAM with global rain-fed and other LULC.

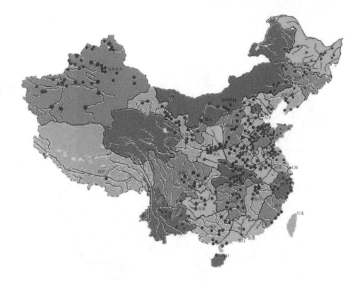

COLOR FIGURE 4.2 Distribution of large irrigation area in China.

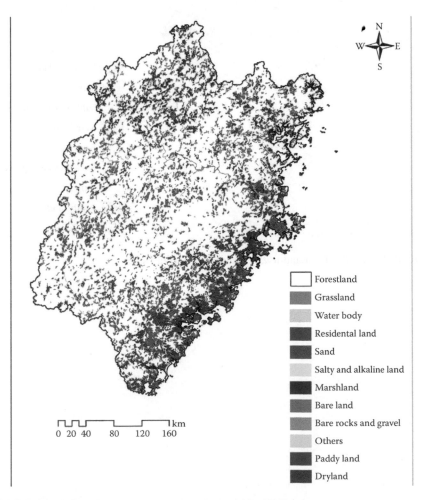

□ Forestland
■ Grassland
■ Water body
■ Residental land
■ Sand
■ Salty and alkaline land
■ Marshland
■ Bare land
■ Bare rocks and gravel
■ Others
■ Paddy land
■ Dryland

COLOR FIGURE 4.5 Land-use map at scale 1:1 M of Fujian.

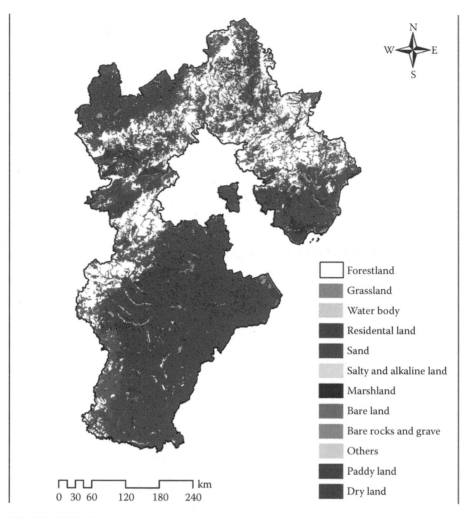

COLOR FIGURE 4.6   Land-use map at scale 1:1 M of Hebei.

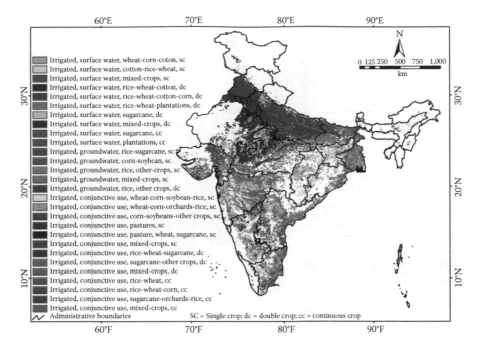

COLOR FIGURE 5.11 Irrigated area map of India derived from NOAA AVHRR 10-km sensor data.

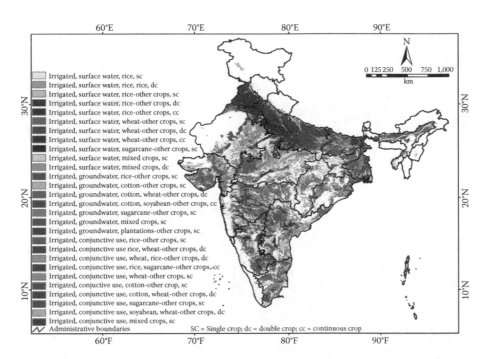

COLOR FIGURE 5.12 Irrigated area map of India derived from MODIS 500-m sensor data.

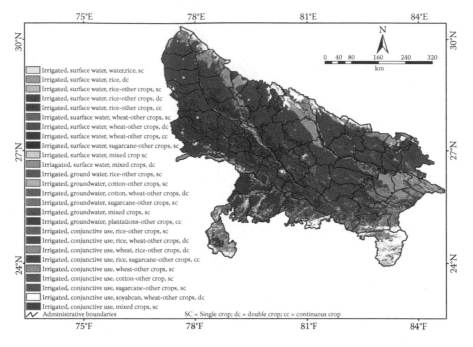

COLOR FIGURE 5.13 Irrigated area map of Uttar Pradesh (GIAM MODIS 500-m).

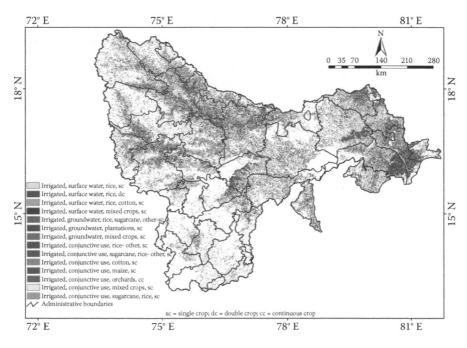

COLOR FIGURE 5.14 Irrigated area map of Krishna basin (GIAM, LANDSAT 30-m).

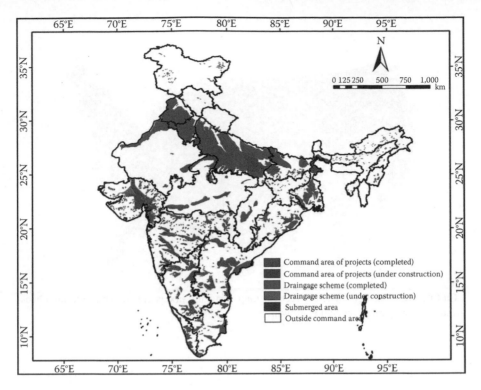

COLOR FIGURE 5.15 Irrigated areas of India mapped by India's CBIP (1994).

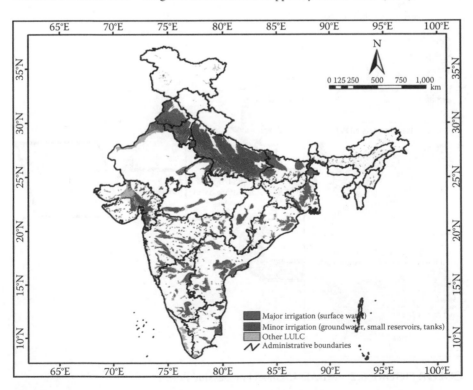

COLOR FIGURE 5.16 Actual irrigated areas of India within India's CBIP (1994) map.

COLOR FIGURE 6.1 The MIrAD-US map. Irrigated pixels are shown in red on a background showing peak vegetation growth.

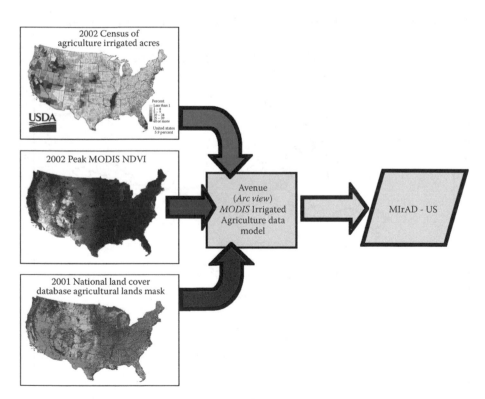

COLOR FIGURE 6.2 MIrAD-US model processing flow.

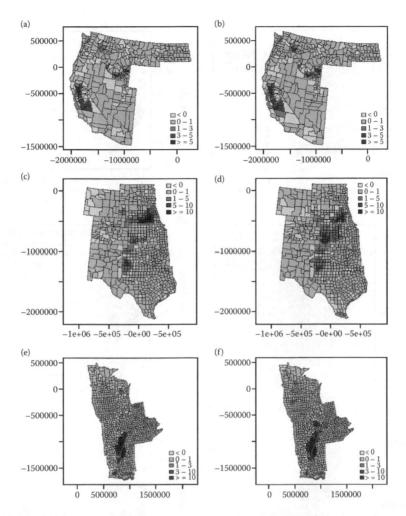

COLOR FIGURE 6.11 The local Moran's statistic maps of LISA for the western region (a) MIrAD-US CIAF, (b) GIAM CIAF; the central region (c) MIrAD-US CIAF, (d) GIAM CIAF; and the MFP region (e) MIrAD-US CIAF, (f) GIAM CIAF.

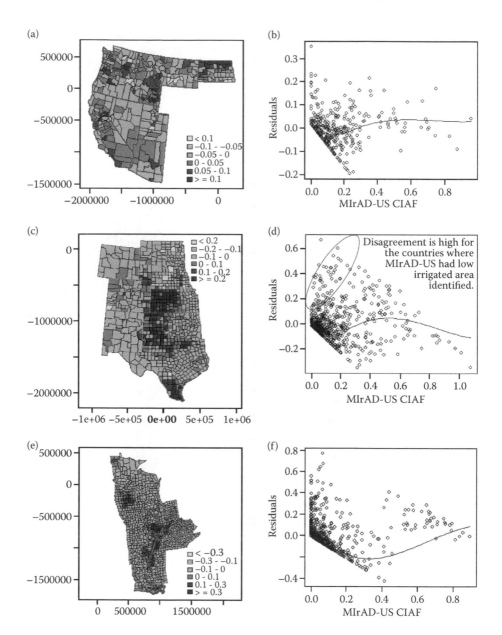

COLOR FIGURE 6.12 (a) The OLS regression residuals map, (b) scatterplot of MIrAD-US CIAF and OLS residuals for the western region; (c) OLS regression residuals map, (d) scatterplot of MIrAD-US CIAF and OLS residuals for the central region; and (e) OLS regression residuals map, (f) scatterplot of MIrAD-US CIAF and OLS residuals for the MFP region.

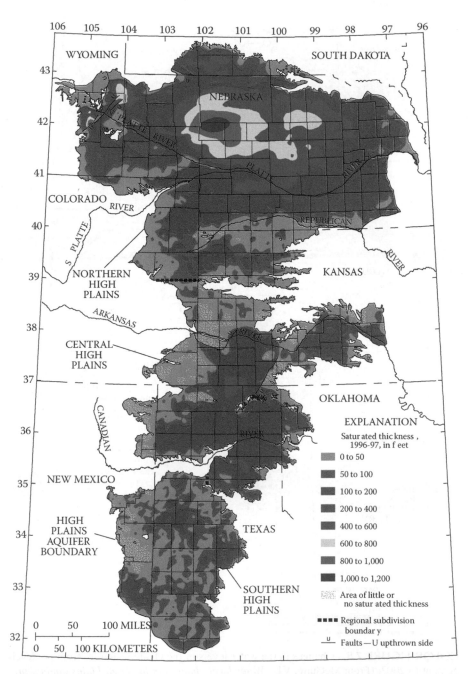

COLOR FIGURE 7.3 Saturated thickness of the Ogallala aquifer. (From McGuire, V.L. and Fischer, B.C., *Water-level changes, 1980 to 1997, and saturated thickness, 1996–97, in the High Plains aquifer*, USGS Fact Sheet 124-99, USGS, Lincoln, NE, 1999.)

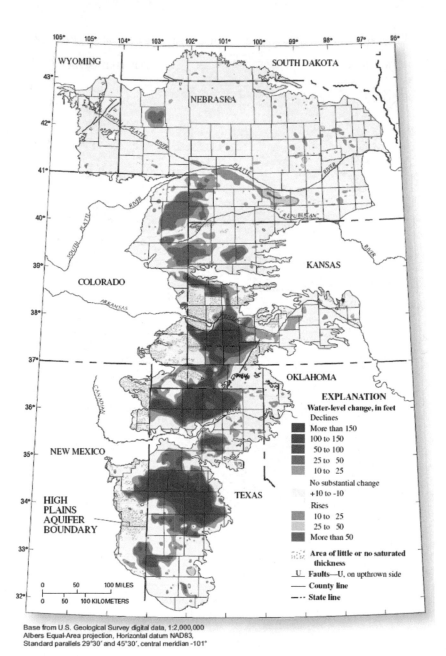

COLOR FIGURE 7.5 Changes in the water levels of the Ogallala aquifer from predevelopment to 2003. (From McGuire, V.L., *Water-level changes in the High Plains aquifer, predevelopment to 2003 and 2002 to 2003*, United States Geological Survey (USGS) Fact Sheet, FS-2004-3097, USGS, Lincoln, NE, 2004.)

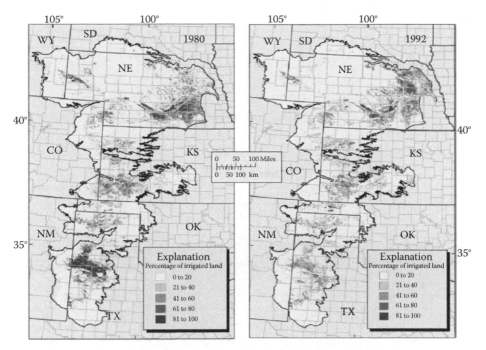

COLOR FIGURE 7.6 Comparison of irrigated areas between 1980 and 1992. (From Zwingle, E. and Richardson, J., Ogallala aquifer: Wellspring of the high plains, *National Geographic Magazine*, 183, 80–109, 1993.)

COLOR FIGURE 7.10 The irrigated area classes identified in GIAM overlying a 2005 30-m Landsat image for an area in Kansas.

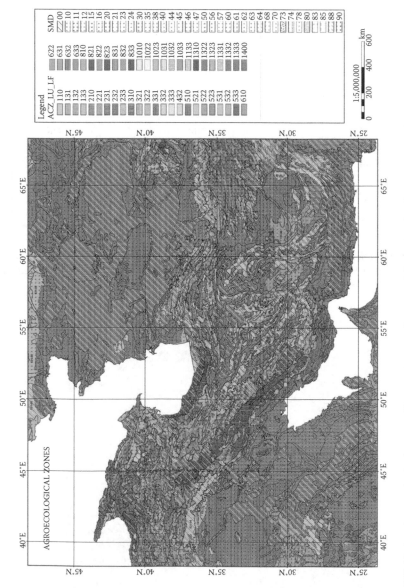

COLOR FIGURE 10.6 Agroecological zones of Iran and parts of West and Central Asia, Caucasus, and Egypt.

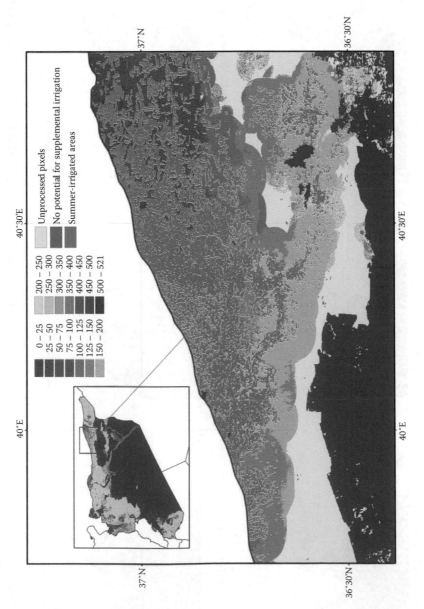

COLOR FIGURE 10.17 Modeled water allocations to rainfed areas, shown for a sample area.

COLOR FIGURE 11.6 Comparison of paddy field cover map from global land cover product based on MODIS and AVHRR: (a) AARS, (b) BU, (c) JRC, (d) USGS, (e) UMD, (f) UT.

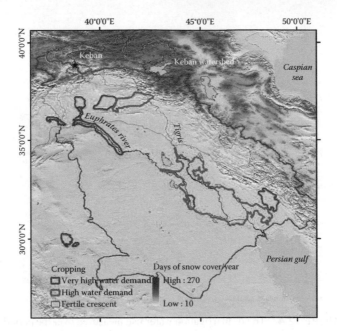

FIGURE 12.1 Distribution of cultivated areas in the Euphrates-Tigris basin (light blue line). Intensely irrigated areas (green) greatly depend on river flow created in the adjacent mountain ranges. More than 80% of river flow of the Euphrates River is created in the Keban watershed (red), forming the main water resource for irrigation in the study area (black box).

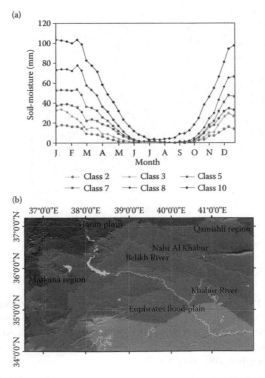

COLOR FIGURE 12.3 Soil moisture zones from an unsupervised classification of modeled mean soil moistures (1982–2000) (a) with major irrigated areas outlined in green and the arable Fertile Crescent in transparent green. Corresponding annual cycles of class mean soil moisture are given in the diagram (b).

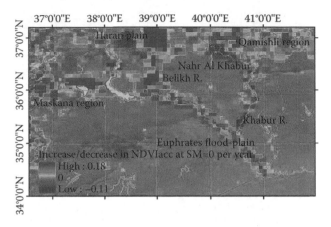

COLOR FIGURE 12.4 Trends in irrigation water use measured over the period 1982–2000.

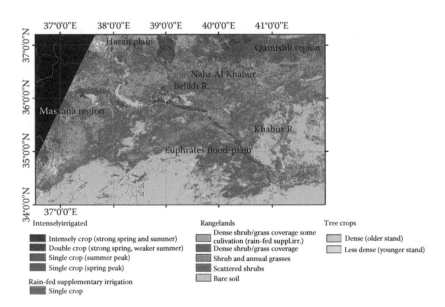

Intensely irrigated	Rangelands	Tree crops
Intensely crop (strong spring and summer)	Dense shrub/grass coverage some culivation (rain-fed suppl.irr.)	Dense (older stand)
Double crop (strong spring, weaker summer)	Dense shrub/grass coverage	Less dense (younger stand)
Single crop (summer peak)	Shrub and annual grasses	
Single crop (spring peak)	Scattered shrubs	
	Bare soil	
Rain-fed supplementary irrigation		
Single crop		

COLOR FIGURE 12.5 Example of a classification of land cover/land use from a 250-m MODIS NDVI time series (year 2005, 16-day intervals) based on an algorithm that uses the shape of NDVI cycles as a primary classifier. (From Geerken, R.A., An algorithm to classify vegetation phenologies and coverage and monitor their inter-annual change, *International Journal of Remote Sensing*, 2007 (In review).)

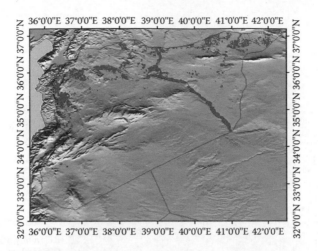

COLOR FIGURE 12.7 Areas under irrigation (green) in Syria during the year 2000. The result is based on 1-km SPOT Vegetation NDVI time series and modeled soil moistures at 1 km. Not included in this analysis are possible irrigated areas in Syria falling west of longitude 36.3 and south of latitude 32.7.

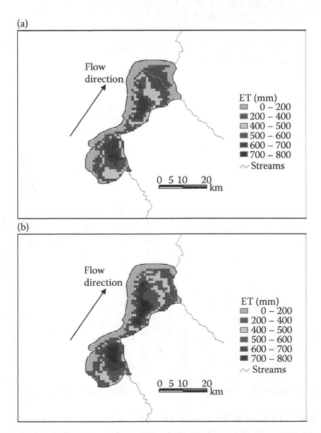

COLOR FIGURE 13.7 (a) Seasonal (June 2 to September 6) actual ET distribution in irrigated fields of the study area in 2002. Arrow indicates the general south-north flow direction of the streams. (b) Seasonal (June 2 to September 6) actual ET distribution in irrigated fields of the study area in 2003. Arrow indicates the general south-north flow direction of the streams.

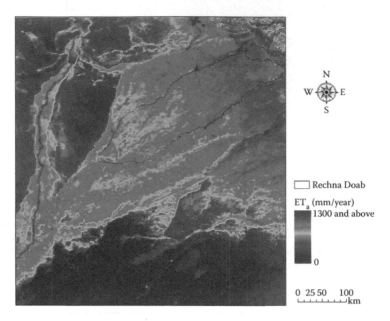

COLOR FIGURE 14.3 Annual ET$_a$ (May 2001–May 2002) in the Rechna Doab and other parts of the upper Indus Basin, Pakistan.

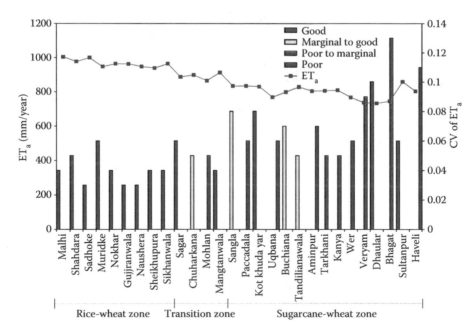

COLOR FIGURE 14.5 Subdivision-level variation in actual evapotranspiration (ET$_a$), coefficient of variation of ET$_a$ and groundwater quality in the Rechna Doab, Pakistan. Groundwater quality is based on farmers' perception. (*Groundwater Quality Data Source: IWMI Socioeconomic survey 2004.*)

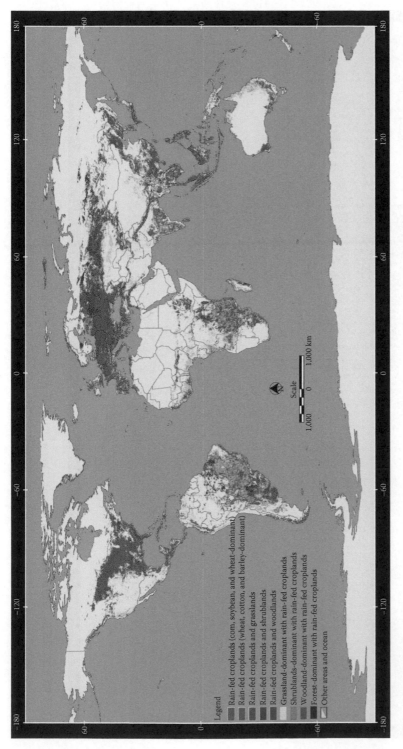

Legend

Rain-fed croplands (corn, soybean, and wheat-dominant)
Rain-fed croplands (wheat, cotton, and barley-dominant)
Rain-fed croplands and grasslands
Rain-fed croplands and shrublands
Rain-fed croplands and woodlands
Grassland-dominant with rain-fed croplands
Shrublands-dominant with rain-fed croplands
Woodland-dominant with rain-fed croplands
Forest-dominant with rain-fed croplands
Other areas and ocean

Scale

1,000   0   1,000 km

COLOR FIGURE 15.4 Global map of rain-fed cropland areas (GMRCA). This is a 9-class GMRCA map at nominal 1-km resolution produced using a fusion of 1-km SPOT VGT, 10-km NOAA AVHRR, and numerous secondary data. The first five classes are dominated by rain-fed classes, the next three are dominated by savannas with significant rain-fed croplands, and class 9 is dominated by forests with significant rain-fed cropland fragments.

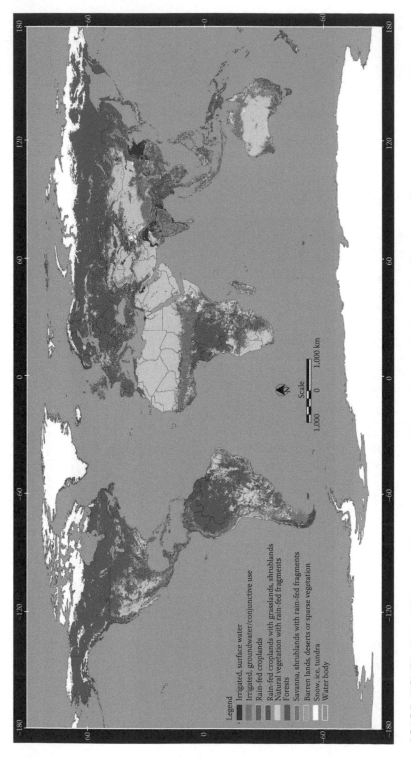

COLOR FIGURE 15.5 Global rain-fed croplands along with irrigated crops and other land use/land cover. Classes 3 and 4 are rain-fed croplands, class 5 is natural vegetation with significant rain-fed fragments, classes 1 and 2 are irrigated croplands, and the rest of the classes are noncroplands.

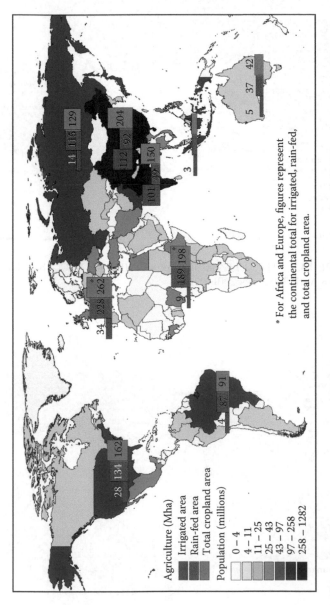

COLOR FIGURE 15.8 Human population in the twentieth century compared with total cropland area. The most populated countries are in dark blue shades, showing the histogram of irrigated, rain-fed, and total cropland area.

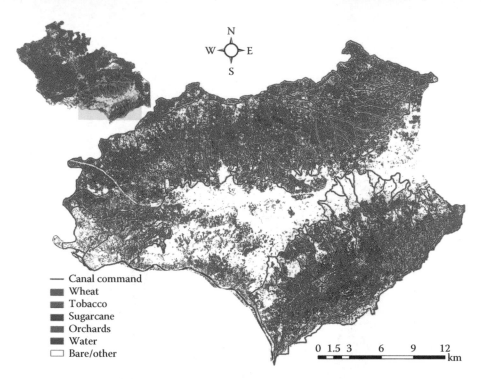

COLOR FIGURE 18.8 The LULC classification of PHLC and USCC, based on Landsat ETM + image of 8 April 2002 (rabi 2001–02).

Legend for Figure 18.8:
- Canal command
- Wheat
- Tobacco
- Sugarcane
- Orchards
- Water
- Bare/other

Scale: 0 1.5 3 6 9 12 km

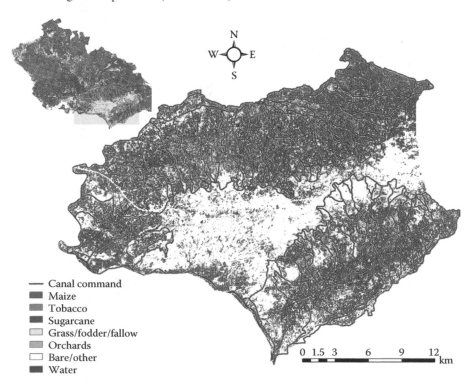

COLOR FIGURE 18.9 The LULC classification of PHLC and USCC based on Landsat ETM + image of 15 September 2002 (kharif 2002).

Legend for Figure 18.9:
- Canal command
- Maize
- Tobacco
- Sugarcane
- Grass/fodder/fallow
- Orchards
- Bare/other
- Water

Scale: 0 1.5 3 6 9 12 km

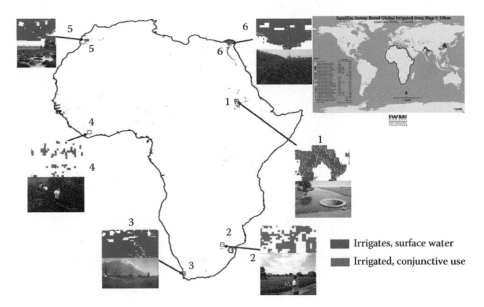

COLOR FIGURE 20.1 Irrigated cropland areas of Africa. The global Annualized Irrigated Area (AIA) in the African continent is only about 2% compared to 14% of the global population. There is a real opportunity to expand irrigated areas in Africa to facilitate green and blue revolutions.

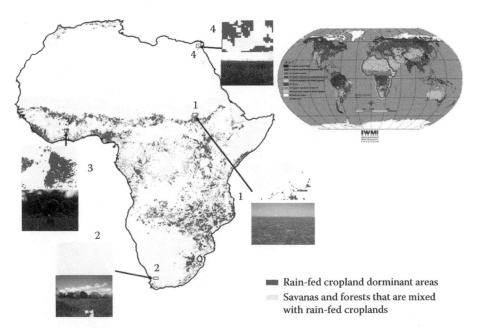

COLOR FIGURE 20.2 Distribution of rain-fed croplands in the African continent. The total global rain-fed cropland area (Global: 1.13 Bha) in Africa is 17% and its productivity can be improved through irrigation and better land management practices to help move towards a green revolution.

1995	667	1,644	190	308	334	25	52	467	90	5	94,354	1,415
1996	757	1,600	194	347	383	28	59	528	106	6	103,552	1,368
1997	804	1,549	199	366	417	22	64	555	121	6	105,583	1,313
1998	886	1,478	202	399	472	15	65	620	135	8	109,902	1,240
1999	880	1,545	206	381	489	11	68	615	141	5	114,764	1,304
2000	877	1,656	210	387	475	16	62	601	156	5	123,716	1,411
2001	908	1,490	212	397	499	12	58	634	163	4	112,856	1,243
Average		1,520	182									1,303

a Information on garage type was not available. Values were obtained by polynomial interpolation of the available data.

of the carports built each year are assumed to the size of one-car garage and half of the size of a two-car garage:

$$P_t = (\text{1-car garage SF}_t * 23 + \text{2-car garage} \times \text{SF}_t * 37$$

$$+ \text{3-car garage SF}_t * 56 + \text{carport} \times \text{SF}_t/2 * 23$$

$$+ \text{carport SF}_t/2 * 37). \tag{8.5}$$

The calculation of D_t (Equation 8.6) assumes driveways to be of the size of 33 m^2 for a one-car garage, 56 m^2 for a two-car garage, and 84 m^2 for a three-car garage. The driveways for the houses with carports are assumed to be half the size of those with one-car garages and half the size of those with two-car garages:

$$D_t = (\text{1-car garage SF}_t * 33 + \text{2-car garage SF}_t * 56 + \text{3-car garage}$$

$$\times \text{SF}_t * 84 + \text{carport SF}_t/2 * 33 + \text{carport SF}_t/2 * 56). \tag{8.6}$$

Total residential turfgrass area based on median values is calculated by multiplying the 1978–2001 average median residential lawn area by the total number of single-family houses from the 2000 US Census. This measure provides a lower limit of the total residential area under turfgrass. An upper limit of the total residential turfgrass area, based on average lot and house sizes, is calculated by multiplying the 1978–2001 average residential lawn area by the total number of single-family houses from the 2000 US Census.

To obtain a total turfgrass area, including the turfgrass found in commercial areas, schools, golf courses, athletic fields, etc., we use available information from turfgrass and green industry surveys on the fraction of turfgrass found in private residences (Table 8.5). Across the 10 states for which recent turfgrass surveys are available, on average, private residences make up 68.3% of the total turfgrass area.

Total turfgrass area is calculated as:

$$A_{\text{tot,med(avg)}} = A_{\text{Res,med(avg)}} / f_{\text{Res}}, \tag{8.7}$$

where, $A_{\text{Res,med(avg)}}$ is the total residential turfgrass area calculated either from median or from average lawn area averaged throughout the 1978–2001 period, and f_{Res} is the average fraction of turfgrass area found in private residences across the surveyed states.

In the conversion of total turfgrass area to urban irrigated area, we are aware that although most of the turfgrass area can be expected to be equipped for irrigation (i.e., through automatic sprinklers or through a rubber hose and manual sprinklers), not every lawn and landscaped area across the country is irrigated. A recent survey of 1005 American adult homeowners indicated that 38% never water their lawns [30]. For example, many eastern states are humid enough during the growing season, so that many homeowners do not water their lawns. As these states are also densely populated, a considerable pressure on water resources for indoor consumptive use contributes to triggering frequent water bans for outdoor water use during dry years. In other parts of the country where the climate is drier and the population is smaller, watering of urban landscapes is more common, as otherwise turfgrasses could not survive or would turn brown and dormant for most of the dry season. Percentages of house owners watering their lawns in different geographic areas are available only for a few regions (Table 8.3). We will assume that 40% of the turfgrass area in the northeastern states, 60% in the southeastern and central states, and 80% in the western states are watered. The assumptions on the prevalence of irrigation for each state are reported in Table 8.5.

8.4 RESULTS AND DISCUSSION

8.4.1 TOTAL TURFGRASS AREA

The estimates of total turfgrass area obtained from the various methods are summarized in Table 8.4.

The estimates of turfgrass area for each state derived from the USGS ISA data set are listed in Table 8.5, along with the estimates derived from the NOAA ISA by

TABLE 8.3
Prevalence of Watering Practices in Various States

State	House Owners Watering their Lawns (%)	Source
Iowa	35	USDA National Agricultural Statistics Service & Iowa Agricultural Statistics Service [17]
New Jersey	75	Govindasamy et al. [21]
New York	15	New York Agricultural Statistics Service [22]
North Carolina	68	Osmond and Hardy [31]
Virginia	50	Aveni [32]
Oregon	74	Nielson and Smith [33]
United States	38	Anonymous [30]

TABLE 8.4
Summary of Total Turfgrass Area and Urban Irrigated Area from Various Approaches

Approach	Total Turfgrass Area (ha)	Urban Irrigated Area (ha)
USGS-based	11,172,171	6,665,332
NOAA-based	16,163,436	9,602,148
US Census-based	7,263,980	4,503,668
US Census-based	14,564,101	9,029,743

TABLE 8.5
Estimates of Total Turfgrass Area Derived from Remote sensing Approaches and Turf Surveys, along with the Proportion Assumed to be Irrigated

State	NOAA-based Turfgrass Area (ha)	USGS-based Turfgrass Area (ha)	Turfgrass Survey (ha)[a]	Turfgrass Area Assumed to be Irrigated (%)
Alabama	314,134	202,119		60
Arizona	256,665	197,615		80
Arkansas	210,678	141,329		60
California	1,118,838	853,814		80
Colorado	247,878	182,113		80
Connecticut	247,614	147,067		40
Delaware	55,255	30,846		40
District of Columbia	5,685	6,227		40
Florida	1,174,029	766,200		60
Georgia	569,874	385,985		60
Idaho	93,541	63,820		80
Illinois	575,386	537,902	624,665 (61)	40
Indiana	385,159	287,210		40
Iowa	228,978	165,595	514,534 (47)	40
Kansas	199,662	167,007	309,423 (21)	60
Kentucky	243,975	158,244		60
Louisiana	336,979	217,969		60
Maine	97,698	54,427		40
Maryland	254,171	151,175	459,237 (83)	40
Massachusetts	429,714	292,310		40
Michigan	461,845	447,658	764,492 (83)	40
Minnesota	318,354	233,235		40
Mississippi	200,147	123,706		60
Missouri	345,604	275,787		60
Montana	73,071	59,004		80
Nebraska	115,201	95,355		80
Nevada	93,124	84,697		80
New Hampshire	111,757	63,736		40
New Jersey	396,434	254,108	360,342 (75)	40
New Mexico	153,904	71,275		80
New York	641,950	360,596	1,387,393 (82)	40
North Carolina	816,036	355,305	864,247 (68)	60
North Dakota	56,676	36,978		80
Ohio	671,807	527,373		40
Oklahoma	269,306	174,423		80
Oregon	198,784	149,597		80
Pennsylvania	728,516	429,059		40
Rhode Island	52,026	42,837		40
South Carolina	405,089	194,811		60
South Dakota	69,352	45,254		80

TABLE 8.5 (Continued)

State	NOAA-based Turfgrass Area (ha)	USGS-based Turfgrass Area (ha)	Turfgrass Survey (ha)[a]	Turfgrass Area Assumed to be Irrigated (%)
Tennessee	419,245	257,523		60
Texas	1,320,628	1,041,741		80
Utah	120,869	96,580		80
Vermont	52,150	26,042		40
Virginia	458,507	261,428	553812 (52)	60
Washington	362,650	309,111		80
West Virginia	149,612	82,070		60
Wisconsin	313,486	233,104	485299 (64)	40
Wyoming	55,527	32,920		80

[a] The fractions (%) of turfgrass found in private residences are given in parentheses.

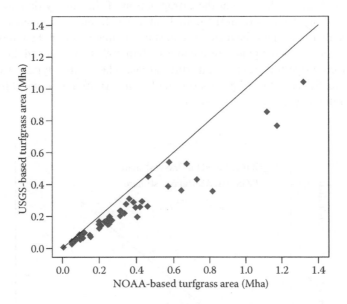

FIGURE 8.2 Scatterplot of the two remote sensing-based estimates of total turfgrass area aggregated by state.

Milesi et al. [9] and, where available, the estimates provided by the turfgrass and green industry surveys. The total turfgrass area estimated from the USGS ISA is 11,172,171 Mha, a surface considerably lower than the previously estimated area of 16,163,436 Mha based on the NOAA ISA.

The USGS-based turfgrass area estimates are smaller than the NOAA-based estimates for all states except the District of Columbia (Figure 8.2), with a median difference of 33%, and differences greater than 50% for the states of New Mexico,

North and South Carolina, and Vermont. The differences between the two turfgrass area estimates are greater than the differences among fractional ISA estimates due to differences in the distribution of fractional ISA values within each data set. For example, the USGS ISA data set has a larger fraction of pixels with less than 10% ISA, for which the lawn estimation model assumes no lawn, as these are assumed to be rural areas.

The statewide estimates of turf area reported in recent turfgrass and green industry surveys available for 10 states are higher than those based on remote sensing data (Figure 8.3). In general, they compare best with the NOAA-based turfgrass estimates, with a Pearson correlation coefficient of .73. For Illinois, New Jersey, and North Carolina, the turf area assessments from the industry and the NOAA-based approach differ by less than 10%. The largest difference is found over the state of New York, where the survey estimate is 50% larger than the NOAA-based estimate. The survey estimates display a weaker correlation with USGS-based turfgrass estimates, with a Pearson correlation coefficient of .5. The USGS-based turfgrass estimates are between 30 and 80% smaller than the areas reported in the surveys. One of the reasons for surveys reporting a greater turfgrass surface than the remote sensing-based approaches is that the extrapolation of the survey data to the whole state is based on averages from the polled data, which may be influenced by larger values. Another reason for the difference in state estimates is that the remote sensing approach assumes no turfgrass present at less than 10% ISA, to avoid the prediction of excessively large turfgrass area in rural areas, where other types of cover may predominate. States with a large proportion of sparse rural development may indeed have a significant surface of turfgrass.

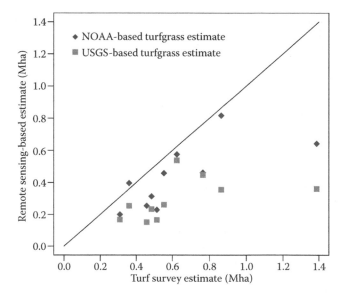

FIGURE 8.3 Scatterplot of turfgrass area derived from surveys and from remote sensing-based approaches.

Turfgrass estimates based on the housing-units approach using median lot and house size are reported in Table 8.2, while the estimates based on average lot and house size are reported in Table 8.3. The data on median lot size, median house size, type of parking, and assumptions of driveway size would indicate that the average single-family house built between 1978 and 2001, averaged a median lawn size of 650 m². The total number of single-family houses in the United States by 2000 was 76,313,410 (66% of all the housing units [34]). Multiplying this lawn size by the total number of single-family houses in the United States, we obtain an estimated 4,961,298 ha of turfgrass area in private residences. Given that turfgrass area can also be found in multifamily housing complexes, commercial lots, around institutional and recreational areas, and other open spaces, the US Census value must be adjusted to account for this portion of nonresidential turfgrasses. The turfgrass industry surveys reported that residential turfgrass area ranges between 20% and 83% of the total turfgrass area in the surveyed states, averaging 68.3%. Based on these data, the total estimate of turfgrass area amounts to 7,263,980 ha. On the other hand, if we repeat the calculations, basing them on average lot and house size, the average lawn size of single-family houses built between 1978 and 2001 would be 1303 m², leading to an estimate of 9,947,281 ha in residential turfgrass area and 14,564,101 ha of total turfgrass area.

While the estimate based on the US Census using median lot and house size was likely to underestimate the total turfgrass area, the method based on average lot and house sizes had a tendency to overestimate it. The housing-unit approach suggests that the estimate of total turfgrass for the United States is closer to the estimate provided by the USGS-based approach. On the other hand, all the industry surveys point to a much larger surface under turfgrass, likely to surpass the NOAA-based estimates of turfgrass area.

8.4.2 Urban Irrigated Area

The estimates of urban irrigated area, obtained by adjusting the total remotely sensed areas for the fraction that is likely to be watered in each state, are reported in Figure 8.4. The irrigated area under cropland is also reported for comparison. The total turfgrass area that is expected to be irrigated at some point during the growing season, according to the assumptions reported in Section 8.3, is 6,665,332 ha according to the USGS-based lawn estimate, and 9,602,148 ha according to the NOAA-based estimates.

An estimate of urban irrigated area can also be obtained from the housing-based turfgrass estimates using the results from a national lawn care poll, indicating that 62% of Americans water their lawns [30]. Using these data, it can be estimated that the surface of urban irrigated area would range between 4,503,668 and 9,029,743 ha.

In comparison, the total cropland area irrigated within the conterminous United States, reported by the Agricultural Census [35], amounts to 22,310,529 ha, with the largest shares represented by forage land (4,156,032 ha) and corn for grain (3,929,445 ha). The estimates of urban irrigated area produced by this study would therefore rank turfgrass as the single largest irrigated crop in the country. Texas, California, and Florida have large irrigated areas, both in croplands and in turfgrass. In many states,

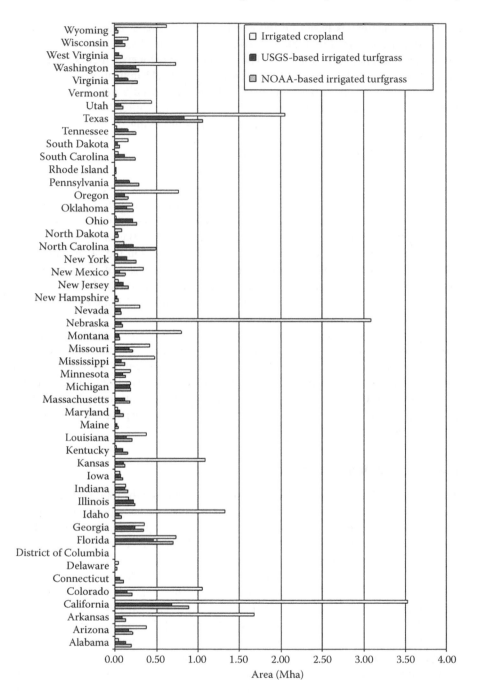

FIGURE 8.4 Distribution of irrigated turfgrass and cropland across the conterminous United States.

such as North Carolina, New York, Ohio, and Virginia, the turfgrass irrigated area is estimated to largely surpass the cropland irrigated area. In many other states, the cropland and turfgrass irrigated areas have comparable surfaces.

8.4.3 RECENT GROWTH TRENDS OF TURFGRASS AREA VS. IRRIGATED CROPLAND AREA

Between 1978 and 2001, 16,366,000 new houses were constructed and sold. Over this period, the yearly addition of these houses to the American landscape has increased the turf grass area by an average of 44,209 ha a year based on data of median lot and house size, and 91,909 ha a year based on average lot and house-size data. This increase does not account for the growth in turfgrass area in nonresidential lawns, which is expected to be considerable, though likely to have proceeded at a slower rate than the expansion in new construction. The expansion of turfgrass in private residences is of the same order of magnitude as that of cropland irrigated areas, which between 1978 and 2002 have increased, on average, by 83,312 ha a year. While the expansion in irrigated cropland has somewhat stabilized over the recent decades, urban growth is expected to continue.

Factors other than urban growth have contributed to the water use by irrigated landscapes. One of these factors is the diffusion of automated sprinklers that, while providing a more uniform cover and therefore improving the efficiency of water application, are often a cause of overwatering, as they are more likely to be left running when meteorological conditions are such that no irrigation is required. A study of water use data from 1129 houses in 14 North American cities [36] indicated that houses in dry and warm climates that rely on automatic sprinklers to water their landscape use, on average, 33.4 inches (84.8 cm) of water (61% more than houses with mobile sprinklers attached to a hose). In cooler and wetter climates, the installation of automatic sprinklers increases water use by 37%, from 6.7 inches (17.0 cm) to 9.2 inches (23.4 cm).

Another factor that has likely contributed to increased water use over the past few decades is a lengthening of the growing season. Although experimental data that link the growing season to increased water use are not available, it can be speculated that a lengthening of the frost-free season in the United States [37] may have triggered an earlier start of the sprinkling season, especially with the expansion of automatic sprinklers, which once programmed at the beginning of the spring tend to be kept running with the same fixed schedule until the fall comes and irrigation is suspended for the winter season.

8.5 CONCLUSION

Irrigation in urban areas makes up a large and growing sector in developed countries like the United States, where a cultural predilection towards well-maintained landscapes predominates. In this study, we attempted estimating the urban irrigated area in the conterminous United States by comparing different approaches. Depending on the method used, the total irrigated turfgrass area ranges from the most conservative 4,503,668 ha, obtained by using time series on the characteristics of new housing collected by the US Census, to 9,602,148 ha, based on a relationship between remote sensing-derived built-up area and turfgrass area. The estimates of urban irrigated area represent an area equivalent to 20–43% of the total cropland irrigated area.

As the methods explored are based on different assumptions, it is not possible to determine the most reliable one. An advantage of the remote sensing-based method is that it provides a level of geographical detail to the estimates of total irrigated turfgrass that is not possible to achieve with US Census-based data.

With increasing population, wealth, and a warming climate, the expansion of irrigated areas in urban settings can be expected to continue, causing increasing pressure on finite freshwater resources not only in urban areas located in dry climates, but eventually also in those located in temperate climates.

ACKNOWLEDGMENTS

This project was funded by NASA's Carbon Cycle research program.

REFERENCES

1. Thenkabail, P.S. et al., *An Irrigated Area Map of the World (1999) Derived From Remote Sensing*, Research Report No. 105, International Water Management Institute, Battaramulla, Sri Lanka, 2006.
2. Döll, P. and Siebert, S., A digital global map of irrigated areas, *ICID Journal*, 49, 2, 55–66, 2000.
3. US Environmental Protection Agency, *Outdoor Water Use in the United States*, 2008, http://www.epa.gov/owm/water-efficiency/pubs/outdoor.htm. (accessed 1 April 2009).
4. Statistics Canada, Canadian lawns and gardens: Where are the 'greenest'? *EnviroStats*, 1, 2, 16-002-XWE, 2007, http://www.statcan.gc.ca/bsolc/olc-cel/olc-cel?catno=16-002-X200700210336&lang=eng. (accessed 1 April 2009).
5. Australian Bureau of Statistics, *Water Account, Australia 2000–01*, 2004, ABS Cat. No. 4610.
6. Beard, J.B., *Turfgrass: Science and Culture*, Prentice Hall, Englewood Cliffs, NJ, 1973.
7. Mayer, P.W. et al., *Residential End Uses of Water*, AWWA Research Foundation and American Water Works Association, Denver, CO, 1999.
8. Liu, H. et al., Enhancing turfgrass nitrogen use under stresses, in *Handbook of Turfgrass Management and Physiology*, M. Pessarakli (Ed.), CRC Press, Taylor & Francis Group, Boca Raton, FL, 2007, 557–601.
9. Milesi, C. et al., Mapping and modeling the biogeochemical cycling of turf grasses in the United States, *Environmental Management*, 36, 426–38, 2005.
10. Bormann, F.H., Balmori, D., and Geballe, G.T., *Redesigning the American Lawn: A Search for Environmental Harmony*, Yale University Press, New Haven and London, 1993.
11. Winston, M.L., *Nature Wars: People vs. Pests*, Harvard University Press, Cambridge, MA, 1997.
12. Stow, D. et al., Irrigated vegetation assessment for urban environments, *Photogrammetric Engineering and Remote Sensing*, 69, 4, 381–90, 2003.

13. Jenkins, V.S., *The Lawn: A History of an American Obsession*, Smithsonian Institution Press, Washington, DC, 1994.
14. Jackson, K.T., *Crabgrass Frontier: The Suburbanization of the United States*, Oxford University Press, New York, 1985.
15. Henry, M.S., Landscaping quality and the price of single-family houses: Further evidence from home sales in Greenville, South Carolina, *Journal of Environmental Horticulture*, 17, 1, 25–30, 1999.
16. Campbell, G.E. et al., *The Illinois Green Industry: Economic Impact, Structure, Characteristics*, University of Illinois at Urbana-Champaign, IL, 2001.
17. USDA National Agricultural Statistics Service & Iowa Agricultural Statistics Service, *Iowa's Turfgrass Industry*, Des Moines, IA, 2001.
18. Kansas Agricultural Statistics, *2006 Kansas Horticultural Survey*, 2007, http://www.nass.usda.gov/Statistics_by_State/Kansas/Publications/Economics_and_Misc/Horticulture/index.asp. (accessed 1 April 2009).
19. USDA National Agricultural Statistics Service, Maryland Field Office, *Turfgrass Survey*, Annapolis, ND, 2006.
20. Michigan Agricultural Statistics Service, *Turfgrass Survey 2002*, Michigan Agricultural Statistics Service, Lansing, MI, 2003.
21. Govindasamy, R. et al., *The New Jersey Turfgrass Industry Economic Survey: Executive Summary*, New Jersey Agricultural Experiment Station, Rutgers, The State University of New Jersey, New Brunswick, NJ, 2007.
22. New York Agricultural Statistics Service, *New York Turfgrass Survey*, New York State Department of Agriculture and Markets, Albany, NY, 2004.
23. North Carolina Department of Agriculture and Consumer Services and Turfgrass Council of North Carolina, *North Carolina Turfgrass Industry*, Raleigh, NC, 1999.
24. USDA National Agricultural Statistics Service Virginia Field Office, *Virginia's Turfgrass Industry*, Richmond, VA, 2006.
25. University of Wisconsin and Wisconsin Agricultural Statistics Service, *1999 Wisconsin Turfgrass Industry Survey*, University of Wisconsin, Madison, UW Extension, and Wisconsin Agricultural Statistics Service, 1999.
26. Vinlove, F.K. and Torla, R.F., Comparative estimations of U.S. home lawn area, *Journal of Turfgrass Management*, 1, 83–97, 1995.
27. Elvidge, C.D. et al., U.S. constructed area approaches the size of Ohio. Eos, Transactions, *American Geophysical Union*, 85, 233, 2004.
28. Yang, L. et al., An approach for mapping large-area impervious surfaces: Synergistic use of Landsat 7 ETM+ and high spatial resolution imagery, *Canadian Journal of Remote Sensing*, 29, 2, 230–40, 2003.
29. US Census Bureau, *Characteristics of New Housing*, http://www.census.gov/const/www/charindex_excel.html. (accessed 1 April 2009)
30. Anonymous, New consumer reports' poll reveals America's take on lawn care, *Consumer Report*, May Issue, 2008, http://www.consumerreports.org/cro/home-garden/resource-center/lawn-care-5-08/yard-survey/lawn-care-yard-survey.htm. (accessed 1 April 2009).
31. Osmond, D.L. and Hardy, D.H., Characterization of turf practices in five North Carolina communities, *Journal of Environmental Quality*, 33, 565–75, 2004.
32. Aveni, M., Homeowner survey reveals lawn management practices in Virginia, *Watershed Protection Techniques*, 1, 2, 85, 1994.
33. Nielson, L. and Smith, C.L., Influences on residential yard care and water quality: Tualatin watershed, Oregon, *Journal of the American Water Resources Association*, 41, 1, 93, 2005.
34. US Census Bureau, *Census 2000 Summary File 3 (SF 3), Matrice H30*, 2000, http://factfinder.census.gov. (accessed 1 April 2009).

35. US Department of Agriculture, *2002 Agricultural Census*, 2002, http://www.agcensus.usda.gov/Publications/2002/index.asp. (accessed 1 April 2009).
36. Maddaus, L.A. and Mayer, P.W., *Splash or Sprinkle? Comparing the Water Use of Swimming Pools and Irrigated Landscapes,* 2001 American Water Works Association Annual Conference, Washington, DC, 2001.
37. Easterling, D.R., Recent changes in frost days and the frost-free season in the United States, *Bulletin of the American Meteorological Society*, 83, 1327–32, 2002.

9 The Challenges of Mapping Irrigated Areas in a Temperate Climate: Experiences from England

Jerry W. Knox, S.A.M. Shamal,
and E.K. Weatherhead
Cranfield University

J.A. Rodriguez-Diaz
University of Cordoba

CONTENTS

9.1   INTRODUCTION

9.1.1   IRRIGATED AREAS IN ENGLAND

In England, irrigated agriculture is an important sector of the agrifood industry, and a small but significant user of water in the drier parts of the country. For many farm businesses, irrigation is an essential component of production, particularly on high-value outdoor horticultural crops, such as potatoes, soft fruit, and field vegetables, where continuous and reliable supplies of premium quality are demanded by the major processors and supermarkets. Although the total area irrigated is small by international standards, typically around 150,000 hectares (ha) in a dry year, the supplemental nature of irrigation, with small depths of water being applied to high-value crops, means that the financial benefit (value) of irrigation water is extremely high [1]. Nationally, irrigated agriculture accounts for only 4% of crop area, but approximately 20% of crop value. It is these characteristics that make water resources for irrigation in England important; without secure and reliable access to water supplies, many agribusinesses would simply not survive; locally, the consequences of water shortages would result in a shift away from intensive high-value production to low-input cereal production, with negative impacts on rural employment.

The use of irrigation in England varies markedly between crop types. Some of the most detailed information available is that collected for the government (Department for Environment, Food and Rural Affairs, DEFRA) through farmer irrigation surveys. These provide data on items including areas irrigated and volumes applied (by crop category), irrigation practices, water resources, and equipment usage. These surveys have been carried out roughly triennially since 1984; the most recent survey was carried out in 2005 (Table 9.1).

Historically, potato and sugar beet were the two most important irrigated crops, but with significant areas of irrigated cereals and grassland for dairy farming. In recent years, however, the composition of the irrigated area has changed considerably, largely in response to agroeconomic policies and shifts in market demand, which have changed the financial benefits of irrigating different crops. By 2005, potato and vegetables together accounted for 70% of the total irrigated area, and the other crop types each had only minor shares. It should be noted that the areas irrigated (and volumes of water applied) vary greatly each year, depending on summer

TABLE 9.1
Irrigated Areas (ha) by Crop Category Derived from Government Surveys, 1984–2005

Crop Category	1984	1987	1990	1992	1995	2001	2005
Potato (early)	7,720	5,360	8,510	8,180	8,730	7,300	6,415
Potato (main crop)	34,610	29,520	43,490	45,290	53,390	69,820	43,140
Sugar beet	25,500	10,100	27,710	10,520	26,820	9,760	8,487
Orchard fruit	3,250	1,330	3,320	2,280	2,910	1,580	1,468
Small fruit	3,560	2,230	3,470	2,750	3,250	3,770	2,631
Vegetables	17,460	11,040	25,250	20,200	27,300	39,180	32,202
Grass	18,940	6,970	15,970	7,240	10,690	3,970	3,671
Cereals	24,700	7,510	28,100	7,160	13,440	4,620	10,979
Other	4,890	2,440	8,650	4,320	9,120	7,280	7,280
Total	140,630	76,500	164,470	107,940	155,650	147,270	116,272

Note: Summing errors due to rounding; data up to 1992 for England and Wales; data from 1995 for England only (Wales represents less than 1% of total).

rainfall and temperature. The survey data must therefore be interpreted in relation to the weather in each year. But it is apparent that irrigation is being increasingly concentrated on the more valuable crops, driven by quality assurance, and those crops that are irrigated receive greater application depths.

Nationally, irrigated agriculture represents only 1–2% of total water abstraction, with most concentrated in eastern England, the driest part of the country where water resources are most constrained. All abstractions for irrigation require a license or permit from the water regulatory authority, the Environment Agency (EA). The EA has records of irrigation abstraction since the Water Resources Act (1963) came into force within their national abstraction licensing database (NALD). Figure 9.1 shows the spatial distribution of agricultural irrigation abstractions in the EA Anglian region in relation to the availability of additional low-flow (summer) water resources. By combining the two data sets, it is estimated that nearly three-quarters of the total volume of water licensed for agricultural irrigation in this region is located within catchments under severe levels of water stress. Some 47% of irrigation is located within "overlicensed" catchments and a further 23% within catchments defined as being "overabstracted."

9.1.2 Applications for Irrigated Area Mapping

The map presented in Figure 9.1 relates to abstraction points for irrigation, rather than to irrigated areas per se, but it provides a good indication of where irrigated areas in the region are concentrated. Clearly, in terms of improving water resources management at both regional and catchment levels, there are a variety of other potential

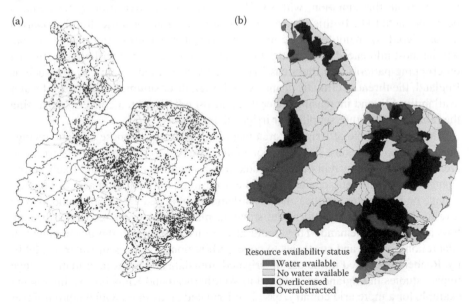

FIGURE 9.1 (a) Spatial distribution of abstraction points for agricultural irrigation, by catchment, in the EA Anglian region in 2006, and (b) additional low-flow water resource availability (*Note*: status estimated for catchments where resource availability has not yet been formally published.)

applications that could be developed, which usefully integrate irrigated area maps and data into the decision-making process [2]. Two selected example applications are described below.

In England and Wales, all water abstractions, including those for irrigation, are licensed and controlled by the EA. The Water Resources Act (1991) allows the EA to impose restrictions on abstraction for spray irrigation in order to avoid or reduce adverse environmental impacts. But imposing total or partial restrictions on irrigation can have significant financial consequences for farm production and farm incomes. The Environment Act (1995) placed a duty on the EA to take account of the likely costs and benefits that may follow from its actions. By taking into account the spatial and temporal variations in irrigation demand (derived from detailed estimates of irrigated areas), a methodology was developed to allow water resources planners to estimate the potential financial impacts associated with imposing alternative restrictions on irrigation abstraction at different stages through the season. Readers interested in a description of the methodology and its application are referred to Knox et al. [1]. However, one limitation identified by the authors was the spatial accuracy of the digital data sets used in the methodology. They noted that the spatial accuracy of the land use data sets and the information relating to irrigated areas would quickly become outdated, and was an inherent problem that needed to be addressed.

Of course, in many arid and semiarid countries, water supplies for irrigation are already under severe pressure, as evidenced by the Comprehensive Assessment of Water Management in Agriculture [3]. But anthropogenic climatic change threatens to exacerbate that situation, with significant increases in irrigation water demand being predicted [4]. In this context, abstraction of water for agricultural irrigation is considered by many to be the most sensitive of water uses, and the sector that will be most influenced by climatic change [5]. Climatic change will directly impact on cropping patterns and the areas irrigated. Even in a temperate country such as England, the threat of climatic change with hotter, drier summers [6], reduced water availability [7], and increasing water demand [8] only heighten concerns regarding the reliability of future supplies for irrigated agriculture.

The potential role of irrigated area mapping for assessing these climatic change impacts appears obvious. The provision of detailed high-resolution maps of irrigated areas, based on the current pattern of land use, and taking into account likely future changes due to climatic change, could be used to quantify the impacts of climatic change on both water supplies and water demand. Irrigated area maps could also be invaluable in helping to inform the strategic planning of water resources through forecasting water demand. They could also be used to inform the development of adaptation policies and strategies to cope with regional changes in water availability. Readers interested in the use of irrigated area data and maps in climatic change impact studies in England are referred to Weatherhead and Knox [9]. An interesting example for a more arid country, based on irrigated areas in the Guadalquivir River Basin in Spain, is given in Rodriguez-Diaz et al. [10].

The difficulty of obtaining accurate data for management purposes is, however, well illustrated by the differences in currently published estimates of total demand,

TABLE 9.2
Estimates of Irrigated Areas for England and Wales Based on Contrasting Methods

Source	Year	Method of Area Estimation	Estimated Irrigated Area (ha)
Government Irrigation Survey	2005	Statistics, based on farm surveys	116,276
DEFRA Irrigation Survey	2001	Statistics, based on farm surveys	147,270
FAO/UF Global Irrigated Area Mapping (GIAM V4.0)	2001	Statistics, based on farm surveys	142,687
IWMI Global Irrigated Area Mapping (GIAM)[a]	1999	Remote sensing and Google Earth Estimate	970,733

[a] Total Area Available for Irrigation (TAAI).

based on different methodologies. Table 9.2 compares the values obtained by surveys with those obtained by remote sensing for the whole of England and Wales; for smaller geographical areas, the agreement is likely to be even lower.

9.2 APPROACHES TO IRRIGATED AREA MAPPING: KEY CHALLENGES

In England, the established methods for estimating and mapping irrigated areas have relied on two main data sources, namely government statistics and industry survey data. Estimates were originally based on spreadsheet analyses combining various government survey data and statistics on the composition of agricultural cropping to estimate irrigated areas, usually at national and regional levels. More recently, Geographic Information System (GIS) spatial analysis techniques have been used to combine government survey (farm) data with digital data sets on soils, land use, and climate (raster) to provide more accurate estimates of irrigated areas, at regional and catchment levels.

Although the application of remote sensing for estimating cropped areas is not new, it still poses significant challenges for estimating irrigated areas, particularly in temperate countries where irrigation is supplemental to rainfall. In the more arid parts of the world, where summer rainfall is negligible, detailed irrigated area maps have already been successfully produced, including, for example, in Italy [10], Turkey [11], Thailand [12], and India [13]. This is in contrast to England, where summer rainfall can easily remove any observable differences in crop reflectance (Normalized Differences Vegetation Index; NDVI) between irrigated and rain-fed crops, thus making it very difficult to accurately estimate irrigated areas.

In this section, the key issues and challenges in moving away from a dependence on government survey data towards approaches based more on a combination of GIS, survey data, and remote-sensing imagery are described.

9.2.1   Using Government Statistics and Industry Survey Data

The main source of national statistics on agricultural irrigation is the government "Irrigation of outdoor crops" surveys. Individual farm responses were previously aggregated into parish, county, and national totals, but due to confidentiality constraints, they could be published only at regional and national levels. This limits the usefulness of the data for GIS modeling, particularly for small-scale irrigation studies, for which parish-level data would provide essential information on the composition and intensity of irrigation. For 2001 and 2005, the data have been aggregated using a GIS at catchment level, which is more useful for water resources planning. The accuracy was still variable, depending on the number of survey questionnaires returned within a selected catchment; not all farms are surveyed and returning the survey questionnaires is voluntary. In 2005, a 43% response rate was achieved, representing 27% of the total irrigated area [14]. The data received by the government were statistically raised to account for missing farms and nonreturns, but this introduced a further degree of error. Despite these reservations, it is believed that the data were broadly correct, particularly with regard to trends, provided they were interpreted in relation to the weather in each year. However, some catchment data cannot be released as yet for the reason of confidentiality, particularly for small catchments and areas outside the main irrigated regions.

Using government survey data for irrigated area mapping has other problems. There was a major issue relating to the integrity of the data supplied, particularly in relation to reported areas and volumes. For example, many UK farmers still interchangeably use metric and imperial units for area (acres and ha) and there was widespread confusion between depth of water applied and volumes (millimeters, inch, acre-inch, meters cubed, and gallons). Incorrect reporting of units in irrigated areas and volumes was a major source of potential error. There was also the issue of farmer sentiment and concerns regarding how the irrigation survey data might be used by the regulatory authorities. For this reason, many farmers choose not to make a return, or may deliberately adjust the reported figures. Finally, an increasing number of farms and agribusinesses now operate at regional or even national level. Farm businesses are no longer confined to their own landholding, but are much more mobile in terms of crop production, often renting distant land to allow longer crop rotations for disease control. This has implications for allocating irrigation survey data to catchments and when farmers fail to report that their land is being irrigated by temporary tenants.

An alternative approach is to base the GIS mapping on the much larger database obtained annually through the government's June Agricultural Survey, which records the cropping pattern of the majority of farms in England and Wales. This is available as a GIS map through the Edinburgh Data Library. Some interpolation and masking have been used to cover missing data and maintain confidentiality, but at catchment scale, the data on the areas of each crop are acceptable. This data set, however, does not identify which fields were irrigated. This was a particular problem in supplemental irrigation, where the same crop can be irrigated, nonirrigated, or partially irrigated. Knox et al. [15] partially resolved this by comparing the June survey and irrigation survey data sets and calculating the proportion of each crop irrigated in each county, and then applying this factor uniformly across the whole county, but this introduced an averaging error to the data. A similar approach could be adopted using maps of crop types based on remote sensing.

In addition to government survey data, some commodity-based industry levy organizations (e.g., potato, horticulture) also collected detailed statistics on crop production, including irrigated areas. For example, growers involved in potato production provided information to the British Potato Council (BPC), a levy-funded organization that represents the interests of the potato industry. Unfortunately, the data held by many of these industrial organizations were usually confidential, although some aggregated statistics could be obtained for research purposes. For example, the BPC collected detailed records of cropped and irrigated potato areas for all their registered growers; these data could be used to estimate total irrigated areas for potato at regional and catchment levels (Figure 9.2).

These maps confirm the high spatial concentration of potato in terms of both irrigated area and water demand (consumption) in eastern England.

It is within this part of the country that the role of remote sensing for accurately estimating potato-cropped areas and differentiating between irrigated and nonirrigated areas is now being investigated.

9.2.2   USING REMOTE-SENSING TECHNIQUES WITH GIS

As part of a broader study investigating the impacts of climatic change on agricultural water demand, a program of research is currently investigating the use of remote-sensing imagery to map potato-irrigated areas in England. This is principally to inform strategic planning by the BPC to prepare the potato industry for changes in future water availability, whether driven by changes in water regulation and/or climatic change. A brief overview of the approaches being applied and the challenges being faced is provided.

Baseline information relating to the spatial distribution of potato production, including cultivar types, cropped areas, and husbandry (dates of planting and harvest) was routinely collected on an annual basis by the BPC from all growers involved in potato production. For selected farm sites across the country, much more comprehensive data were also collated, including information relating to crop development (emergence, full cover), agronomic management practices (irrigated, nonirrigated, fertilizer inputs, etc.), and final yields. These data sets will provide a highly valuable source of information for calibrating and validating estimates of potato-irrigated areas derived from remote sensing. For the purposes of this study, the EA Anglian region has been chosen to test the application of remote sensing, since it includes a significant proportion (33%) of the total potato-cropped area in England.

The first stage in the methodology was to examine the spectral differences observed between specific irrigated and rain-fed potato-cropped fields. The BPC database was used to identify the most important potato cultivars grown in the region, and ones that were known to be responsive to drought stress. The Maris Piper and King Edward cultivars were chosen as they represented 35% of the total potato-cropped area in the EA Anglian region. Both these cultivars are used for processing (e.g., for manufacturing crisps) and prepackaging (e.g., for sale in supermarkets) and

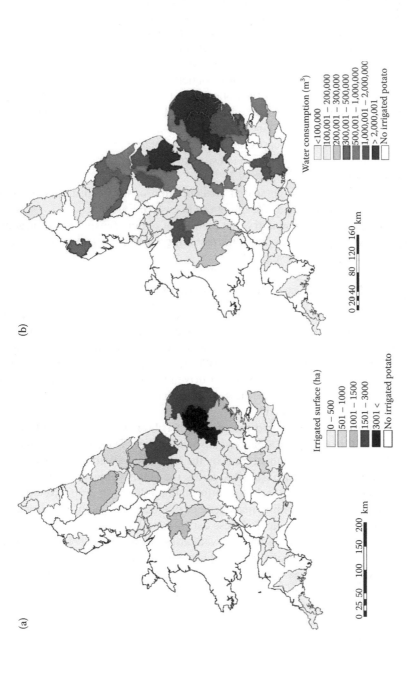

FIGURE 9.2 (a) Estimated areas of irrigated potato (ha) and (b) volumetric irrigation water demand for potato (m³), by catchment, in England and Wales in 2006.

have high agronomic demands for irrigation. Using the BPC dataset, four paired sites were identified where irrigated fields were close to nonirrigated fields of the same cultivar (Figure 9.3).

For this study, Moderate Resolution Imaging Spectroradiometer (MODIS) remote-sensed data were obtained. MODIS was available at a high temporal resolution (1–2 days) and varying spatial resolutions (250, 500, and 1,000 m), and is available free of charge for 36 spectral bands, ranging from 0.4 to 14.4 μm. The relevant MODIS 250-m data were downloaded from the Earth Observing System Data Gateway.

The fields had to be selected such that each field contained a whole (or nearly whole) MODIS 250-m pixel. This dimension was relatively large compared to typical plots of irrigated potato in the UK, which would become a major constraint to applying the methodology generally. Various approaches were evaluated; first, based on MODIS 16-day composite data, then using daily MODIS data, and last, based on other remote-sensed data.

9.2.2.1 Using MODIS 16-day NDVI Composite Data

The MODIS NDVI was then used for spectral comparison between the irrigated and rain-fed field sites. The use of NDVI has been tested in previous irrigation mapping studies, for example, by Thenkabail et al. [16], Ozdogan et al. [11], and Wardlow et al. [17]. The NDVI is the ratio between near-infrared (NIR) and visible (Red) bands, where NDVI=(NIR–Red)/(NIR+Red). For each field site, a series of 16-day NDVI composite images (MOD13Q1), spanning the entire potato-growing season of 2006, were obtained.

A multitemporal analysis was then conducted using 10 images covering the period from preplanting (18 February 2006) until harvest (14 September 2006) (Figure 9.4). This analysis showed that there was no distinct trend separation between the irrigated and rain-fed fields across the four paired sites during the growing season.

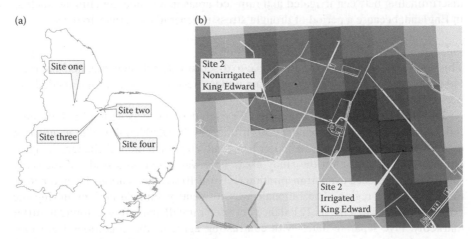

FIGURE 9.3   (a) Geographic location of the four reference farm sites in the EA Anglian region, (b) together with a MODIS 250 pixel overlay for Site 2, showing the location of the irrigated and nonirrigated (rain-fed) potato fields.

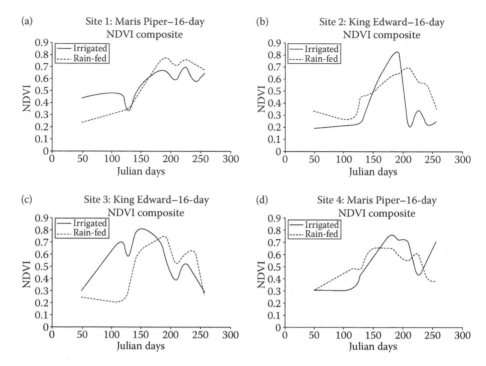

FIGURE 9.4 (a)–(d) Temporal analysis of the 16-day MODIS 250-m NDVI composite for each paired site, showing the trends between irrigated and rain-fed fields during 2006.

This was mainly due to the effects of summer rainfall, and the high variation in the observed NDVI values during the 16-day periods. This suggested that a multitemporal analysis of NDVI composite imagery using 16-day imagery was unsuitable for discriminating between irrigated and rain-fed areas in a temperate climate, such as in England, because a period of drought stress is generally less than 16 days.

9.2.2.2 Using MODIS Daily NDVI Data

This approach has been shown to be able to discriminate between irrigated and rain-fed areas when nonirrigated crops come under stress at the end of an extended dry period without rainfall [11,12,18].

The NDVI was, therefore, compared for each paired site using the MODIS daily reflectance (Red and NIR bands) 250-m imagery (MOD09GQK). The acquisition date for each image was selected after first analyzing the temporal changes in agroclimatic conditions within the study area. This was based on a study of the daily rainfall and reference evapotranspiration (ETo) at 10 weather stations located across the EA Anglian region. The rationale was to identify suitable periods during the growing season when the daily balance between rainfall and ETo was found to differ markedly. Five separate dates with contrasting agroclimatic conditions were ultimately selected and the NDVI was then calculated, working from the Red and NIR bands for each field.

The results based on the daily NDVI (Figure 9.5) showed an improved separation and suggested that it was possible to distinguish between irrigated and rain-fed fields, particularly after a short period where no rainfall was recorded. The NDVI differences on 17 July 2006 showed a higher NDVI for the irrigated fields than for the rain-fed fields, except at one site where the reverse situation was observed. Further investigation is therefore required to assess these NDVI differences, and

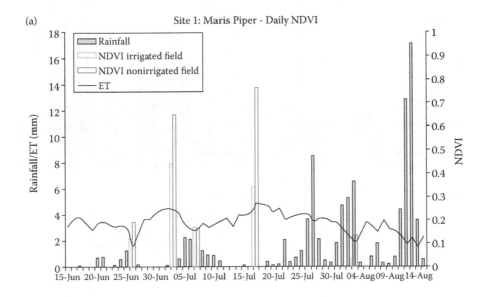

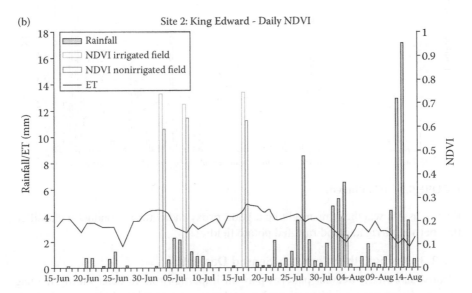

FIGURE 9.5 (a)–(d) Temporal analysis of daily NDVI on five selected dates for the four paired field sites in 2006; daily rainfall and evapotranspiration shown in the background.

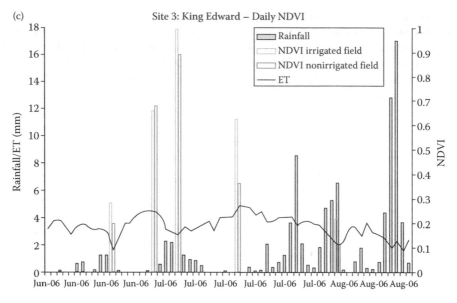

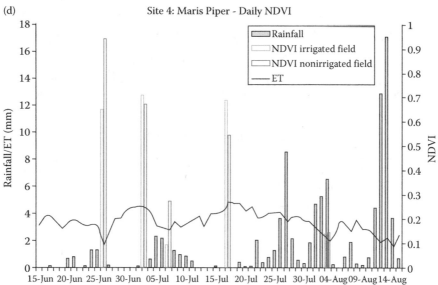

FIGURE 9.5 (Continued)

then to assess whether this might be a good proxy indicator to conduct a classification between irrigated and rain-fed potato fields.

9.2.2.3 Using Other Remotely Sensed Datasets

Clearly, short-term weather patterns play an important role in determining if imagery can be separated with statistical confidence to determine differences between irrigated and rain-fed areas; a data set with frequent acquisition dates is therefore essential. Using the MODIS data also clearly revealed the problems of the large pixel size compared to the typical field size. Future research will, therefore, be focused on signature separation analyses using Landsat TM imagery at 30-m resolution (available every 16 days). Statistical analyses will be carried out for all visible bands, using separability methods such as the Jeffries–Matusita Distance (JM) and Transformed Divergence (TD) techniques. It may then be possible to apply an image classification to map irrigated potato across the EA Anglian region, using data from the BPC data set for validation, and a land mask based on the CORINE land cover data set (100 m).

An alternative approach would be to compare ET values between irrigated and nonirrigated areas. Spatial estimates of actual ET (ETa) can be derived from commercial data sets such as those obtained through the Surface Energy Balance Algorithm for Land (SEBAL) developed by Bastiaanssen et al. [19]. SEBAL is an algorithm model that can be used to estimate ETa empirically, but its application raises significant cost issues and constraints for research purposes. Another alternative would be to use the Simplified Surface Energy Balance (SSEB) approach developed by Senay et al. [20] to calculate ET. It is understood that neither the SEBAL nor the SSEB approach has yet been tested for application in the UK.

9.3 SUMMARY AND CONCLUSIONS

Production of maps showing the areas of land irrigated for different crops would be extremely valuable for irrigation water resources planning and industrial uses. By combining the maps with GIS data on soil types and climate, maps of irrigation demand and climatic change impacts can be produced, and the spatial benefits of suggested adaptations studied, for example. The use of remote-sensing data, if accurate, would be a clear improvement and/or addition to existing methods based on postal surveys.

Recent studies have confirmed the limitations of using MODIS data for discriminating between irrigated and nonirrigated cultivation of the same crop in areas of supplemental irrigation such as England. The largely unpredictable rainfall events mask the development of water stress, making it important to time the data acquisition to the end of a dry period. Unfortunately, delays between time of image acquisition and the frequency with which cloud or mist obscures the target area makes this difficult and unreliable.

Furthermore, the large pixel size of the data does not correspond well with the typical field size. With crop rotations typically limiting irrigation in given fields to one year in three or more, new fields would have to be identified each year. This becomes an increasing problem for studying crops grown under sequential planting, such as lettuces, where individual plots would be even smaller.

By using Landsat TM data, with smaller pixels and different wave bands, some of these problems may be overcome and this is being investigated. The authors acknowledge the British Potato Council (BPC) for provision fo field survey data.

REFERENCES

1. Knox, J.W. et al., Mapping the financial benefits of spray irrigation and potential financial impact of restrictions on abstraction: A case study in Anglian region, *Journal of Environmental Management*, 58, 45–59, 2000.

2. Knox, J.W. and Weatherhead, E.K., The application of GIS to irrigation water resource management in England and Wales, *Geographical Journal*, 165, 1, 90–8, 1999.
3. Comprehensive Assessment of Water Management in Agriculture, *Water for Food, Water for Life: A Comprehensive Assessment of Water Management in Agriculture*, Earthscan, London and International Water Management Institute, Colombo, 2007.
4. Döll, P., Impact of climate change and variability on irrigation requirements: A global perspective, *Climatic Change*, 54, 269–93, 2002.
5. Frederick, K.D. and Major, D.C., Climate change and water resources, *Climatic Change*, 37, 7–23, 1997.
6. Hulme, M. et al., *Climate change scenarios for the United Kingdom: The UKCIP02 scientific report*, Tyndall Centre for Climate Change Research, School of Environmental Sciences, University of East Anglia, Norwich, UK, 2002.
7. Arnell, N.W., Relative effects of multi-decadal climatic variability and changes in the mean and variability of climate due to global warming: Future stream flows in Britain, *Journal of Hydrology*, 270, 195–213, 2003.
8. Weatherhead, E.K. et al., *Sustainable water resources: A framework for assessing adaptation options in the rural sector*, Technical Report 44, Tyndall Centre for Climate Change Research, Norwich, UK, 2005.
9. Weatherhead, E.K. and Knox, J.W., Predicting and mapping the future demand for irrigation water in England and Wales, *Agricultural Water Management,* 43, 203–18, 1999.
10. Rodriguez-Diaz, J.A. et al., Climate change impacts on irrigation water requirements in the Guidalquivir river basin in Spain, *Regional Environmental Change*, 7, 3, 149–59, 2007.
11. Baruffi, F. et al., Crop classification and crop water need estimation of Piave river basin by using MIVIS, Landsat-TM/ETM+ and ground-climatological data, *Proceedings of SPIE – The International Society for Optical Engineering*, Vol. 5976, 2005.
12. Ozdogan, M., Woodcock, C.E., and Salvucci, G.D., Monitoring changes in irrigated lands in southeastern Turkey with remote sensing, Geoscience and Remote Sensing Symposium (IGARSS) Proceedings, *IEEE International Volume 3*, 1570–72, 2003.
13. Kamthonkiat, D. et al., Discrimination of irrigated and rainfed rice in a tropical agricultural system using SPOT VEGETATION NDVI and rainfall data, *International Journal of Remote Sensing*, 26, 12, 2527–47, 2005.
14. Thenkabail, P.S., Schull, M., and Turral, H., Ganges and Indus river basin land use/land cover (LULC) and irrigated area mapping using continuous streams of MODIS data, *Remote Sensing of Environment*, 95, 3, 317–41, 2005.
15. Weatherhead, E.K., *Survey of irrigation of outdoor crops in 2005 – England and Wales*, Cranfield University, UK, 2006.
16. Knox, J.W., Weatherhead, E.K., and Bradley, R.I., Mapping the total volumetric irrigation water requirements in England and Wales, *Agricultural Water Management*, 33, 1–19, 1997.
17. Thenkabail, P.S. et al., *An irrigated area map of the world (1999) derived from remote sensing*, Research Report 105, International Water Management Institute, Colombo, Sri Lanka, 2006.
18. Wardlow, B.D., Egbert, S.L., and Kastens, J.H., Analysis of time-series MODIS 250 m vegetation index data for crop classification in the US Central Great Plains, *Remote Sensing of Environment*, 108, 3, 290–310, 2007.
19. Biggs, T.W. et al., Irrigated area mapping in heterogeneous landscapes with MODIS time series, ground truth and census data, Krishna Basin, India, *International Journal of Remote Sensing*, 27, 19, 4245–66, 2006.
20. Bastiaanssen, W.G.M. et al., A remote sensing surface energy balance algorithm for land (SEBAL) 1 formulation, *Journal of Hydrology*, 212–13, 198–212, 1998.
21. Senay, G.B. et al., A coupled remote sensing and simplified surface energy balance approach to estimate actual evapotranspiration from irrigated fields, *MDPI/Sensors*, 7, 979–1000, 2007.

10 Irrigated Area Mapping in the CWANA Region and Its Use in Spatial Applications for Land Use Planning, Poverty Mapping, and Water Resources Management

Eddy De Pauw
International Center for Agricultural
Research in the Dry Areas

CONTENTS

10.1 INTRODUCTION

The International Center for Agricultural Research in the Dry Areas (ICARDA) deals with germ plasm enhancement and natural resource management issues, including land degradation in drylands. Its current mandate region includes North

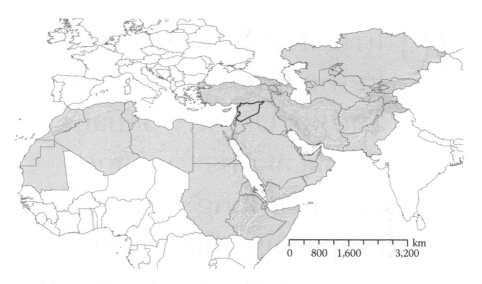

FIGURE 10.1   CWANA region (gray with hill shading); Syria in thick black line.

Africa, West Asia, Central and South Asia, and the Horn of Africa (Figure 10.1, area in gray). For convenience, this area is abbreviated as Central Asia+West Asia+North Africa (CWANA).

In a dryland region, as huge and diverse as CWANA, with limited reliable ground-based resource inventories and monitoring systems, remote sensing presents a highly valued tool for providing a spatially explicit dimension to the highly complex issues of land use/land cover (LULC) identification, change detection, land degradation, and their relationships to socioeconomic systems. Moreover, in this region, the growth, decline, or degradation of the land and water resources in the irrigated areas has a particular bearing on the livelihoods of rural populations, since in most cases, water is the key limiting resource for intensification and diversification of crop- and livestock-based agricultural systems. For this reason, the mapping and monitoring of changes in the irrigated areas are of particular relevance.

In this chapter, case studies are presented at the regional (CWANA) and national (Syria) scale that look at irrigated area mapping from both a producer and a user perspective. They describe first how such maps were derived from either low-resolution or medium-resolution satellite platforms and how, on the other hand, the products were used, in conjunction with other thematic data sets, in regional and national studies related to biodiversity assessments, crop diversification and targeting, water management, and poverty research.

10.2   IRRIGATED AREA MAPPING OF THE CWANA REGION

Since 1989, ICARDA has made a considerable investment in an institutional project dedicated to the agroecological characterization of the CWANA region, with the objective of developing spatial frameworks for the extrapolation of site-specific research. In 2000, it was realized that LULC data for the entire region and at a relatively fine resolution were indispensable for a wide range of applications and that the maps available at the time had serious limitations in the CWANA region [1]. For this reason, the same year it was decided to embark on an LULC mapping project, which resulted, one year later, in an LULC map for the CWANA region at 1-km resolution, based on the classification of Normalized Difference Vegetation Index (NDVI) patterns obtained from the National Oceanic and Atmospheric Administration (NOAA) Advanced Very High Resolution Radiometer (AVHRR) satellite platform. The NDVI is a contrast-stretch ratio calculated from the red band (R) and the near-infrared band (NIR) that are covered by AVHRR bands 1 and 2. NDVI can be considered a proxy but an integrated indicator for major phenological vegetation characteristics (green biomass, ground cover, leaf area) and photosynthetic activity, and its correlation with vegetation productivity has been well established [2].

Thirty-six 10-day composites of AVHRR-NDVI data with 1-km resolution were downloaded from the USGS Eros Data Center's website (http://edcdaac.usgs.gov/1km/comp10d.html). They covered a 1-year period from 1 April 1992 until 31 March 1993. The selected area spans latitudes –2°–59° and longitudes –23°–91°, covering the entire CWANA region. Radiometric calibrations and atmospheric corrections were already done by the EROS Data Center. Data-inherent noise was reduced by creating 12 monthly composites, after noisy pixels had been flagged. These images were reprojected from the Goode's Homolosine projection into the Plate-Carree latitude/longitude projection with 30" resolution, using the software ERMapper. Using a 30"-grid slope map, derived from the DEM GTOPO30 [3], a slope correction was made by multiplying each pixel's NDVI value with the cosine of its slope angle.

The LULC patterns are usually derived from vegetation indices through statistical techniques such as *unsupervised* and *supervised classifications*. An unsupervised classification automatically groups pixels together in a predefined number of classes based on their spectral similarity. Supervised classification starts with assigning well-known training regions, whose spectral information is used to calculate class means. Class membership is then attributed based on the spectral similarity with these class means.

The LULC mapping approach uses key NDVI parameters such as onset, peak, and end-of-greenness period, which are grouped into a limited number of patterns with similar NDVI phenology. The classes established statistically are then interpreted and translated into LULC types using a limited number of groundtruthing points. In principle, in addition to NDVI, other parameters can be used in these automated classifications, such as individual band reflectance values.

In order to develop a new LULC map for CWANA, it was decided not to use statistical classification methods, as these methods assume that only the LULC type (LCLUT) determines the NDVI pattern. In reality, LULC is itself adapted to different agroclimatic conditions. In a very large area, such as CWANA, it is therefore quite possible that in different agroclimatic zones (ACZs) *similar NDVI patterns could correspond to different LCLUTs, and also that different NDVI patterns could represent the same LCLUT*. For this reason, the classification of NDVI patterns into LCLUTs was based on the use of an *empirical hierarchical decision tree* that takes due consideration of differences in agroclimatic conditions.

A decision-tree algorithm does not assign all classes in one step, but groups pixels together based on stepwise conditions. This way, the data are split up into smaller subgroups, which allow better control of the classification process. The classes that are easier to identify are selected first. Major LULC categories were established on the basis of expert opinion, field surveys, together with thematic maps, climatic data, and Landsat satellite imagery (Landsat) for Syria as a particular form of groundtruth. For these categories, the NDVI patterns were analyzed and empirical relationships established with agroclimatic conditions. On the basis of these tests, a final classification scheme was established that contained fewer classes than other land-cover classifications, but could be identified with a higher degree of accuracy.

The key characteristic of this decision tree is that the thresholds of NDVI used for allocating a pixel into a particular LULC class are variable or "sliding," depending on the particular aridity regime at the location. In order to set these thresholds, use was made of the aridity index (AI), a concept used by UNESCO [4] in its mapping of the arid zones of the world. The AI is defined by UNESCO as the ratio of the annual precipitation to the annual potential evapotranspiration (PET) according to the Penman method and the aridity zones distinguished are the following:

- Hyperarid: AI < 0.03
- Arid: $0.03 \leq AI < 0.2$
- Semiarid: $0.2 \leq AI < 0.5$
- Subhumid: $0.5 \leq AI < 0.75$
- Humid: AI > 0.75

Figure 10.2 shows the decision tree used and it contained the following classes:

- Forests/tree crops/closed shrublands
- Dry-season irrigated field crops
- Rain-fed/supplementary irrigated field crops
- Woodland savannahs
- Open shrublands/grasslands
- Barren/sparsely vegetated

The classes that contain fully or partially irrigated land use are shown in gray boxes.

To separate croplands, the NDVI pattern during the dry season was an important indicator. The addition of latitude as a criterion in the decision tree enabled the inclusion of the typical vegetation patterns of each latitudinal zone and the setting of different values on the timing of the dry season.

Some of the classes in the decision tree are not homogeneous. Due to the large pixel size, greenness and density characteristics could not always be separated consistently, which is why the categories "forests," "tree crops," and "closed shrublands (maquis)" were grouped into a single class. "Open shrublands" and "grasslands" were grouped together because they form natural intergrades with different proportions of the pure categories. "Rain-fed agriculture" in many regions is a pure class, but in other regions is supplemented with irrigation water from wells. If that

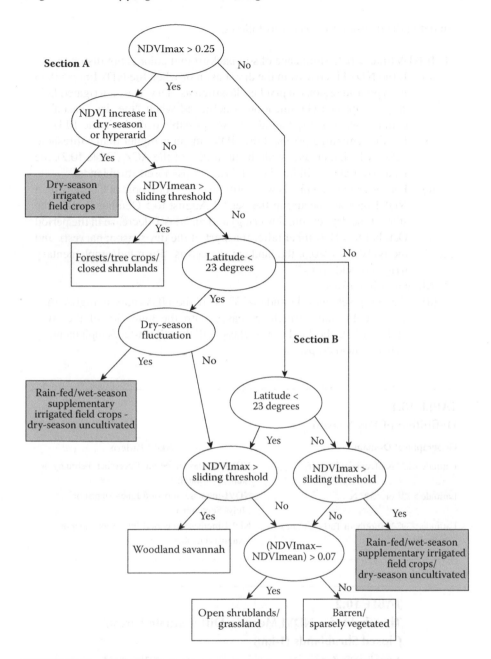

FIGURE 10.2 Hierarchical decision tree for land use/land cover assessment.

irrigation simply serves to stabilize yields of rain-fed crops and does not extend into the dry season, it is very difficult to separate the phenology of purely rain-fed crops from that of rain-fed crops with some addition of irrigation water; hence, they were also grouped into a single class.

In detail the decision tree works as follows:

1. If NDVImax >0.25 (influence of soil background color is not distinctive):
   (a)   If the NDVI increases in the dry season or when the NDVImax >0.25 in hyperarid regions, a pixel is classified as "Dry-season irrigated field crops." Optimal classification is achieved when "the dry season" is defined, according to latitude, as two possible periods (Table 10.1).
   (b)   For the remaining pixels, if the NDVI mean is higher than a threshold value, which increases with the humidity of the ACZ (Table 10.2), the land cover class will be "Forests/tree crops/closed shrublands."
   (c)   For the remaining pixels with latitude <23° N (Tropic of Cancer), if an NDVI-increase occurs in the period August to >October (before the start of the dry season: flowering) and an NDVI-decrease in the period October to >December (after the start of the dry season: harvest), and the NDVImax >0.5, the land cover class is "Rain-fed/supplementary irrigated field crops."
2. All remaining pixels:
   (d)   For the pixels with latitude >23° N, if the NDVImax is higher than a threshold value, which increases with the humidity of the ACZ (Table 10.3), the land cover class will be "Rain-fed/supplementary irrigated field crops."

TABLE 10.1
Definition of Dry Season

Geographical Domain	NDVI Pattern
Latitude <23° N (Tropic of Cancer)	NDVI-increase in period December–February or January–March
Latitude >23° or <39° N	NDVI-increase in period June–August or July–September
Latitude >39° N (northern Turkey, Caucasus, Central Asia)	NDVI-increase in period July–September or August–October

TABLE 10.2
Thresholds for NDVI Mean to Differentiate Forest/ Closed Shrublands (1-km)

Agroclimatic Zone	Threshold
All arid zones	0.20
All semiarid zones	0.25
Subhumid zones with mild summer and mild winter	0.35
Other subhumid zones	0.30
All humid zones	0.40

TABLE 10.3
Thresholds for NDVImax to Differentiate Rain-Fed Field Crops (1-km)

Agroclimatic Zone	Threshold
All arid zones	0.22
All semiarid zones except with cold winter	0.35
All subhumid zones except with cold winter, semiarid zones with cold winter	0.40
All humid zones, subhumid zones with cold winter	0.50

TABLE 10.4
Thresholds for NDVImax to Differentiate Woodland Savannah (1-km)

Agroclimatic Zone	Threshold
All arid zones	0.25
All semiarid zones	0.35
All subhumid zones	0.40
All humid zones	0.50

(e) For the pixels with latitude < 23° N (Tropic of Cancer), if the NDVImax is higher than a threshold value, which increases with the humidity of the agroclimatic zone (ACZ) (Table 10.4), the land cover class will be "Woodland savannahs."

(f) For the remaining pixels, if the difference (NDVImax–NDVImean) > 0.07 (reflecting a distinctive vegetation change throughout the year), the land cover class will be "Open shrublands/grasslands." All the remaining pixels are classified as "Barren, sparsely vegetated."

(g) The classes Forests/tree crops/closed shrublands and Woodland savannahs could be further differentiated on the basis of greenness or density characteristics. If the NDVImin > 0.2, these classes got the specification of "High density, evergreen," which resulted in the following LCLUTs:

High density and evergreen forests/tree crops/closed shrublands or
High density and evergreen woodland savannahs.

Similarly, field crops could be differentiated on the basis of the same greenness/density characteristics. The criterion for subdivision is the NDVImax.

If NDVImax > 0.6, the following LCLUTs are differentiated:

High density and high-yield dry-season irrigated field crops or
High density and high-yield rain-fed/supplementary irrigated field crops.

TABLE 10.5
Existing LULC Maps Based on AVHRR Information

LULC Map	Geographical Scope	Derived From	Method for Classification	KHAT
Asian Association for Remote Sensing [7]	Asia	NDVI 4, 5	Unsupervised statistical+decision tree	0.2667
University of Maryland [8]	Global	All 5 AVHRR	Supervised statistical+decision tree	0.2373
International Global Biosphere Program [9]	Global	NDVI	Unsupervised statistical+postclassification refinement	0.1976
ICARDA	CWANA		Decision tree	0.5533

The performance of this classification based on the decision tree was compared to three regional or global land classifications based on statistical procedures (Table 10.5). The accuracy of the four classifications was assessed by comparison with a higher resolution LULC map for Syria [5] using the KHAT Accuracy Index [6]. The LULC map of Syria was obtained from a visual interpretation of 1:200,000 scale false-color prints of Landsat 5 scenes from the period 1989/1990, followed by digitizing and integration in a Geographic Information System (GIS), and verified through fieldwork. Additional accuracy tests were undertaken using groundtruth samples collected by Tateishi [7] in Central Asia during the summers of 1996 and 1997 and the spring of 1998.

The full map of LULC for the CWANA region is described by Celis et al. [1]. Figure 10.3 shows the fully irrigated areas (in black) and the rain-fed/supplemental irrigated areas (in gray), derived from the map. Given the small size of many irrigated or rain-fed areas, this scale tends to underestimate the extent of irrigated or rain-fed land in this region. Figure 10.4 better represents the true extent in the inset map, covering the western part of Turkey.

10.3   IRRIGATED AREAS AND AEZs OF CWANA

The CWANA region constitutes the largest contiguous area of drylands in the world. Often overlooked is the fact that, within its overall dryland context, this region, the current focal area of ICARDA's research, also has a very high diversity in agro-ecologies and agricultural systems. Climate is the first determinant of agroecological diversity. Using the 1979 UNESCO classification system for arid zones [4], by the simple combination of major aridity and temperature regimes, 84 different ACZ can be differentiated in CWANA. Further subdivision is possible, based on the distribution of precipitation in relation to the interannual temperature pattern. Within this highly heterogeneous climatic setting, more diversity in biophysical environments is created by differences in landscapes, soils, geological substrata, and surface water and groundwater resources. Irrigation development is the single most important

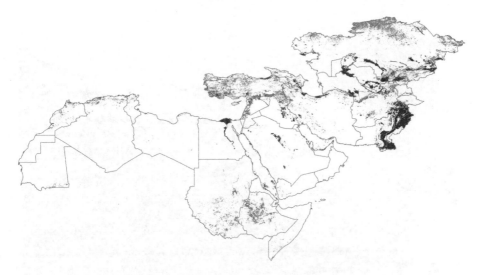

FIGURE 10.3 Irrigated (black) and rainfed/supplemental irrigated areas (gray) in CWANA. The inset map covering western Turkey is expanded in Figure 10.4.

factor in creating artificial agroecological niches, areas where natural conditions and production systems show abrupt differences with their surroundings [10].

Ignoring the complex nature of such diverse agricultural environments has been one of the main reasons why it has been remarkably difficult to transfer research results or "lessons-learnt," whether they concern biodiversity management, intensification or diversification of crop production, land use optimization, or combating land degradation, from one dryland area to another.

In the early days of ICARDA, it was already realized that in order to make headway with the "outscaling" challenge at the level of the CWANA region, there was a need for an integrated approach using an *AEZ framework*. There is nothing new about the AEZ concept. AEZs have been used in different regions or countries, and at different scales, for a variety of agricultural purposes, including the identification of agricultural production zones, fertilizer use recommendations, rationalizing the location of research stations, targeting of new technologies and crop cultivars, and regional comparative advantage and crop-subsidy planning.

At the level of the CWANA region, the main use of an AEZ framework is as a tool for rapid identification of different biophysical environments and characterization of resources and constraints to assist agricultural research planning and policy development. For this purpose, a map of the AEZs of the CWANA region was created at 1-km resolution using the following six-step procedure:

- Converting point climatic data into basic climatic "surfaces" through spatial interpolation.
- Generating a spatial framework of ACZ by combining the basic climatic surfaces into more integrated variables that provide a synthesis of climatic conditions.

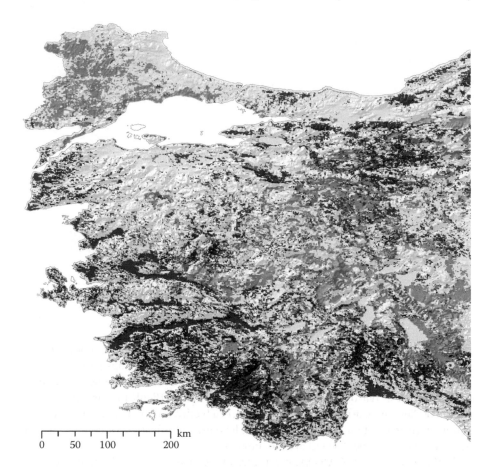

FIGURE 10.4   Expanded inset map of Figure 10.3 showing irrigated (black) and rainfed/supplemental irrigated areas (gray) in western Turkey on a hill-shaded light gray background.

- Generating a spatial framework of land systems, which are land-based mapping units, created by the combinations of major LULC, landscape, and soil categories.
- Integrating the frameworks for ACZ and land systems by overlaying in GIS.
- Removing redundancies, inconsistencies, and spurious mapping units generated by the overlaying process.
- Characterizing the AEZ in terms of other relevant themes.

A climatic database containing monthly precipitation and temperature averages for more than 1900 stations in the CWANA region was compiled. The main sources of climatic data were mainly the FAOCLIM database [11], and national meteorological data sets from the Central Asian and Caucasus countries and Iran. PET estimates, according to the Penman–Monteith method [12], were obtained through a two-step disaggregation regression with temperature. This point database was then converted into raster layers with 30 arc-second spatial resolution (corresponding with roughly 1-km pixel size), by spatial interpolation using the "thin-plate smoothing spline" method of Hutchinson [13], as implemented in the ANUSPLIN software [14], and elevation data obtained from the GTOPO30 global digital elevation model (DEM) as a covariable. The spatial framework of ACZ was then created by transforming these monthly "climate surfaces" of precipitation, temperature, and PET into classes of the UNESCO classification for arid zones, using the criteria *AI, summer temperature,* and *winter temperature* and the class limits of this climatic classification.

To generate a framework of land systems required a more complex approach. First of all, there was a need for a regional or global LULC classification at the same spatial resolution as the ACZ framework. In view of its good accuracy, the CWANA LULC map for the year 1993 [1] was selected for this purpose and reclassified into three categories: (i) irrigated crops, (ii) rain-fed crops, and (iii) nonagricultural land use. To establish broad landform categories, the GTOPO30 DEM was reclassified into three terrain classes based on the difference in elevation between neighboring pixels: (i) plains and plateaus, (ii) hills, and (iii) mountains.

Soil patterns can vary significantly over small distances and were, unsurprisingly, the most difficult to integrate into a land systems framework at the regional scale. Using the FAO Soil Map of the World [15] as the data source, it was found that within CWANA, 1047 soil associations occur, as determined by varying combinations of 112 FAO soil types. Reducing this vast variability by regrouping was necessary in order to establish ecosystems that were not overfragmented. This simplification was done in two steps:

(1) Regrouping the FAO soil types into broader classes that are relevant to their general management properties ("soil management groups").
(2) Mapping the major combinations of these soil management groups ("soil management domains").

The 26 FAO soil units were reduced to the following nine soil management groups: (i) agricultural soils; (ii) soils of wetlands, poorly drained areas and floodplains; (iii) sandy soils; (iv) sodic and saline soils; (v) rock outcrops and shallow soils; (vi) semidesert soils; (vii) desert soils; (viii) nonagricultural soils; and (ix) soils with high-acidity and/or low-nutrient status. Using these new soil groupings, the units of the Soil Map of the World were converted by reclassifying the 1047 FAO soil associations into 60 soil management domains, combinations of the main soil management groups inside an FAO soil association. Figure 10.5 outlines the entire process. Figure 10.6 shows Iran and parts of Central Asia.

The AEZ methodology, as implemented in this regional approach, at the national level, has been taken up in Turkey, Syria and at the subnational level in Iran in the Karkhe, Aras, and Daryacheh–Uromieh River basins through collaborative projects between national agricultural research systems and ICARDA. Irrespective of the scale of the assessment, from regional to subnational, irrigated land use is an indispensable factor for differentiating AEZs.

10.4 IRRIGATED AREAS AND RESOURCE POVERTY IN CWANA

The CWANA may well be the region with the highest proportion of marginal land for agriculture, and hence of "agricultural resource poverty." This concept can be used

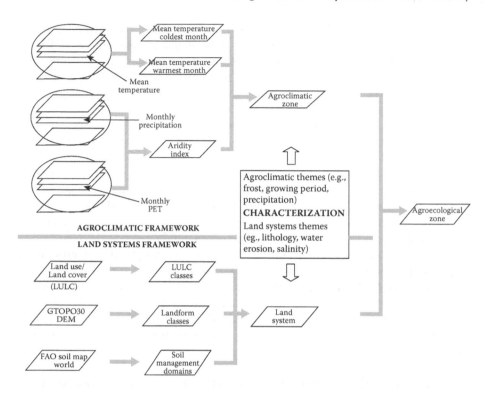

FIGURE 10.5 Flowchart for compilation of the CWANA agroecological zones map.

to quantify the presence of an unfavorable agricultural climate, or inadequate water and soil resources and topography. Of these, moisture availability in the CWANA region is the most limiting factor and is alleviated, where possible, by irrigation. Also, low temperatures can be a major limitation, particularly in the many mountainous areas of CWANA, where topography presents an additional constraint due to steep slopes and dissected topography, coupled with problems of accessibility, which strongly reduce the area of good-quality farmland. The main soil limitations, which are physically not possible to correct, or only at excessive cost, are salinity, sodicity, shallow depth, high stoniness, and very coarse texture.

A number of recent studies arising from poverty mapping research indicate the strong linkages that exist between rural poverty, access to resources (e.g., owned land, water, animals, machinery), and agroecological variables (e.g., climate, soil, availability of water for irrigation) and suggest a potential for integrating current and emerging GIS-based data on environmental characteristics with socioeconomic data, to analyze the interaction between poverty and natural resources. Yet, perhaps because the CWANA region is in aggregate and not the poorest region in the world, the constraints of the agricultural resource base, in this region, have not received sufficient attention as a key factor for explaining human poverty. To address this shortcoming, first, agricultural resource poverty needed to be quantified. To overcome the challenge of the complexity and interaction of the biophysical factors, an index-based method was developed for quantifying resource poverty,

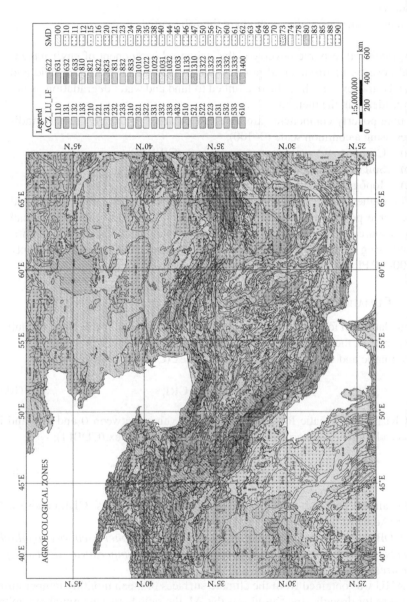

FIGURE 10.6 (See color insert following page 224.) Agroecological zones of Iran and parts of West and Central Asia, Caucasus, and Egypt.

which is simple, yet considers all relevant biophysical factors, and allows consistent comparison between different locations. The method incorporates an audit trail to assess the contribution of individual environmental factors toward agricultural resource poverty; it is scale-independent and can be applied using currently available data sets.

The main principles of the method are the following:

1. Agricultural resource poverty is the structural component of environmental poverty. Environmental poverty due to managerial factors at field and landscape level, such as the one linked to land and water degradation, is not considered in the method.
2. Three poverty components are assessed separately through thematic indices using a common scale (0–100):
   (a) Climatic and water resources poverty (CWRPI)
   (b) Soil resource poverty (SRPI)
   (c) Topographic resource poverty (TRPI)
3. The highest of these thematic indices sets the value for the agricultural resource poverty index (ARPI), also on the same scale of 0–100. Instead of resource poverty, agricultural resource endowment can also be quantified using the agricultural resource poverty index (AREI), which is simply 100-ARPI.

10.4.1 Climatic and Water Resource Poverty

The input data used for the mapping of climatic resource poverty are 1-km resolution climate surfaces, the same as discussed in Section 10.3.

The climatic and water resources poverty index (CWRPI) was defined as:

$$CWRPI = 100 - CRI, \tag{10.1}$$

where CRI is the Climatic Resource Index with values between 0 and 100 and is based on a climatically determined biomass productivity index (CDBPI):

$$CDBPI = AHU \times AI_{adj}, \tag{10.2}$$

where the annual heat unit (AHU)=annual growing-degree-days (GDD) above 0°C and AI_{adj}=Adjusted AI.

The CDBPI can be considered a proxy for the *atmospheric energy available for biomass production, as expressed by accumulated temperature, adjusted for drought stress.*

The AHU was obtained from the climate surfaces of mean monthly temperature. The measure for drought stress used was the AI, the ratio between annual precipitation and annual PET, calculated by the Penman–Monteith method. If the ratio was higher than 1, it was adjusted to 1, to ensure that very cold areas (with low AHU) would not get an excessive CDBPI through compensation from a high AI (as is often the case in cold areas because of very low evapotranspiration rates).

The CDBPI can also be used in irrigated areas by overlaying a mask of the irrigated areas on top of the rain-fed BPI map. For these areas, AI = 1 (no moisture limitation) and

$$CDBPI = AHU. \qquad (10.3)$$

In this case, the irrigated areas were also derived from the LULC map of the CWANA region [1].

The process of setting an adjusted value for AI is summarized in Figure 10.7.

This range was determined empirically from CDBPI values in parts of Central and West Asia and North Africa with well-known very high agricultural and biomass productivity [16]. Examples are given in Table 10.6.

On the basis of Table 10.6, the optimum range for CDBPI was set at 7000 (opt1) to 9000 (opt2). GDD and CRI were calculated as:

- If CDBPI < opt1, then CRI=CDBPI/opt1*100
- If CDBPI is between opt1 and opt2, then CRI=100
- If CDBPI > opt2, then CRI=100×[1−(CDBPI−opt2)/opt2] (Figure 10.8)

The combined map of CDBPI for both rain-fed and irrigated areas made it possible to calculate a CWRPI that considers both climatic and water resources.

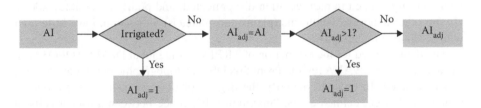

FIGURE 10.7   Adjustment of the Aridity Index.

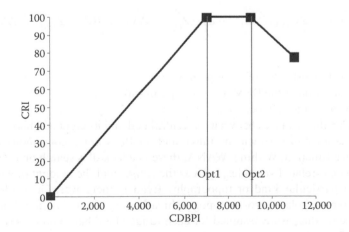

FIGURE 10.8   Empirical relationship between CRI and CDBPI.

TABLE 10.6

Examples of CDBPI for High-Potential Agricultural Areas

Area	CDBPI
Georgia, valley	5,340
Azerbaijan, irrigated areas	5,360
Iran, Caspian seacoast	5,790
Uzbekistan, Fergana valley	5,380
Iraq, Tigris-Euphrates, irrigated	8,350
Saudi Arabia, irrigated central part	8,760
Turkey, Ceyhan plain	6,900
Egypt, Nile delta	7,680
Pakistan, Indus basin, upper reaches	8,880

10.4.2 TOPOGRAPHIC RESOURCE POVERTY

The topographic resource poverty index (TRPI) is the *percentage of land that is problematic for agriculture due to topographic limitations*, related to slopes and dissected landscapes. Strongly dissected landscapes contain little land with agricultural value, mostly located in narrow, often disconnected, and poorly accessible valleys. On the other hand, flat landforms (plains and plateaus), in general, have little land with unsuitable topography.

The data set used for the estimation of TRPI was the global DEM GTOPO30 [3], extracted for the CWANA region. From this DEM, a topographic parameter "range" was calculated. In order to quantify the degree of dissection, the *relief intensity* concept was used, defined as the "maximum difference between adjacent cells in the DEM," expressed as meter per kilometer and was calculated according to the following formula:

$$S = \max[((Z_{ij} - Z_{i-1,j-1}) / \Delta X_1), \ ((Z_{ij} - Z_{i-1,j}) / \Delta X_2),\ldots,((Z_{ij} - Z_{i+1,j+1}) / \Delta X_8)],$$

(10.4)

where:

max=the highest value of the 8-point vector,

Z=the elevation of the DEM at a given point (m),

i and j=index the cell location,

$\Delta X_1 - \Delta X_8$=the distances between the central cell and its eight neighbor cells (km).

The relationship between the range and TRPI is *not* a continuous or even a quantitative function. Within CWANA, there is a lack of research on relationships between slope-related variables, such as the range, and the amount of agricultural land on a particular kind of topography. Even if there were, it would be very difficult to extrapolate across a topographically very diverse region. For this reason, blanket values were assigned for each range class, based on expert judgment (Table 10.7).

TABLE 10.7
Relationship Between Topographic Properties and TRPI

Degree of Dissection	TRPI
Flat (range 0–20 m)	2
Rolling plains (range 20–50 m)	10
Low hills (range 50–100 m)	80
Steep hills (range 100–300 m)	90
Medium-gradient mountains (range 300–600 m)	95
High-gradient mountains (range >600 m)	99

10.4.3 SOIL RESOURCE POVERTY

The soil resource poverty index (SRPI) is the percentage of each grid cell occupied by *problem soils*. For the purpose of this study, problem soils were defined as soils having limitations that are too severe for sustained use.

The data set used for developing the SRPI was the Digital Soil Map of the World [15]. The following soils were considered problem soils and their classification in the FAO system is as follows:

- Soils with Al-toxicity: all Ferralsols (F) and Acrisols (A), except "humic" subgroups.
- Saline soils: all Solonchaks (Z)+salt flats (SALT).
- Soils with high sodium content: all Solonetz (S)+Solodic Planosols (Ws).
- Shallow soils: all Lithosols (I), Rendzinas (E), and Rankers (U)+Rock Outcrops (ROCK).
- Coarse-textured soils: dunes and shifting sands (D/SS) and the presence of textural class 1.

Under the decision rules for defining the SRPI, soils having specific management needs but are productive under appropriate management were not considered problem soils. Examples of soils with special management needs are Vertisols (requiring practices promoting good drainage), acid soils (requiring adapted fertilization), erosion-prone soils (requiring soil-conservation measures), or stony soils (requiring stone clearance, as is already standard practice in many parts of the region).

On the other hand, highly acid soils, with associated Al-toxicity, were considered problem soils, since the required use of lime may not be economically feasible or even physically possible. Soils that are coarse, saline, or sodic were also considered problem soils, although they can obviously be reclaimed, but at great expense. Very shallow soils are an example of soils with very severe limitations precluding sustained use, although in special cases stone and bedrock removal may also deepen them.

In some cases, soils that are problematic in temperate climates are not considered problem soils in dry areas. For example, the poorly drained Gleysols in drylands should be considered a resource rather than a problem due to their scarcity value, e.g., as wetlands for biodiversity.

10.4.4 INTEGRATION OF THEMATIC RESOURCE POVERTY INDICES

By combining these three thematic indices, it is possible to estimate an index that expresses the agricultural resource base in a given place on a scale of 0–100 in terms of either poverty (ARPI) or endowment (AREI). These indices are calculated as follows:

$$ARPI = \max(CRPI, TRPI, SRPI), \text{ the highest of the three indices} \quad (10.5)$$

$$AREI = 100 - ARPI \quad (10.6)$$

The entire procedure for developing the resource poverty or endowment indices at regional scale is shown in Figure 10.9.

10.5 IRRIGATED AREAS AND AGRICULTURAL INCOME DISTRIBUTION IN SYRIA

Following the above approach for calculating thematic agricultural resource poverty or endowment indices, a study was undertaken in Syria to assess the possibility of using these fine-resolution indices in combination with agricultural statistics to more accurately map agricultural income, and in the process detect hot spots of rural poverty. Using 1-km resolution climate surfaces for Syria, a national soil map [17], and the SRTM DEM, the agricultural resource potential (or endowment) index (ARI)

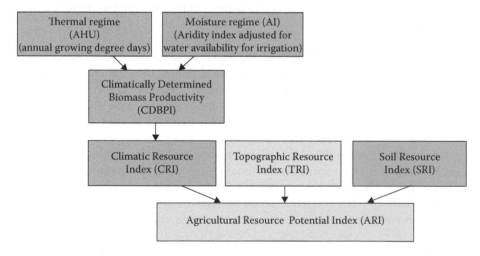

FIGURE 10.9 Main steps in developing the Agricultural Resource Poverty and Endowment Indices.

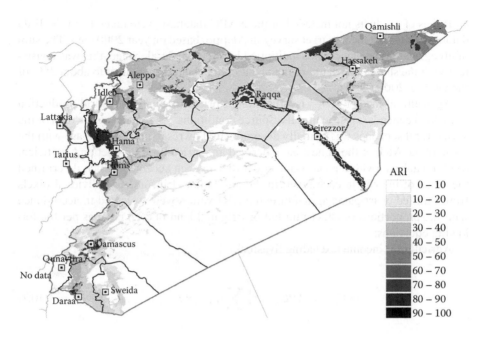

FIGURE 10.10   Agricultural resource poverty index (ARPI) in Syria.

was calculated by integrating the climatic, topographic, and soil resource indices (Figure 10.10) as follows:

- CRI: an index scaled between 0 and 100 for capturing the climatic potential for biomass production.
- Soil resource index (SRI): the proportion of the pixel without problematic soil types.
- Topographic resource index (TRI): the percentage of SRTM pixels with a slope of 15% or above.

The thematic indices (CRI, SRI, TRI) were combined as raster themes in GIS, with the same spatial scope and resolution, into the ARI, which was calculated for rain-fed areas as the lowest value of the indices CRI, SRI, and TRI. If, however, a fraction of the pixel is irrigated, only the CRI is considered for the irrigated fraction in the ARI calculation, assuming that irrigation takes place where soil and topographic conditions are not severely constraining. The presence of irrigation water was inferred from an earlier LULC map of Syria [5].

The Agricultural Production Database of Syria [18] provides production and price data for crops, fruits, vegetables, and animal products for "agricultural subzones," which are the spatial units resulting from the intersections of the provinces and the "Agricultural Stability Zones," an agricultural zoning system with five classes based on the quantity and reliability of annual precipitation [19].

Prices of products not included in the NAPC database were taken from the FAO database or from a local market survey in Aleppo, based on year 2000 data. The sum of all agricultural products was multiplied by their wholesale value; this was consistent with the share of agriculture in the national account, equivalent to about 25% of the GDP in 2000 (NAPC).

Agricultural income (INC) is a function of prices (p) and agricultural production (q), which depends on agricultural resource conditions. Aggregate census data of the agricultural subzones was spatially disaggregated by rescaling the database with the mean of the ARI for the same area. It derived an Agricultural Production Coefficient (APC) database (with a mean value equal to 1 in each subzone), which determined the value of income in each pixel. In the allocation of income to individual pixels from rain-fed or irrigated agriculture, the ARI values were weighted in accordance with the proportions of either rain-fed or irrigated land in each pixel as per the following equations:

Agricultural income (excluding livestock):

$$INC_{w,j,z} = APC_{w,j,z} * \left(\sum_i q_{i,w,z} * p_{i,w,z} \right) \Big/ n_z. \tag{10.7}$$

Agricultural Production Coefficient:

$$APC_{w,j,z} = \frac{ARI_{w,j,z} * M_{w,j}}{\left(\sum ARI_{w,j,z} \right) / n} \quad \text{and} \quad \left(\sum_j APC_{w,j,c} \right) \Big/ n = 1, \tag{10.8}$$

where:

i = products (e.g., crops, fruits, vegetables, livestock),
w = water availability ("r" rain-fed, "ir" irrigated),
j = pixels (~1 km²),
z = subzones,
M = percent of irrigated "ir" and rain-fed "r" areas in pixel j,
n = number of agricultural regions inside a nahia (county).

On the basis of the above equations, agricultural income derived from large statistical units could be disaggregated to pixel level. Figure 10.11 provides a comparison of the agricultural income from irrigation, using the aggregated and disaggregated (modeled) data.

Using the same equations, and following for livestock a somewhat different procedure [16], the incomes from rain-fed agriculture and livestock were calculated and added to the income from irrigation, to generate the total income from agriculture (Figure 10.12).

A last step consisted in relating this total income from agriculture to population density, itself downscaled from the nahia to the pixel level [16]. The result is a map of modeled per-capita in agricultural income (Figure 10.13).

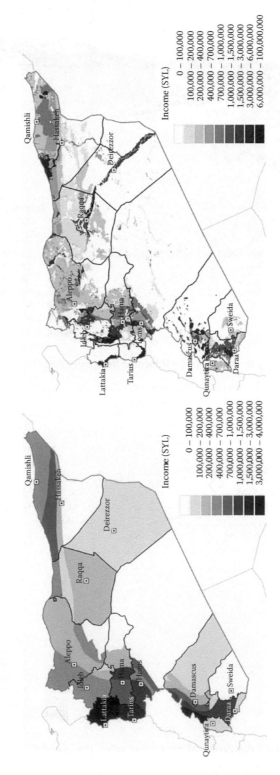

FIGURE 10.11 Aggregated (left) and disaggregated (right) income distribution from irrigation.

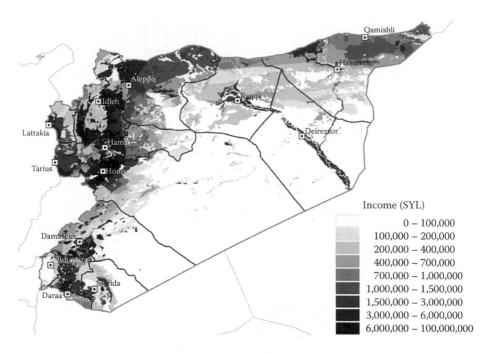

FIGURE 10.12 Spatially disaggregated total income from agriculture.

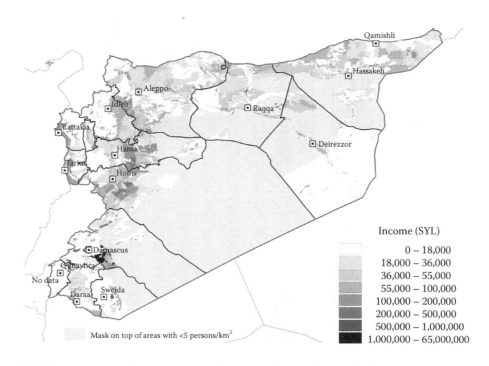

FIGURE 10.13 Spatially disaggregated per-capita income from agriculture.

The overall pattern of the ARI-modelled income distribution clearly reflects the gradients in precipitation and the distribution of irrigated land. Ultimately, these are the key resources of agriculture in Syria, since even livestock income depends on them. Within the better-income areas, pockets with shallow soils or unfavorable topography appear as lower-income spots. One exception is the coastal mountains in the west of Syria, where income is better than expected from its agricultural resource potential. This is explained by large-scale investments in land improvement by the government and the private sector, chiefly through terracing and rock removal, which have significantly improved the earning capacity of local agriculture.

10.6 SPATIAL MODELING OF THE BIOPHYSICAL POTENTIAL FOR SUPPLEMENTAL IRRIGATION: A CASE STUDY IN SYRIA

Supplemental irrigation, in contrast to full irrigation, makes optimal use of natural precipitation, resulting in substantial water savings in the irrigated areas, which can then be used to irrigate previously rain-fed areas and reduce vulnerability to drought. In order to evaluate *where* supplemental irrigation could have the highest impact, a method was developed that combines the use of GIS, remote sensing, and modeling.

The method is based on the combination of a simple model to calculate the possible water savings made by the shift from spring/summer fully irrigated crops to supplemental-irrigated winter/spring crops, with a water allocation procedure for the surrounding rain-fed areas based on land suitability criteria. In their simplest concept, areas suitable for irrigation but currently nonirrigated would be characterized by the presence of arable soils, nonconstraining slopes, agricultural land use, and within a distance of existing irrigation schemes that do not impose uneconomical costs of water conveyance or pumping. Critical in the application of this method is, therefore, a knowledge about the location of the irrigated areas, since these act not only as the reservoirs of potential water savings, but also as the location of the target areas (rain-fed, and within a reasonable distance from the irrigated areas).

Twelve rectified and UTM-projected Landsat ETM scenes, taken in two different seasons (summer 2002 and spring 2003) covering the whole area of Syria, were used to extract the land cover features required in the spatial modeling. Image processing was undertaken using ERDAS Imagine software and some functions of the ArcGIS 8.3. ESRI software. The Principal components analysis (PCA) was central in the identification of most land cover categories. The PCA is often used as a method of data compression. It allows redundant information to be compacted into fewer bands, in the case of Landsat from eight bands to three to four.

After identification of water bodies, forest areas, and arable land [20], the irrigated areas were easy to recognize in summer satellite images due to their presence as green agricultural fields in the dry season and their regular patterns. The NDVI was, therefore, suitable for identifying and mapping irrigated fields.

Figure 10.14 illustrates the use of the NDVI technique: Figure 10.14a shows the Landsat image in color combination RGB 741, Figure 10.14b shows the NDVI, and Figure 10.14c shows the areas where the NDVI is above the necessary threshold to

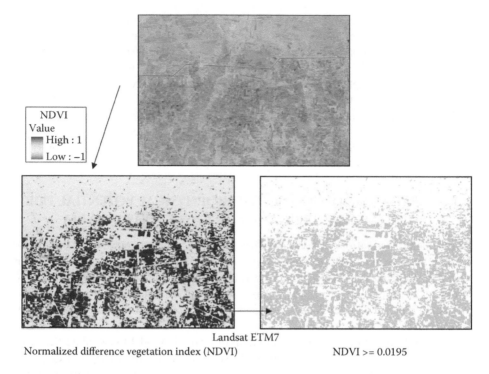

Landsat ETM7
Normalized difference vegetation index (NDVI)   NDVI >= 0.0195

FIGURE 10.14 Use of NDVI threshold for extracting summer-irrigated areas.

be considered irrigated for the particular scene. After screening of the forested areas, different NDVI thresholds were used to detect the irrigated areas.

The water savings within the demarcated irrigated areas (the "water bank") were obtained as the difference in total crop water need between a standard fully irrigated crop (cotton) and a standard partially irrigated crop (wheat), calculated by textbook methods for crop water requirements, summed for all pixels inside the irrigated area (Figure 10.15). These potential water savings from irrigated areas were then allocated to neighboring rain-fed areas, using an allocation procedure that reflects the suitability of areas surrounding an irrigation perimeter to benefit from a possible water allocation (Figure 10.16). The criteria used were based on distance from the irrigated perimeter, slope, soils, and presence of forests. The scores obtained against these four criteria were combined in a multicriteria evaluation using the "most limiting factor method." The results of the multicriteria evaluation were retained in an Allocation priority layer (APL) specific to each irrigated perimeter. Water was then allocated to the rain-fed areas in an iterative process, first to the pixels with the highest scores in the APL, and as these filled up, to pixels with lower priority scores until the water bank for each perimeter was exhausted. An example of the modeled water allocations to rain-fed areas is shown for a sample area in northern Syria in Figure 10.17.

Cotton						
Date	Kc	ETo	Etc	Prec	Etc$_{irr}$	Potential water savings
15 Apr	0.35	123	43	34		
15 May	0.64	193	124	19		
15 Jun	1.17	267	312	2		
15 Jul	1.17	308	360	0		
15 Aug	1.08	274	296	0		
15 Sep	0.74	178	131	1		
Total			1,266	56	1,210	1,056
Wheat						
Date	Kc	ETo	Etc	Prec	Etc$_{irr}$	Water allocation needs
15 Nov	0.70	40	28	36		
15 Dec	0.76	21	16	59		
15 Jan	0.89	22	20	60		
15 Feb	1.02	37	38	51		
15 Mar	1.13	72	81	45		
15 Apr	1.15	123	141	34		
15 May	0.70	193	134	19		
Total			458	304	154	154

FIGURE 10.15 Example of potential water savings (for a single pixel) by shifting from cotton to wheat.

10.7 CONCLUSIONS

Invariably in drylands, agriculture is a major and sensitive sector of the economy, and unavailability of water is the main constraint. Agricultural production of major grain crops is strongly affected by precipitation fluctuations [21], and crop and live-stock losses due to drought can have very severe repercussions on both the balance of payments and the livelihoods of individual producers of the countries. For this reason, irrigation plays a critical role in the widening of agricultural options and has a stabilizing influence on rain-fed agriculture and farmer livelihoods that far exceeds its areal extent.

The few case studies presented in this chapter offer an inkling of the diversity of spatial applications in agricultural research that can benefit from information on irrigated areas. These applications may cover a wide range of scales, from the regional to the subnational, and themes, including zoning, land-use planning, land suitability and degradation assessments, poverty mapping, and water management.

The case studies undertaken in Syria can be easily extended and adapted to the needs of other countries, using widely available medium-resolution satellite imagery. On the other hand, there is a particular need for a 1-km resolution global map of the irrigated areas, in order to outscale the regional agroecological zoning and resource poverty studies beyond the CWANA region.

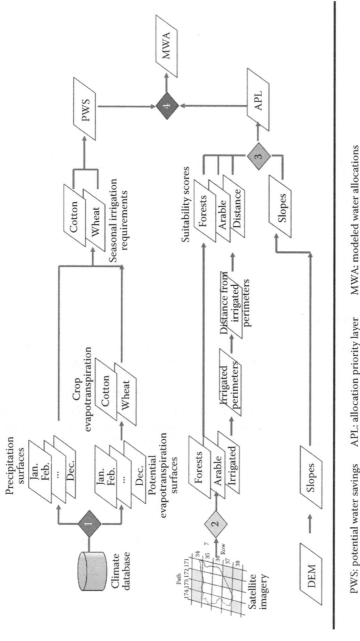

FIGURE 10.16 Outline of the method and data requirements.

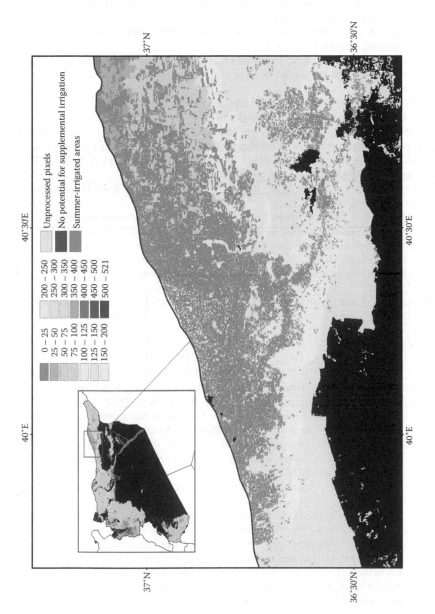

FIGURE 10.17 (See color insert following page **224**.) Modeled water allocations to rainfed areas, shown for a sample area.

REFERENCES

1. Celis, D., De Pauw, E., and Geerken, R., *Assessment of Land Cover/Land Use in the CWANA Region Using AVHRR Imagery and Agroclimatic Data, Part 1, Land Cover/ Land Use – Base Year 1993*, International Center for Agricultural Research in the Dry Areas (ICARDA), Aleppo, Syria, 2007.

2. Myneni, R.B. et al., Increased plant growth in the northern high latitudes from 1981 to 1991, *Nature*, 386, 698–702, 1997.

3. Gesch, D.B. and Larson, K.S., Techniques for development of global 1-kilometer digital elevation models, in *Human Interactions with the Environment, Pecora Thirteen* publication 1996.

4. UNESCO, *Map of the World Distribution of Arid Regions*, Map at Scale 1:25,000,000 with Explanatory Notes, UNESCO, Paris, 1979.

5. De Pauw, E., Oberle, A., and Zöbisch, M., *Land Cover and Land Use in Syria – An Overview*, Asian Institute of Technology (AIT), International Center for Agricultural Research in the Dry Areas (ICARDA), and the World Association of Soil and Water Conservation (WASWC), Aleppo, Syria, 2004.

6. Lillesand, T.M. and Kiefer, R.W., *Remote Sensing and Image Interpretation*, 3rd ed, Wiley, New York, 1994.

7. Tateishi, R., *AARS Asia 30" Land Cover Data Set with Ground Truth Information by the Land Cover Working Group of the Asian Association on Remote Sensing and CEReS*, Center for Environmental Remote Sensing of Chiba University, Japan, 1999.

8. Hansen, M. et al., Global land cover classification at 1 km spatial resolution using a classification tree approach, *International Journal of Remote Sensing*, 21, 6, 1331–64, 2000.

9. Belward, A., Estes, J., and Kline, K., The IGBP-DIS global 1-km land-cover data set DISCover: A project overview, *Photogrammetric Engineering & Remote Sensing*, 65, 9, 1013–20, 1999.

10. De Pauw, E., Goebel, W., and Adam, H., Agrometeorological aspects of agriculture and forestry in the arid zones, *Agricultural and Forest Meteorology*, 2793, 1–16, 2000.

11. FAO, FAOCLIM, *A CD-ROM with World-Wide Agroclimatic Data, version 2*, Environment and Natural Resources Service (SDRN) Working Paper No. 5, Food and Agriculture Organization of the United Nations, Rome, Italy, 2001.

12. Allen R.G. et al., *Crop Evapotranspiration, Guidelines for Computing Crop Water Requirements*, FAO Irrigation and Drainage Paper 56, FAO, Rome, 1998.

13. Hutchinson, M.F., Interpolating mean rainfall using thin plate smoothing splines, *International Journal of Geographic Information Systems*, 9, 385–403, 1995.

14. Hutchinson, M.F., *ANUSPLIN Version 4.1. User Guide*, Center for Resource and Environmental Studies, Australian National University, Canberra, 2000, http://cres.anu.edu.au.

15. FAO (Food and Agriculture Organization of the United Nations), *Digital Soil Map of the World and Derived Soil Properties*, CD-ROM, version 3.5, FAO, Rome, Italy, 1995.

16. Szõnyi, J.A. et al., *Mapping Agricultural Income Distribution in Rural Syria: A Case Study in Linking Poverty to Resource Endowment*, International Center for Agricultural Research in the Dry Areas (ICARDA), Aleppo, Syria, 2005.

17. Louis Berger International, Inc., *Reconnaissance Soil Survey, (1:500,000 Scale)*, Land Classification/Soil Survey Project of the Syrian Arab Republic, South Dakota University, SD, 1982.

18. NAPC (National Agricultural Policy Center), *Syrian Agricultural Database, Assistance in Agricultural Planning, Policy Analysis and Statistics*, National Agricultural Policy Center, Syrian Arab Republic, 2003.

19. MAAR (Ministry of Agriculture and Agrarian Reform), *Agricultural Statistical Abstract* (several years), Ministry of Agriculture and Agrarian Reform Statistics Department, Syrian Arab Republic, 2000.

20. De Pauw, E. Oweis, T., and Youssef, J., *Integrating Expert Knowledge in GIS to Locate Biophysical Potential for Water Harvesting: Methodology and a Case Study for Syria*, ICARDA, Aleppo, Syria, 2008.

21. Keatinge, J.D.H., Dennett, M.D., and Rogers, J., The influence of precipitation regime on the crop management in dry areas in northern Syria, *Field Crops Research*, 12, 239–49, 1986.

11 Subpixel Mapping of Rice Paddy Fields over Asia Using MODIS Time Series

Wataru Takeuchi
University of Tokyo

Yoshifumi Yasuoka
National Institute for Environmental Studies

CONTENTS

11.1 INTRODUCTION

11.1.1 THE NEED FOR BETTER INFORMATION

Two-thirds of the rice-growing areas in the world are in Asian countries and hundreds of millions of people depend on rice as their staple food source. Paddy fields have been considered one of the likely and most important sources of atmospheric methane since the rapid increase in atmospheric methane was recognized in the early 1980s (Wassmann, Lantin, and Neue, 2003). Paddy field cover is an important variable for modeling regional biogeochemical cycles and climate (Dickinson, 1995). Agricultural water use by irrigation accounted for 70% of global freshwater withdrawals, and most of the Asian rice agriculture is irrigated, especially in eastern Asia, including Japan, Korea, and China (Huke and Huke, 1997). The improved understanding of paddy field distribution at large spatial scales has increased the interest in deriving crop yield, methane emission, and water use estimations.

Since a large amount of water is required for rice production, precipitation is the most limiting factor in rice cultivation, particularly in regions of rain-fed rice culture. On the other hand, rice productivity largely depends on the temperature and solar radiation where irrigation systems are available. It is difficult to define a simple formula between the climate and the yield of rice. In most of the Asian countries, there is no suitable infrastructure for the implementation of programs to monitor rice production. The conventional method of compiling statistics on the rice crop depends on the ground surveys by the government ministry or village officials. This local information is then spatially extrapolated to generate data and predictions on a regional basis. The major difficulty associated with the collection of ground data in this way is the remoteness of many of the croplands, especially for the upland areas. This means that the traditional surveys are both time consuming and expensive. In addition, the information derived from ground surveys is sometimes inaccurate and unreliable, which leads to inaccurate forecasts on crop yield.

11.1.2 THE BENEFITS OF SPACE-BASED MONITORING

In this situation, remotely sensed data from satellite images provide an alternative means of obtaining paddy field distribution for inexpensive and synoptic information on rice crop areas. The usefulness of optical satellite imagery with high spatial resolution data from the Landsat, Advanced Spaceborne Thermal Emission and Reflection Radiometer (ASTER), and Système Pour l'Observation de la Terre (SPOT) missions for rice monitoring has an advantage in spatial resolution and has yielded a good rice paddy map estimation (Okamoto and Fukuhara, 1996; Samarasinghe, 2003). However, using high spatial resolution data over a continental scale has proven to be limited, as the main rice growth takes place during the rainy season when there is typically 90% cloud cover. Another objection to monitoring rice paddies only with high spatial resolution data over a continental scale is the cost and logistics of data volume.

Valuable efforts have been made on global land cover characteristics by the international science community (Hill and Donald, 2003). The most recently developed and the only global land cover map including paddy fields is that by the International Geosphere-Biosphere Programme, Data and Information Systems (IGBP-DIS) initiative. The IGBP DISCover global land cover product includes 94 general land cover classes defined by Olson global ecosystems. However, current continental paddy field cover data are inadequate to meet either the research needs of the continental change science community, or the operational requirements of national and international resource management organizations (Loveland et al., 2000). During the rice-transplanting period and the early part of the rice-growing period over Asia, because of a long warm season, paddy fields are a mixture of green rice plants and open water (Xiao et al., 2002b).

The Moderate Resolution Imaging Spectroradiometer (MODIS) sensors on Aqua and Terra have the capability of acquiring data regularly in monsoonal regions, since they can acquire the radiometric data over the range of viewing angles in 36 discrete channels in 250-m to 1-km ground resolutions (Justice et al., 1998). The newly launched SPOT VEGETATION (VEG) provide shortwave infrared (SWIR) bands that are sensitive to vegetation moisture and soil water conditions. The availability of an additional SWIR spectral band on MODIS or VEG provides the capability of developing and generating improved indices, sensitive to equivalent water thickness (Xiao et al., 2002a).

On the other hand, MODIS data may not detect fine spatial structures in mixtures of vegetation, soil, and water in paddy fields over Asia because of its coarse spatial resolution. In this sense, it is necessary to devise a method for the fusion of the data with different spatial resolutions. By depicting each pixel as a percent coverage, areas of complexity or heterogeneity are better represented. Discrete classes do not allow for the variety of variability for spatially complex areas. Since the scale of paddy fields is typically finer than 1 km, subpixel cover mapping from MODIS data can yield a usable product for crop yield and methane estimation (Hansen et al., 2002).

11.1.3 OBJECTIVE OF THIS STUDY

This chapter proposes an improved methodology for deriving percent paddy field cover estimates over the previous methodologies. The procedure is presented together with MODIS data over Asia supplemented by ASTER-based training data and groundtruth data. A continuous map of paddy field cover offers advantages over traditional discrete classifications.

11.2 METHODOLOGY

11.2.1 A FLOWCHART OF RICE PADDY MAPPING

Figure 11.1 shows a flowchart of rice paddy mapping proposed in this study. The methodology presented here mainly consists of three parts, including MODIS processing, training data, and paddy-mapping model. The following sections describe them accordingly.

Terra MODIS data used in this study were daily level 1b records from 2001 July to 2007 June, received on a direct broadcasting system installed at the Institute

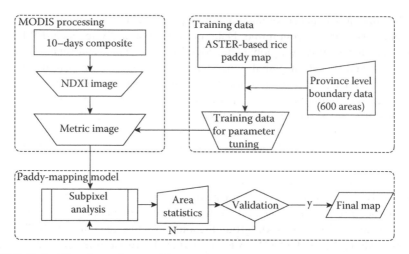

FIGURE 11.1   Flowchart of rice paddy mapping.

of Industrial Science, University of Tokyo in Tokyo and the Asian Institute of Technology in Bangkok. The original level 0 data were converted to level 1b data by the International MODIS/AIRS Processing Package software (version 1.5) developed by the University of Wisconsin. Level 1b data were in hierarchical data format-EOS format, the standard data format of Terra and Aqua MODIS sensors. MODIS data were in the form of 12-bit precision brightness counts and coded to a 16-bit scale, freely available on the Web at http://webmodis.iis.u-tokyo.ac.jp/. Then a cloud-screening process was carried out to produce 10-days composite imagery (Takeuchi and Yasuoka, 2006).

11.2.2   Normalized Difference Vegetation, Soil, and Water Indices 4

In order to incorporate SWIR channels that were sensitive to vegetation moisture and soil water conditions, we defined normalized difference vegetation, soil, and water indices, namely, NDVI, NDSI, and NDWI, collectively called NDXI as defined below:

$$NDVI = (NIR - VIS) / (NIR + VIS), \tag{11.1}$$

$$NDSI = (SWIR - NIR) / (SWIR + NIR), \tag{11.2}$$

$$NDWI = (VIS - SWIR) / (VIS + SWIR). \tag{11.3}$$

In Equations 11.1 through 11.3, VIS represents the reflectance value of visible channel, near-infrared (NIR), and SWIR, respectively. NDXI was defined by extending the idea of NDVI with the combination of VIS, NIR, and SWIR channels, and each index had the symmetric expression as a function of VIS, NIR, and SWIR. The reflectance values in VIS, NIR, and SWIR channel, σ, were calculated by an integration of the spectral response of the sensor $\Phi(\lambda)$, and the reflectance value of the target $\sigma(\lambda)$ as a function of wavelength λ, as shown in Equation 11.4, where $\lambda 1$ and $\lambda 2$, respectively represented the upper and lower boundary wavelength of the sensor's response function.

$$\overline{\sigma} = \int_{\lambda_1}^{\lambda_2} \Phi(\lambda)\sigma(\lambda). \tag{11.4}$$

11.2.3   MULTITEMPORAL METRICS METHOD

The approach presented for mapping continuous paddy field cover followed the annual metrics approach (DeFries et al., 2000). Multitemporal metrics captured the salient points of phonological variation by calculating annual means, maxima, minima, and amplitudes of spectral information. The value of metric generation vs. using a series of 10-days values was that the metrics were not sensitive to the time of year or the seasonal cycle and could limit the inclusion of atmospheric contamination (Hansen et al., 2000). The metrics to be tested mimicked those shown in Table 11.1. The minimum of channel 1 reflectance was negatively correlated with paddy field cover, since the surface of paddy fields was covered with soil and water at the initial stage of rice growth. Maximum NDVI reflectance was positively correlated with paddy fields, since the surface of paddy fields was covered with rice plant vegetation at the maturing stage of rice growth. Taking these rice-growth conditions into consideration, channel 1–7 and NDXI were ranked individually. From these rankings, a set of metrics was derived. Table 11.1 shows metrics as an example, using channel 1.

11.2.4   TRAINING DATA FROM ASTER

Training data were created by classifying and interpreting 54 pairs of ASTER to identify homogeneous areas over the Asian region. The training data supporting system comprised around 2000 scenes of ASTER database shown in Figure 11.2 and they are available at http://webpanda.iis.u-tokyo.ac.jp/ASTER/. Training data were prepared based on provincial-level administrative boundaries (Figure 11.3), assuming that the same types of spatiotemporal cropping patterns were carried out in the same administrative boundary region. To distinguish paddy fields from other categories (croplands, forests, urban areas), the scene acquisition date and the spatial resolutions were important. As for scene acquisition dates, two stages were important; the early planting stage of rice growing (early May to late May in Japan) and the midgrowing stages (late July to mid-August in Japan).

TABLE 11.1

An Example of Metrics Derived for MODIS Channel 1 Where the Same Metrics are Calculated for Channel 2–7 and NDXI

Metric type	Ranking criteria: Each channel is individually ranked
Monthly values	Minimum, median, maximum red values
Means	Mean of three, six, and nine darkest reflectance in 10-days values
Amplitudes	Amplitude of reflectance for minimum, median, and maximum values

FIGURE 11.2 A set of ASTER database around 2000 scenes over Asia. Dots correspond to footprints of ASTER imagery.

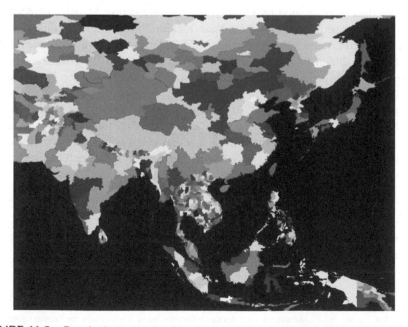

FIGURE 11.3 Provincial-level administrative boundaries consisting of 600 areas of interests.

At the early planting stage, the surface of rice paddy is covered with soil and water, whose spectral response is similar to inland water, such as lakes, small ponds, and rivers, whereas it is covered with vegetation at the midgrowing stage, whose spectral signature seems similar to other croplands, including potato, soybean, and so on. As for the spatial resolution, the traditionally used Landsat 30-m spatial resolution was not enough to classify paddy fields (Okamoto and Fukuhara, 1996). The patch size of paddy rice in Asian countries is finer than 30 m, and linear structures, such as farm roads, irrigation, and drainage canals, are several meters wide and difficult to distinguish. To overcome this problem, three bands from visible to infrared with 15 m spatial resolution were used. These areas were then aggregated to develop a coarse resolution training data sets for a discrete classification system.

11.2.5 Paddy-mapping Model with Subpixel Characterization

The approach was based on a linear mixture model to derive continuous fields of paddy cover maps. The linear mixture model was based on the assumption that the reflectance at a pixel was the sum of the reflectance of each component within the pixel weighted by the respective proportional covers. It was based on the following relationship:

$$R_i = \sum_{j=1}^{k} r_{ij} x_j + e_j, \qquad (11.5)$$

$$\sum_{j=1}^{k} x_j = 1, \qquad (11.6)$$

where R_i is the reflectance in band i, r_{ij} is the reflectance of component j in band i, x_j is the fractional cover 3 of component j, e_j is the error term, and i is the number of components. In addition, the model should have the constraints shown in Equation 11.6.

We derived linear discriminants to combine MODIS metrics into a smaller number of variations of variables with a maximum ratio of the separation of class means to within-class variance. Weightings for each metric were derived to maximize this ratio, and linear discriminants were the linear combination of these weighted metrics, shown in Figure 11.4. To derive the linear discriminants, training data obtained from ASTER were used to calculate the class means and within-class means covariance matrices.

11.3 RESULTS AND DISCUSSION

11.3.1 Comparison of 500-m Paddy Cover Map from MODIS and Advanced Very High Resolution Radiometer (AVHRR) 1-km Global Products

Figure 11.5 shows the annual spectral patterns of paddy fields in single (a) and double cropping (b), (c), and (d). A single cropping pattern was characterized in a peak of NDWI in June, followed by a peak of NDVI in October. In the same way, a double cropping pattern was identified by a peak of NDVI in March followed by a peak of NDWI in April, then a peak of NDVI return in July followed by an NDWI peak in

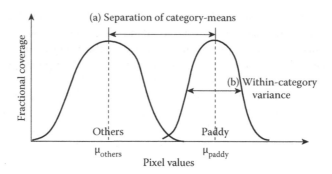

FIGURE 11.4 Conceptual description of separation of category-means and variance between paddy fields and others.

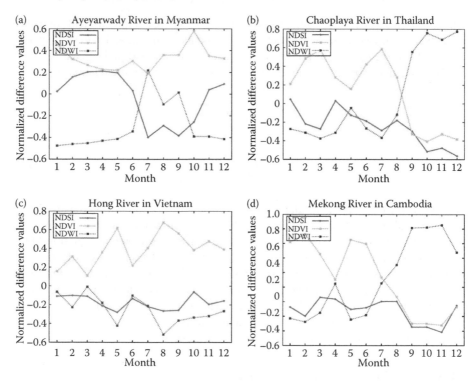

FIGURE 11.5 Annual spectral patterns of paddy fields. (a) Single cropping of the Ayeyarwady River in Myanmar, (b) double croppings of the Chaoplaya River in Thailand, (c) Hong River in Vietnam, and (d) Mekong River in Cambodia.

October. The time periods used captured the seasonality for paddy fields. To assess the ability of the continuous paddy fields map, we compared the results with the other land cover data sets, including AARS (Tateishi et al., 1995), BU (Friedl et al., 2002), JRC (JRC, 2003), UMD (Hansen et al., 2002), and USGS (Loveland et al., 2000), shown in Figure 11.6. Bright values indicated areas of high fractional abundance of paddy fields in a given pixel. They were mainly developed as the 1-km global land cover

FIGURE 11.6 (See color insert following page 224.) Comparison of paddy field cover map from global land cover product based on MODIS and AVHRR: (a) AARS, (b) BU, (c) JRC, (d) USGS, (e) UMD, (f) UT.

database through a continent-by-continent unsupervised classification using AVHRR NDVI composites covering 1992–1993. Among these five products, AARS and USGS databases included the legend "paddy field;" however, the other three examples did not. Since a visual interpretation indicated that the "cropland" class represented or included paddy fields, cropland classes were used as a comparison for the following studies. Although AARS, BU, JRC, USGS, and UMD land cover classifications do not provide estimates of proportional cover, we can compare the MODIS 1-km product with them to determine consistency in geographic distributions of paddy fields. The area of interest covered South Asia, Southeast Asia, and East Asian countries. The MODIS 1-km product was compared with the International Rice Research Institute (IRRI) statistics of paddy area obtained from the web site (IRRI, 2007).

Figure 11.7 shows the comparison of estimates of paddy field area over major rice paddy crop areas in Thailand. In a qualitative sense, the MODIS continuous field data reproduced the geographic patterns of paddy field distributions that would be expected. Therefore, it was important to bridge the finer spatial and spectral resolution of MODIS. Figure 11.4 shows the statistical analysis of MODIS-derived paddy area in Southeast Asia and the other paddy field coverage based on AVHRR estimation with the IRRI statistical report. The accuracy assessment was conducted by comparing the paddy field area based on the classification result with the IRRI

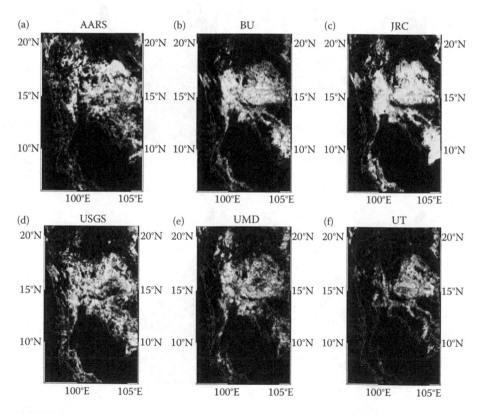

FIGURE 11.7 Comparison of MODIS-derived paddy field map in Thailand with other studies: (a) AARS, (b) BU, (c) JRC, (d) USGS, (e) UMD, (f) UT.

statistical report. According to the correlation coefficients, our estimation (UT) had the highest value (0.987), followed by JRC (0.811), BU (0.752), USGS (0.747), AARS (0.672), and UMD (0.654). One possible reason for the discrepancy between the satellite-monitoring-based estimation and the statistical values was that the definition of paddy field cover was different between the satellite-based product and IRRI statistics. The IRRI statistical paddy field included area without crops. On the other hand, the satellite-monitoring-derived product defines area that was planted and harvested. This implies that the product estimation was smaller than the statistical planting area. For the crop estimation, the latter should be the definition of paddy fields of that year. Taking these facts into account, the AVHRR product seems to overestimate the paddy fields, which points to the usefulness of percent cover and fractional estimation using MODIS finer resolution.

Figure 11.8 shows the comparison of this product with the IGBP-DIS over Seoul city in Korea. In past work, AVHRR sensor's resolution of 1 km did not allow for the depiction of detailed land surfaces, which resulted in such a potential dramatic loss of information. That was because the original resolution and sensor characteristics of the AVHRR captured an image that was too coarse to view many of the ground features visible with MODIS.

11.3.2   VALIDATION OF THE 500-M PADDY COVER
MAP WITH THE ASTER 15-M MAP

The 500-m MODIS paddy field cover map was compared to the 15-m ASTER map for the five areas where the rice crop was common in Japan. We selected five regions over Japan's main island from north to south, which covered different climatic regions and cultivation practices. Figure 11.9 shows the comparison of percent paddy field cover estimated between ASTER, MODIS, and MODIS NDVI over Miyagi in northeastern

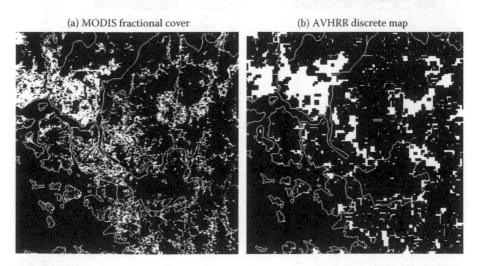

(a) MODIS fractional cover   (b) AVHRR discrete map

FIGURE 11.8   Comparison of paddy field cover map over Seoul city (37.58 N, 126.93 E) in Korea: (a) MODIS estimation in 500 m in 2007, (b) AVHRR estimation in 1 km in 1992–1993. The image covers approximately 100×100 km^2.

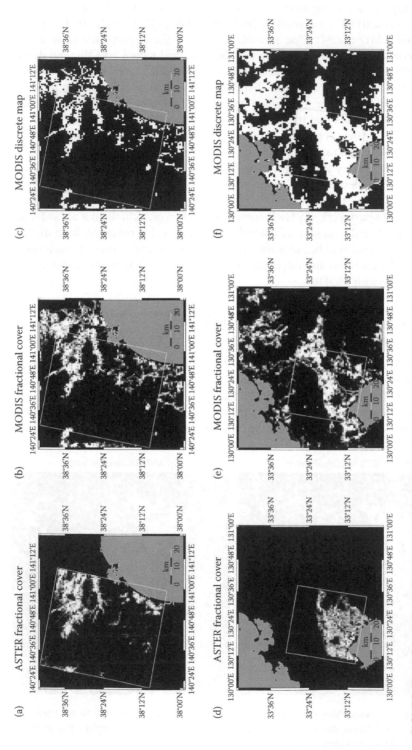

FIGURE 11.9 Comparison of percent paddy field cover estimates among ASTER fractional cover, MODIS fractional cover, and MODIS discrete map over Miyagi in northeastern Japan (38.56 N, 141.15 E) shown in (a), (b), and (c), and in Saga in southwestern Japan (33.25 N, 130.29 E) shown in (d), (e), and (f).

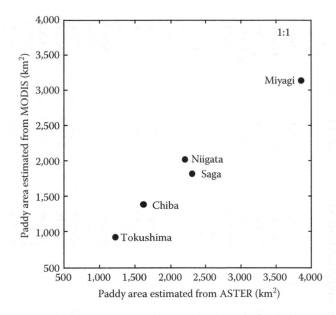

FIGURE 11.10   Comparison of estimates of paddy field area between ASTER and MODIS over five pairs of validation site: Miyagi (38.56 N, 141.15 E), Niigata (37.81 N, 139.08 E), Chiba (35.86 N, 140.03 E), Tokushima (34.10 N, 134.48 E), and Saga (33.25 N, 130.29 E).

Japan and Saga in southwestern Japan. ASTER images of 15 m were aggregated to 500 m resolution using simple area average calculation. The 500-m MODIS product seems to have the fine geographical distribution compared with the ASTER result.

Figure 11.10 shows the comparison of estimates of paddy field area between ASTER and MODIS over five major rice paddy crop areas in Japan, including Miyagi, Niigata, Chiba, Tokushima, and Saga. The comparison shows that the estimation of MODIS was 15% lower than the ASTER estimation. This result shows that features finer than 500 m were partly lost due to the coarse resolution of MODIS. However, the overall estimates show the linear relationship between ASTER and MODIS. In a qualitative sense, the MODIS continuous field reproduced the geographic patterns of paddy field distributions that would be expected. It was therefore important to bridge the coarse resolution MODIS and the finer resolution ASTER using the nonmixing approach.

11.3.3   COMPARISON OF SATELLITE-BASED ESTIMATION WITH DATA DERIVED FROM THE IRRI STATISTICAL REPORT

Although AARS, BU, JRC, UMD, and USGS land cover classifications did not provide estimates of proportional cover, we can compare the MODIS 500-m product with them to determine consistency in geographic distributions of paddy fields. The area of interest covers 18 countries, including Bangladesh, Bhutan, Cambodia, China, India, Indonesia, Japan, Korea DPR, Korea Rep., Lao PDR, Malaysia, Myanmar, Nepal, Pakistan, Philippines, Sri Lanka, Thailand, and Vietnam. The MODIS 500-m product was compared with statistics of the IRRI paddy area obtained from the web site (IRRI, 2007). Table 11.2 shows the country-level comparison of MODIS and

TABLE 11.2
The Country-Level Comparison of MODIS and AVHRR Estimation with the Statistical Data Derived from the IRRI (2007) Report

Country	AARS	BU	JRC	UMD	USGS	UT	IRRI
Bangladesh	1,714	96,497	125,184	74,478	83,589	115,784	110,000
Bhutan	21	169	2,284	534	691	86	300
Cambodia	807	42,385	53,529	17,281	63,901	23,633	23,000
China	236,758	2,084,015	2,025,062	1,787,372	1,153,760	309,050	294,200
India	2,913	1,989,304	2,028,661	541,795	598,964	435,923	425,000
Indonesia	19,703	86,015	175,746	41,572	132,372	109,666	117,530
Japan	13,355	22,080	118,268	25,626	31,006	17,279	16,500
Korea DPR	0	45,165	27,202	15,430	14,111	6,035	5,830
Korea Rep.	938	13,923	50,442	17,997	15,641	9,435	9,900
Lao DPR	1,645	7,365	17,714	12,338	17,810	8,177	8,200
Malaysia	11,657	16,355	33,379	6,528	65,310	6,380	6,700
Myanmar	1,607	117,730	169,593	64,292	100,093	62,843	60,000
Nepal	2	34,456	60,495	34,481	21,887	15,989	15,500
Pakistan	453	223,839	331,688	89,888	28,699	22,431	22,100
Philippines	3,726	55,920	70,875	24,672	42,748	40,227	40,000
Sri Lanka	1,128	14,590	36,444	7,610	26,666	9,023	8,525
Thailand	4,919	222,730	291,202	190,740	234,789	101,318	98,000
Vietnam	8,482	72,985	125,065	82,298	101,779	73,858	74,000

Note:　Countries are listed in alphabetical order and the area represented in kilometers.

AVHRR estimation with the statistical data derived from the IRRI report (IRRI, 2007). Countries are listed in alphabetical order and the area is represented in kilometers squared. The classification result of the UT product had a reasonable estimation in all countries compared with the IRRI report, whereas the definition of paddy field cover was different with the product and IRRI. The IRRI statistical paddy field data included areas without crops. On the other hand, the product defines area that was planted and harvested. This implied that the product estimation was smaller than the statistical planting area. For the crop estimation, the latter should be the definition of paddy fields of that year. Taking these facts into account, the MODIS and AVHRR product seems to overestimate the paddy field mapping, which points to the usefulness of percent cover and fractional estimation using MODIS 500-m.

11.4   DISCUSSION AND CONCLUSIONS

This new procedure for depicting a continuous field of paddy cover map using MODIS-derived metrics is an improvement over past efforts using AVHRR data. The metrics were not sensitive to the time of year or the seasonal cycle, and helped limit the inclusion of atmospheric contamination. The 500-m MODIS paddy cover product shown here was compared to 1-km AARS, BU, JRC, UMD, USGS classification results over 18 countries, including Bangladesh, Bhutan, Cambodia, China, India, Indonesia, Japan, Korea DPR, Korea Rep., Lao PDR, Malaysia, Myanmar, Nepal, Pakistan, Philippines, Sri Lanka, Thailand, and Vietnam, and statistical data reported by IRRI. The result shows that the 1-km product tends to overestimate the cover map area and points to the usefulness of finer resolution MODIS data. The main advantage was that the higher resolution (500 m) could distinguish finer surface features that were too coarse to distinguish AVHRR's 1-km resolution. The finer ground features were lost in MODIS imagery, which implies that the continuous field training data using ASTER may play a crucial role by containing signatures across a wide range of spatial and spectral mixtures. A combination of low- and finer-resolution images can offer an improved solution to derive continental-scaled paddy cover maps. The assessment of thematic accuracy of map products and data sets derived from the processing of remote-sensing data has a long way to reach scientific consensus. In order to overcome that problem, the validation efforts should overcome the difficult logistics challenge. Since the derived paddy field maps created in this study were distributed over a large area, a considerable number of scientists or their information on the local land use will be indispensable to obtain assessments of accuracy and to reach a better scientific consensus. It is only through discussions between the intimately familiar with operationally organizational needs, and those with a considerable background in land cover characterization strategies and capabilities, that acceptable levels of accuracy of results can be achieved.

ACKNOWLEDGMENTS

This study was financially supported by the Japan Science and Technology Agency (JST) under the research project "Development of network monitoring system for environment and disasters in Asian region." The authors would like to thank JST for their support. We would also like to thank Dr. Xiangming Xiao with the New Hampshire University and Dr. Robert J. Hijmans with the International Rice Research Institute for their fruitful discussions and useful suggestions for our studies.

REFERENCES

DeFries, R.S. et al., A new global 1-km dataset of percentage tree cover derived from remote sensing, *Global Change Biology*, 6, 247–54, 2000.

Dickinson, R.E., Land processes in climate models, *Remote Sensing of Environment*, 51, 27–38, 1995.

Friedl, M.A. et al., Global land cover mapping from MODIS: Algorithms and early results, *Remote Sensing of Environment*, 83, 1–2, 287–302, 2002.

Hansen, M.C. et al., Global land cover classification at 1km spatial resolution using a classification tree approach, *International Journal of Remote Sensing*, 21, 1331–64, 2000.

Hansen, M.C. et al., Towards an operational MODIS continuous field of percent tree cover algorithm: Examples using AVHRR and MODIS data, *Remote Sensing of Environment*, 83, 303–19, 2002.

Hill, M.J. and Donald, G.E., Estimating spatio-temporal patterns of agricultural productivity in fragmented landscapes using AVHRR time series, *Remote Sensing of Environment*, 84, 367–84, 2003.

Huke, R.E. and Huke, E.H., Rice area by type of culture: South, Southeast, and East Asia, a revised and updated data base, International Rice Research Institute (IRRI) 2007, Los Banos, Laguna, Philippines, 1997, http://www.irri.org/science/ricestat/ (accessed 29 January 2008).

JRC (Joint Research Centre), Global land cover 2000 database, European Commission, Joint Research Centre, 2003, http://www-gvm.jrc.it/glc2000/ (accessed 29 January 2008).

Justice, C.O. et al., The moderate resolution imaging spectroradiometer (MODIS): Land remote sensing for global change research, *IEEE Transactions on Geoscience and Remote Sensing*, 36, 4, 1228–49, 1998.

Loveland, T.R. et al., Development of a global land cover characteristics database and IGBP DISCover from 1 km AVHRR data, *International Journal of Remote Sensing*, 21, 6&7, 1303–30, 2000.

Okamoto, K. and Fukuhara, M., Estimation of paddy field area using the area ratio of categories in each pixel of Landsat TM, *International Journal of Remote Sensing*, 17, 9, 1735–49, 1996.

Samarasinghe, G.B., Growth and yields of Sri Lanka's major crops interpreted from public domain satellites, *Agricultural Water Management*, 58, 145–57, 2003.

Takeuchi, W. and Yasuoka, Y., Development of cloud and shadow free compositing technique with MODIS QKM in *Proceedings of the American Society of Photogrammetry and Remote Sensing Spring Meeting 2006 (ASPRS)*, Reno, NV, May 5, 2006.

Tateishi, R. et al., Land cover classification system for continental/global applications, in *Proceedings of the 15th Asian Conference on Remote Sensing*, 23–27, 1995.

Wassmann, R., Lantin, R.S., and Neue, H.U., Methane emissions from major rice ecosystems, Kluwer Academic, Dordrecht, 2003.

Xiao, X. et al., Observation of flooding and rice transplanting of paddy rice fields at the site to landscape scales in China using VEGETATION sensor data, *International Journal of Remote Sensing*, 23, 3009–22, 2002a.

Xiao, X. et al., Quantitative relationships between field-measured leaf area index and vegetation index derived from VEGETATION images for paddy rice fields, *International Journal of Remote Sensing*, 23, 3595–3604, 2002b.

Section IV

Evapotranspiration Models, Water Use, and Irrigated Areas

12 Assessment of Water Resources and Demands of Agriculture in the Semiarid Middle East

Roland A. Geerken
German Development Cooperation

Ronald B. Smith
Yale University

Z. Masri and Eddy De Pauw
International Center for Agricultural
Research in the Dry Areas

CONTENTS

12.1 INTRODUCTION

Except for its most northwestern part, arable land in the northeastern Fertile Crescent (precipitation ~300 mm/year) falls entirely into climatic zones, classified as semiarid or arid [1]. The need to satisfy the rising food demand of a rapidly growing population forces authorities to press for higher agricultural yields and for the development of new agricultural areas [2,3]. Demands, however, can only be met by expanding agriculture into less fertile rangeland areas [4], which requires additional dam constructions and the expansion of an already dense irrigation network [5,6]. An increasing dependence of agriculture and, consequently, of food supply on irrigation water is the result [7,8]. Neglecting climate change, recent developments in irrigation projects and in dam constructions will lead to a water deficit in Syria by 2020 at the latest [9]. These developments, however, fall into a time where climate models predict substantial temperature increases in this region [10] with adverse impacts on the water budget, which may cause water shortages to occur even earlier. Irrigation water is either pumped from renewable and nonrenewable aquifers or diverted from river water [3] originating in the Taurus Mountains (Turkey) and the Zagros Mountains (Iraq) [11]. Due to lower transportation costs (pumping vs. mostly gravity flow), farmers give preference to river water wherever possible. Snowfields in the Taurus Mountains are the main source of river water for the Euphrates. This water is gradually released during springtime (March to May) and ensures a continuous water flow through the dry season from May to October. The predicted climatic changes are likely to change this balance, with impacts on downstream agriculture (Figure 12.1).

In this study, we monitored and analyzed the development of irrigated agriculture in southern Turkey and in Syria (Figure 12.4) over the period from 1982–2000. Annual, semiquantitative calculations of irrigation water demands highlight temporal trends and spatial distribution in irrigated agriculture. To assess the sustainability of these trends, we modeled various hydrologic parameters (soil moisture (SM), runoff, subsurface flow, snow pack) for the Euphrates–Tigris watershed under the assumption of different climatic change scenarios. Natural river discharge was modeled for the upper Euphrates (Keban watershed), the main source of river water (more than 80%) for agricultural areas in southern Turkey and in Syria.

12.2 METHODS

12.2.1 HYDROLOGIC MODELING

For the quantification of hydrological variables in a data-sparse region like the Middle East, we developed a simple linear hydrologic model that required a minimum of input variables, including temperature (minimum, maximum, mean), precipitation, and topography. Potential evapotranspiration (PET) is calculated after Hargreaves and converted to more meaningful Penman PET, using look-up tables specified for different climatic zones [12]. Further details and equations to our model are given in the Appendix.

12.2.2 SEMIQUANTITATIVE ASSESSMENT OF IRRIGATION

To analyze spatiotemporal changes in irrigation patterns, we developed an approach that uses ten-day intervals of Advanced Very High Resolution Radiometer (AVHRR) Normalized Difference Vegetation Index (NDVI) data as a proxy for

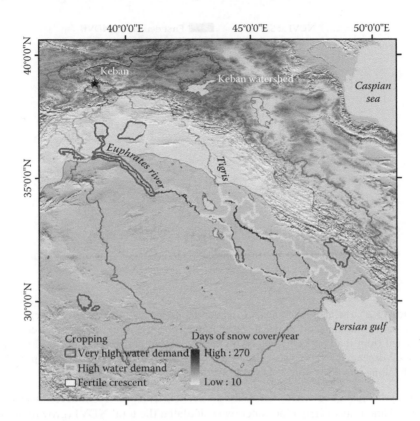

FIGURE 12.1   (See color insert following page 224.) Distribution of cultivated areas in the Euphrates–Tigris basin (light blue line). Intensely irrigated areas (green) greatly depend on river flow created in the adjacent mountain ranges. More than 80% of river flow of the Euphrates River is created in the Keban watershed (red), forming the main water resource for irrigation in the study area (black box).

seasonal/interannual variations in vegetation vigor and modeled SM. We did not calculate the actual amount of irrigation water used, but rather the relative changes in irrigation intensity during the period from 1982–2000, as explained in the following.

While natural vegetation and rain-fed crops will follow the seasonal SM cycle, areas under irrigation will continue to maintain a high NDVI value even when SM drops below a critical minimum. Accordingly, an area under irrigation is defined by two criteria: first, SM must drop below a defined threshold value, and secondly, vegetation must be in a growing stage. For the first criterion, we assumed a minimum SM of 5 mm (assumed wilting point), and for the second criterion, we used the first derivative of the NDVI cycle, where positive values that exceed a threshold of 0.01 (slope) during at least three consecutive compositing periods indicated growth; negative values indicated senescence. For consistent results in the calculation of derivatives, we used Fourier filtered NDVI cycles. This is explained in Figure 12.2, showing an annual NDVI cycle of a double crop together with its SM cycle: while SM conditions during springtime ensure continuous growth of the winter crop, the growth period of the summer crop falls entirely into the dry season. Such growth

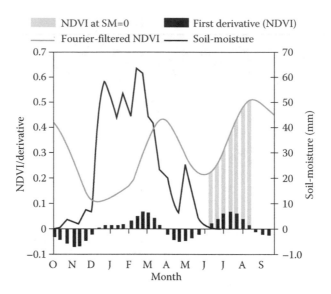

FIGURE 12.2 Estimation of annual irrigation requirements using ten-day interval NDVI cycles and ten-day interval modeled soil moisture: positive values for the first derivative, calculated from the Fourier filtered NDVI cycle, indicate periods of growth. The annual, irrigation-dependant NDVI was calculated as the accumulated NDVI during periods of growth falling into below-threshold soil moisture (SM) periods.

can only be sustained where crops are under irrigation. As an indirect measure of required amounts of irrigation water, we calculated the total NDVI (growth period) during times when SM was less than 5 mm. Irrigation was assumed to end once the NDVI cycle reached its peak and greenness began to decline. This stage is indicated by a change from positive to negative values for the first derivative. The interannual comparison of annual total NDVIs under irrigation then reveals spatiotemporal dynamics in irrigation activities.

12.2.3 DATA

Station data from the National Climate Data Center (NCDC) of daily rainfall and temperature were interpolated to create gridded layers covering the period from 1982–2000. The temporal sampling interval was chosen to match the ten-day AVHRR compositing scheme (data from 1982–2000). For data interpolation, we used a Gaussian weighting function with a flexible radius, where the total weight of stations contributing to the calculation of each value was set to 1.1.

Gridded data (1 km² resolution) of ten-day interval temperatures, precipitation, and of calculated PET, together with topography (90-m data from the Shuttle Radar Topographic Mission [SRTM], degraded to 1000 m), and an assumed mean water-holding capacity of 110 mm, form the input to our model. Model outputs include SM, surface runoff, subsurface flow, and snowpack for the entire Euphrates–Tigris basin and river discharge calculated for the Keban gauging station (Figure 12.1). For more meaningful estimates on the area under irrigation, instead of 8-km AVHRR data, we used one year (2000) of ten-day NDVI data from the Système Pour l'Observation de la Terre (SPOT) Vegetation satellite with a spatial resolution of 1 km.

12.3 RESULTS AND DISCUSSION

While locally available SM defines the general suitability of an area for agricultural activities (Figure 12.3), it is the access to irrigation water that makes an area highly productive. As illustrated in Figure 12.1, it is the adjacent mountain areas and their snowpack that feed the major rivers, providing the necessary irrigation water. Mean annual cycles of modeled SM (Figure 12.3) reveal the importance of a timely release of mountain water, nowadays regulated by six large reservoirs, to compensate for SM shortages during the dry season from about May to October.

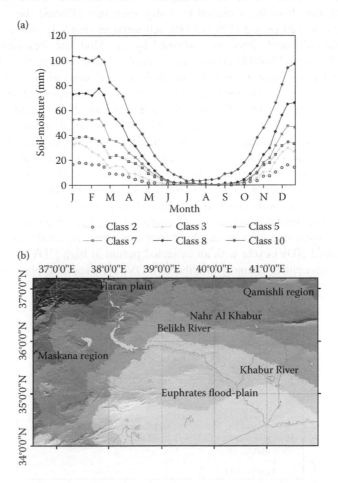

FIGURE 12.3 (See color insert following page 224.) Soil moisture zones from an unsupervised classification of modeled mean soil moistures (1982–2000) (a) with major irrigated areas outlined in green and the arable Fertile Crescent in transparent green. Corresponding annual cycles of class mean soil moisture are given in the diagram (b).

12.3.1 Trends in Irrigated Agriculture

The calculation of temporal trends in annual NDVI under irrigation, using the above-described method, revealed the spatiotemporal distribution and trends in irrigation water demands (Figure 12.4). Annual rates of increase are highest in newly established irrigation areas, like the Haran Plain (Turkey) and the Maskana Region (Syria) (Figures 12.4, 12.5, and 12.6). Areas with a long tradition in irrigated agriculture, such as the Euphrates flood-plain, show lower rates of increase. There, increases are lowest in more downstream areas, which we attribute to more widespread salinization processes, probably triggered by river water, at this point, heavily loaded with soluble salts as a consequence of its repeated use for irrigation. Slightly negative trends are visible along the Khabur River. Additionally, strong negative trends are visible west of Haran. Some weaker negative trends are also visible further to the east, which may have been caused by a dry spell that affected the region at the end of the observed period (1999–2000). All stronger trends (positive or negative) fall into the cultivated areas, as confirmed by our Moderate Resolution Imaging Spectroradiometer (MODIS) classification (Figure 12.5). As expected, there are no trends in temporal changes of water needs in natural vegetation covers (compare Figures 12.4 and 12.5). The significance of trends was not important, since we were only interested in the general trends in water demands. Some trends might actually be of very low significance, e.g., where changes in irrigation were linked to the completion of dams causing a rather sporadic than a continuous change.

Based on defined irrigation criteria, vegetation covers, such as deep-rooting trees or evergreen trees, may pretend to be under irrigation since they can preserve their greenness into the dry period, by drawing water from deeper layers. The distinction of such vegetation covers from truly irrigated crops could be achieved through a shape classifier that takes advantage of characteristic differences in the plant phenology. A typical tree (deciduous, deciduous evergreen, or coniferous) produces a plateau-shaped NDVI cycle with an extended period of high NDVI values. In contrast, crops have either normally distributed NDVI cycles or multiple peaks (multiple

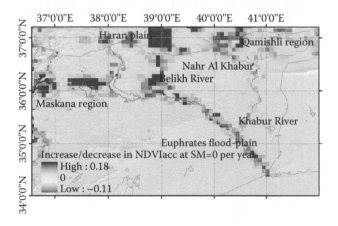

FIGURE 12.4   (See color insert following page 224.) Trends in irrigation water use measured over the period 1982–2000.

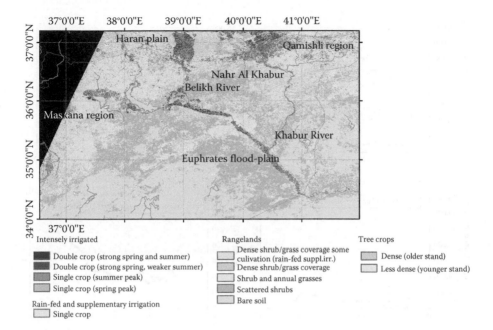

FIGURE 12.5 (See color insert following page 224.) Example of a classification of land cover/land use from a 250-m MODIS NDVI time series (year 2005, 16-day intervals) based on an algorithm that uses the shape of NDVI cycles as a primary classifier (From Geerken, R.A., An algorithm to classify vegetation phenologies and coverage and monitor their inter-annual change, *International Journal of Remote Sensing*, 2007 (In review).)

crops) with a narrow peaking period. A separation of individual cycles based on their NDVI shapes could be consistently achieved using the classification algorithm of Geerken [13] (Figure 12.5). The algorithm uses one year of NDVI data, acquired at regular temporal intervals (e.g., SPOT or MODIS), and identifies identical shapes independent from coverage variations, phenological shifts, and background. A pixel is assigned to one of several user-defined shape references based on maximum shape similarity, which is described using the Fourier components magnitude (magnitude ratio) and Phase (phase difference) [13,14]. The level of accuracy (shape fidelity) can be set according to individual user needs or data quality.

The result of the shape classification could be used not only to eliminate tree covers mistakenly identified as irrigated, but also to confirm the trend analysis in Figure 12.4, with intensely irrigated areas precisely lining up with intensely cultivated areas in Figure 12.5. All strong positive trends between 1982 and 2000 in Figure 12.4 (blue areas, marking growing irrigation demand) fall into areas where double crops could be identified from the year 2000 NDVI cycle classification (SPOT Vegetation, 1 km).

To get a better insight into the temporal development of irrigation water needs, we summarized the results of our trend analyses subdivided into various large agricultural areas, as indicated in Figure 12.1. The rising demand for irrigation water in the Haran Plain, in Maskana, and in the Euphrates flood-plain is covered by diverting water from the Euphrates River. The sharp increase in water use that we see in

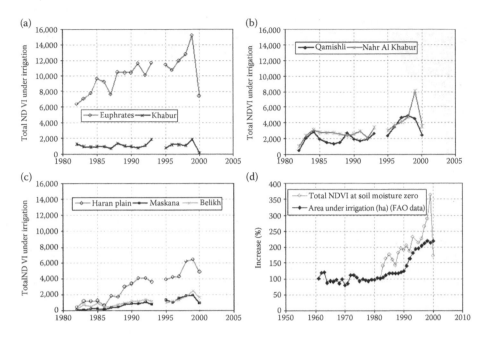

FIGURE 12.6 Development of the annual total NDVI at SM zero (here used as a surrogate for irrigation water consumption) in selected agricultural areas of southern Turkey and Syria (a, b, and c). Comparison of temporal percent changes in "Area under irrigation" (Syria) as reported by FAO with percent changes measured in "NDVI under irrigation" (d). The situation in 1982 was set to 100% for both measures. Measurements in NDVI under irrigation do not include irrigated areas in the coastal mountains and in southwest Syria.

the Haran Plain after 1997 coincides with the completion of the Karakamis dam. Qamishli and the Nahr Al Khabur areas not only draw water from tributaries of the Khabur, but also receive water through pumping from local aquifers. In both areas, irrigation has been significantly expanded. The apparent drop in irrigation water use at the end of the century in the Khabur flood-plain temporally coincides with the damming of an upstream reservoir, leaving the river dry and cutting fields off from the necessary water supply. To compensate for losses in the Khabur, farmers intensified cultivation in the Steppe, west of the Khabur, marked by weak positive trends (indicated by cyan colors in Figure 12.4) that are caused by well irrigation. According to these analyses, water demands have at least doubled or tripled over a period of 20 years in most areas. The construction of dams and irrigation canals appears to have the strongest impact on changes in irrigation activities.

12.3.2 IRRIGATED AREA ESTIMATES

When calculating percent changes in total NDVI under irrigation for the area of Syria, the result compares fairly well with percent changes in the area under irrigation as reported by the *Food and Agriculture Organization* of the United Nations (FAO). While the area under irrigation (FAO) has continuously increased since 1982, the NDVI under irrigation shows a comparable general trend, but is characterized by ups and downs, as a result of interannual climatic variations that dictate the need for the amount of irrigation . Other than climatic variations, fluctuations in productivity or cropping cycles can impact the curve of "NDVI under irrigation."

These influences are the likely cause for differences between FAO data and our model outputs that we see at the beginning of the eighties and particularly in 2000. Rare April/May precipitation in 2000 created sufficient SM that required less irrigation to produce similar amounts of green biomass compared to previous years. The total area under irrigation, of course, was not reduced, but the irrigation need according to the model was not as high. Though of only minor influence in Syria, we are reminded that a change to more water efficient breeds or to a different crop type may also be responsible for some of the variations.

While our product describes the spatial distribution of irrigated fields, we considered an 8×8 km pixel size as too coarse for realistic area estimates. Instead, we used 1-km SPOT Vegetation data to calculate the area under irrigation for the year 2000 (Figure 12.7). The actual area under irrigation can be calculated as the total number of NDVI pixels that fulfill the criteria "crop in growing stage during SM below threshold," as described in Figure 12.2. In a simplified approach, where no SMs are available, the ten-day interval SM information can be replaced by a file indicating dry and wet periods.

To prevent the result from being falsified by other vegetation covers such as trees—as explained above—any other vegetation type than field crops was eliminated using the classification of NDVI shapes. The calculated area under irrigation

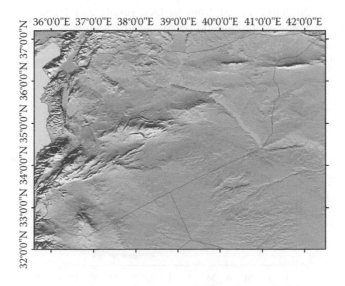

FIGURE 12.7   (See color insert following page 224.) Areas under irrigation (green) in Syria during the year 2000. The result is based on 1-km SPOT Vegetation NDVI time series and modeled soil moistures at 1 km. Not included in this analysis are possible irrigated areas in Syria falling west of longitude 36.3 and south of latitude 32.7.

for Syria (excluding some smaller areas west of longitude 36.3 and south of latitude 32.7) amounts to 1,026,920 ha compared to 1,211,000 ha for the entire Syria reported by FAO. Other than the uncovered areas, irrigated agriculture in narrow river flood plains and wadis that is difficult to detect with 1 km data, and the nearly abandoned irrigation in the Khabur flood plain during 2000, may be reasons for the lower number.

12.3.3 River Discharge

With SM dropping below a threshold of 5 mm at the latest with the beginning of June (Figure 12.3), summer crops in the Haran Plain, Maskana, and the Euphrates flood plain completely depend on river water that, to a vast majority (>80%), is created in the Keban watershed (Figure 12.1). The Keban watershed, located in the Taurus Mountains, is under a thick snow cover for several months each year. The snow-melt, typically starting in April and ending in May, creates high discharge rates that quickly drop to a much lower but continuous flow at the beginning of the summer. Any changes affecting this resource will affect its downstream users. To assess the vulnerability of current developments in irrigated agriculture, we studied the natural flow of the upper Euphrates and how it may change under the assumption of different climate change scenarios.

For modeling the current situation of river discharge, we used a set of mean climatic parameters (temperature, precipitation) calculated from data covering the period 1994–2004. Due to a series of large dam constructions started after 1972, a mean river discharge measured prior to 1972 at the Keban station was assumed to reflect the natural flow and was used to verify modeled discharge amounts (Figure 12.8). Modeled river discharge is fed from surface runoff and from subsurface flow (Appendix). Differences in temporal discharge distribution between modeled and observed data during springtime (Figure 12.8) are due to the ten-day sampling interval that we used in our model; however, temperature changes that occurred

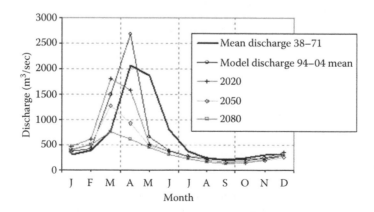

FIGURE 12.8 Measured and modeled natural river discharge at the Keban station under present climatic conditions and projected climate changes (B2a scenario).

since 1971 may also be a major cause (about 0.5°C increase according to records at the NCDC).

For modeling future changes in river discharge, we used outputs from the General Circulation Model (GCM) of the Canadian Center for Climate Modelling and Analysis (CCCma) (minimum-, maximum-, mean temperature, precipitation). The B2a scenario that we chose in the presented analysis assumed a moderate economic development, a continuously increasing global population, and reduced CO_2 emissions [10].

Modeled discharges for 2020, 2050, and 2080 (Figure 12.8) show dramatic declines in annual total discharge amounts. Since precipitation in the CCCma model runs does not change too much, but mean temperature may rise as much as 5°C by 2080, most of the discharge losses must be attributed to evapotranspiration. While temporal shifts in water availability can be regulated by dam constructions, they will be ineffective in evaporation losses. Total annual reductions in river discharge under a B2a scenario are around 47%. For the irrigation intensive period from March to July (snowmelt), river discharge will be reduced by 60% from its current flow. In a modeled A2a scenario (continuous increase in CO_2 emissions), these figures will increase to 55 and 68%, respectively.

In addition to water losses stemming from river discharge, downstream cultivated areas will suffer substantial losses in local SM caused by rising temperatures. Figure 12.9 shows mean ten-day SM losses by 2080 for selected agricultural areas. As Figure 12.9b suggests, with a near 0% change in SM during the summer months, the summer crop remains unaffected by SM changes, since these crops are already growing under near-dry conditions. However, rising evapotranspiration rates will

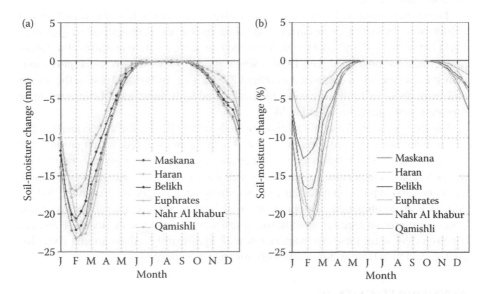

FIGURE 12.9 Modeled changes in soil moisture for a B2a-scenario in the year 2080. Changes are shown as absolute soil moisture loss (left) and as percent change (right) with respect to the current mean (1994–2004) soil moisture.

increase the demand for irrigation water during a time when river discharge is low. More dramatic are SM losses during springtime, which may trigger an increasing demand for supplementary irrigation of winter crops, putting additional pressure on depleted rivers.

A possible compensation for the projected losses in SM and irrigation water could be a shift in the cropping calendar. With rising temperatures, winter frost may no longer be a limitation and, all other parameters permitting (cloud cover, sunshine hours), the main growing period could be shifted from the spring/summer period to the winter/spring period, a time when evapotranspiration rates are still low. How much of the water losses can be compensated for by simply shifting crop calendars have to be assessed through growth models. A final conclusion for alternative crop management practices, however, depends on future interannual climate variability, which cannot be deduced from the CCCma outputs. Interannual climate variability is the ultimate control in dryland farming, defining the sustainability of any land use.

12.4 CONCLUSIONS

The spatial and temporal description of hydrologic parameters by our model proved to be of good quality and sufficiently accurate for regional to subregional assessments. This could be verified through local SM measurements (Appendix), measured river discharge, and satellite image analysis (snow cover). Major points of uncertainty include the quality of the climate data and their interpolation. More accurate information on actual water-holding capacities instead of a single default value, as used in our model, will certainly result in local modifications that may turn out to be quite substantial.

These inaccuracies, however, will not affect the trends that we found in our model runs. Evidence comes from FAO data on the development of irrigation in Syria between 1982 and 2000, which matches quite well with our results. For future projections, we only used data from a single model (CCCma) whose outputs range near the average when compared to other GCMs. To gain further trust in the results and learn about extreme scenarios, the performed hydrologic model runs should be repeated with data from other GCMs.

But even without the confirmation through other GCMs, the consequences due to rising temperatures are apparent and will force farmers to reduce irrigation water consumption. This can be done by more efficient irrigation techniques (gravity flow in open canals and surface irrigation are still most common), changes in crop management (crop type, crop calendar), or by cutting back irrigated agriculture. Which approach or combination of approaches will help to sufficiently reduce irrigation water consumption requires quantitative assessments using growth models. However, the projected water losses to evaporation will significantly reduce annual river discharge, and thus limit water availability for irrigation purposes.

ACKNOWLEDGMENTS

This study was carried out as part of the research project "The Water Cycle of the Tigris–Euphrates Watershed: Natural Processes and Human Impacts," funded through NASA (NNG05GB36G). We want to thank the reviewers for their effort and their helpful comments.

REFERENCES

1. UNEP (United Nations Environmental Programme), *World Atlas of Desertification*, Edward Arnold, London, 1992.
2. Perrin de Brichanbaut, G. and Wallen, C.C., *A Study of Agroclimatology in Semi-Arid and Arid Zones of the Near East*, Technical Note No. 56, WMO No. 141, Tp66, World Meteorological Organization, Rome, 1963.
3. Beaumont, P., Water and armed conflict in the Middle East–Fantasy or reality? in *Conflict and the Environment*, N.P. Gleditsch (Ed.), Kluwer, Dordrecht, 1977, 355–74.
4. Zaitchik, B.F. et al., Climate and vegetation in the Middle East: Inter-annual variability and drought feedbacks, *Journal of Climate*, 20, 15, 3924–41, 2007.
5. Gruen, G.E., Turkish waters regional conflict or catalyst for peace? *Water, Air, and Soil Pollution*, 123, 565–79, 2000.
6. Beaumont, P., Water policies for the Middle East in the 21st century: The new economic realities, *International Journal of Water Resources Development*, 18, 2, 315–34, 2002.
7. Allen, S.J., Measurement and estimation of evaporation from soil under sparsely barley crops in Northern Syria, *Agricultural and Forest Meteorology*, 49, 291–309, 1990.
8. Kolars, J., Water resources of the Middle East, *Canadian Journal of Development Studies*, Special Issue, Sustainable Water Resources Management in Arid Countries, 1992.
9. Wakil, M., Analysis of future water needs for different sectors in Syria, *Water International*, 18, 1, 18–22, 1993.
10. IPCC (International Panel on Climate Change), Solomon, S. et al. (Eds.), *Climate Change 2007: The Physical Science Basis*, 2007, http://ipcc-wg1.ucar.edu/wg1/wg1-report.html.
11. Evans, J.P. and Smith, R.B., Water vapor transport and the production of precipitation in the eastern Fertile Crescent, *Journal of Hydrometeorology*, 7, 6, 1295–1307, 2006.
12. De Pauw, E., *Agroecological Methods*, ICARDA, Aleppo, Syria, 2007 (In preparation).
13. Geerken, R.A., An algorithm to classify vegetation phenologies and coverage and monitor their inter-annual change, *International Journal of Remote Sensing*, 2007 (Accepted).
14. Evans, J.P. and Geerken, R., Classifying rangeland vegetation type and coverage using a Fourier component based similarity measure, *Remote Sensing of Environment*, 105, 1–8, 2006.
15. Kumar, L., Skidmore, A.K., and Knowles, E., Modelling topographic variation in solar radiation in a GIS environment, *International Journal of Geographical Information Science*, 11, 5, 475–97, 1997.
16. Eggleston, E.I. and Riley, J.J., *Hybrid Computer Simulation of the Accumulation and Melt Processes in a Snowpack*, Technical Report PRWG65-1, Utah State University, Logan, UT, 1971.
17. Caltrans, *Caltrans Storm Water Quality Handbooks,* State of California, Department of Transportation, Sacramento, CA, 2003.

APPENDIX

PET after Hargreaves (PET_{HG}) is controlled by air temperature and extraterrestrial radiation:

$$\text{PET}_{\text{HG}} = 0.0023 * I_0 * (T_{\text{mean}} + 17.8) * \sqrt{T_{\text{max}} - T_{\text{min}}} \quad (\text{mm}), \qquad (12.1)$$

where:

I_0 = top of atmosphere radiation (extraterrestrial radiation) (mm/day),
T_{mean} = mean temperature (°C),
T_{max} = maximum temperature (°C),
T_{min} = minimum temperature (°C).

The daily total top of atmosphere (TOA) radiation was calculated for each 10-day interval, where each period is represented by its mean day. TOA was corrected for topographic features (slope and aspect) using topographic data from the Shuttle Radar Topographic Mission (SRTM) degraded to 1,000-m spatial resolution. Equations for TOA radiation calculations and for radiation on inclined surfaces are given by various authors, e.g., Kumar et al. [15]. For the conversion of PET$_{(HG)}$ to more meaningful PET$_{(PENMAN)}$, we used regression equations from selected stations and those given by DePauw [12], measured for different climatic regions (Koeppen classification) (Figure 12.10).

The general equation for calculating SM is

$$\text{SM}(t) = P(t) + M(t) - \text{AET}(t) - R(t), \qquad (12.2)$$

where precipitation (P), snowmelt (M), actual evapotranspiration (AET), and runoff (R) are measured in millimeters. AET is assumed to be proportional to PET and the available SM according to

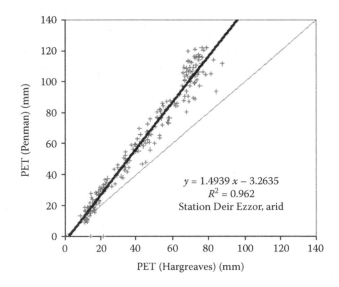

FIGURE 12.10 Example of a transfer function used to convert PET (Hargreaves) to PET (Penman) in arid climates.

$$\text{AET}(t) = \text{PET}(t)\left(\frac{\text{SM}(t)}{\text{MSM}}\right). \tag{12.3}$$

Equation 12.3 indicates that with constant PET, the AET will decrease as the soil becomes drier. The ten-day precipitation, snowmelt, and PET are assumed uniform during a ten-day interval. The time development of SM can then be written as

$$\frac{d\text{SM}}{dt} = P + M - R - F * \text{SM}, \tag{12.4}$$

where

$$F(T) = \frac{\text{PET}}{\text{MSM}}, \tag{12.5}$$

is the factor for evaporation. The term "excess" removes water in excess of the maximum soil moisture (MSM) and becomes runoff. The SM at the end of a ten-day period ($\text{SM}_{(10)}$) is given by

$$\text{SM}_{(10)} = \text{SM}_{\text{SS}} - (\text{SM}_{\text{SS}} - \text{SM}_{(0)}) * \text{EXP}(-F * t), \tag{12.6}$$

where SM_{SS} is a steady soil moisture state,

$$\text{SM}_{\text{SS}} = (P + M) / F, \tag{12.7}$$

where $\text{SM}_{(0)}$ is the SM at the beginning of the ten-day period, and t marks the end of the time interval, e.g., ten for a period of ten days. Values for P in Equation 12.7 and for PET in Equation 12.5 are a matter of chosen time steps, days, hours, seconds, or other.

A quantitative comparison of temporal soil moisture variations shows good agreement between field-measured SM and modeled SM (Figure 12.11).

The amount of subsurface flow is calculated as a percentage from the total soil moisture (for the Euphrates/Tigris watershed, 7% of modeled soil moisture was assumed to be transported to the rivers as subsurface flow).

If the temperature drops below T_{crit}, precipitation is assumed to accumulate as snow. If the temperature is above T_{crit}, accumulated snow is reduced by melting (M), using the equation of Eggleston and Riley [16].

$$M = C_{\text{m}} * (T_{\text{mean}} - T_{\text{crit}}). \tag{12.8}$$

Without considering vegetation cover as done in the original equation, the melt rate factor C_{m} (mm/°C per day) is determined by

$$C_{\text{m}} = k_m * R_{\text{I}}(1 - A), \tag{12.9}$$

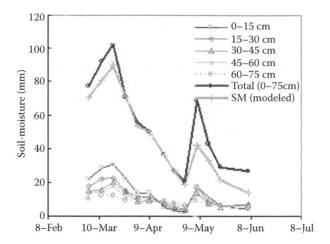

FIGURE 12.11 Comparison between modeled soil moisture and observed soil moistures. The "Total" is the sum of observed soil moisture measured at different depths (Site: Aleppo Steppe/Syria).

where k_m is a proportional constant, R_I the solar radiation index, and A the snow albedo. Albedo changes with time t (in days) described by

$$A = 0.4 * (1 = e^{-k_e t}),\qquad(12.10)$$

where the time constant k_e is 0.2. R_I is the ratio between TOA radiation on a surface to TOA radiation on this surface if it was horizontal. For periods of rain, C_m is adjusted according to

$$C_m(\text{rain}) = C_m + 0.00126 * P,\qquad(12.11)$$

where P is precipitation in millimeters.

In addition to the excess water (Equation 12.4), the amount of precipitation that becomes runoff (R) is triggered by slope according to:

$$R = C_{RO} * P.\qquad(12.12)$$

Values for C_{RO} range between 0.08 and 0.35, and depend on slope ranges as defined in *Caltrans Storm Water Quality Handbook* [17].

To calculate the time it takes surface runoff to reach a defined pour point, we used the equations from Manning for sheet flow:

$$T_t = \frac{0.007(n * L)^{0.8}}{(P_2)^{0.5} s^{0.4}},\qquad(12.13)$$

where:

T_t=travel time (h),

n=Manning's roughness coefficient,

L=flow length (ft),

P_2 =2-year, 24-hour rainfall (in.),

s=slope of hydraulic grade line (ft/ft),

and for open channel flow

$$V = \frac{k}{n}\left(\frac{A}{P}\right)^{2/3} S^{1/2},$$ (12.14)

where:

V=velocity,

k=1 when measured in meters and seconds, 1.49 when measured in feet and seconds,

S=channel slope,

n=Manning coefficient; differs for different materials,

A=area of channel (ft^2),

P=perimeter of channel (ft) (wetted perimeter),

A/P=also called hydraulic radius (Rh).

Knowing pixel size and flow direction, the calculated velocities can be converted to travel times. The criterion for open channel is a minimum number of pixels that flow into a pixel that is defined by the user.

Travel time was calculated for each pixel. Total travel time from a specific location is the sum of travel times along the flow path from that pixel to the pour point.

The calculation of subsurface flow velocities V_F in a porous medium of a distinct transmissivity k was done after Darcy

$$V_F = k\frac{h}{l},$$ (12.15)

where h is the head gradient given by terrain slope, and l is flow distance given by pixel size and flow direction. Total travel times from each pixel to the pour point are calculated as described for the surface flow.

13 Estimating Actual Evapotranspiration from Irrigated Fields Using a Simplified Surface Energy Balance Approach

Gabriel B. Senay, Michael E. Budde,
James P. Verdin, and James Rowland
U.S. Geological Survey Earth Resources
Observation and Science Centre

CONTENTS

13.1   INTRODUCTION

Food security assessment in many developing countries, such as Afghanistan, is vital because the early identification of populations at risk can enable the timely and appropriate actions needed to avert widespread hunger, destitution, or even famine. The assessment is complex, requiring the simultaneous consideration of multiple socioeconomic and environmental variables. Since large and widely dispersed populations depend on rain-fed and irrigated agriculture and pastoralism, large-area weather monitoring and forecasting are important inputs to food security assessments. The Famine Early Warning Systems Network (FEWS NET), an activity funded by the United States Agency for International Development (USAID), employs a crop water balance model (based on the water demand and supply at a given location) to monitor the performance of rain-fed agriculture and forecast relative production before the end of the crop-growing season. While a crop water balance approach appears to be effective in rain-fed agriculture [1,2], irrigated agriculture is best monitored by other methods, since the supply (water used for irrigation) is usually generated from upstream areas, farther away from the demand location.

13.2   BACKGROUND

Several researchers [3–7] have successfully applied the surface energy balance method to estimate crop water use in irrigated areas. Their approach requires solving the energy balance equation at the surface (Equation 13.1) where the actual evapotranspiration (ETa) is calculated as the residual of the difference between the net radiation to the surface and losses due to the sensible heat flux (energy used to heat the air) and ground heat flux (energy stored in the soil and vegetation).

$$LE = Rn - G - H, \tag{13.1}$$

where:

   LE = latent heat flux (energy consumed by ETa) (W/m^2),
   Rn = net radiation at the surface (W/m^2),
   G = ground heat flux (W/m^2),
   H = sensible heat flux (W/m^2).

The estimation of each of these terms from remotely sensed imagery requires quality data sets. Allen et al. [6] described well the various steps required to estimate ETa using the surface energy balance method, which employs the *hot* and *cold* pixel approach by Bastiaanssen et al. [3]. In summary, for the net radiation, data on incoming and outgoing radiation and the associated surface albedo and emissivity fractions for shortwave and longwave bands are required. The ground heat flux is estimated using surface temperature, albedo, and Normalized Difference Vegetation Index (NDVI). The sensible heat flux is estimated as a function of the temperature gradient above the surface, surface roughness, and wind speed.

13.3   RATIONALE

While solving the full energy-balance equation has been shown to give good results in many parts of the world, the data and skill requirements to solve the various terms in the equation are prohibitive for operational applications in large regions where anomalies are more useful than absolute values. In this study, we implemented a Simplified surface energy balance (SSEB) approach [8] to estimate ETa, which maintains and extends the major assumptions in the Surface Energy Balance Algorithm for Land (SEBAL [3]) and the Mapping ETa at High Resolution using Internalized Calibration (METRIC [6]). Both SEBAL and METRIC assume that the temperature difference between the land surface and the air (near-surface temperature difference) varies linearly with land surface temperature (LST). They derive this relationship based on two anchor pixels, known as the hot and cold pixels, representing dry and bare agricultural fields and wet and well-vegetated fields, respectively. SEBAL and METRIC methods use the linear relationship between the near-surface temperature difference and the LST to estimate the sensible heat flux, which varies as a function of the near-surface temperature difference, by assuming that the hot pixel experiences no latent heat, i.e., ET=0.0, whereas the cold pixel achieves maximum ET.

The SSEB modeling approach extends this assumption with a further simplification by stating that the latent heat flux (ETa) also varies linearly between the hot and cold pixels. This assumption is based on the logic that temperature difference between soil surface and air is linearly related to soil moisture [9]. On the other hand, crop soil water balance methods estimate ETa using a linear reduction from the potential ET, depending on soil moisture [2,10]. Therefore, we argue that ETa can be estimated by the near-surface temperature difference, which in turn is estimated from the land surface temperatures (LSTs) of the hot and cold pixels in the study area. In other words, while the hot pixel of a bare agricultural area experiences little ET and the cold pixel of a well-watered irrigated field experiences maximum ET, the remaining pixels in the study area will experience ET in proportion to their LST in relation to the hot and cold pixels. This approach can be compared to the Crop Water Stress Index (CWSI) first developed by Jackson [11]. The CWSI is derived from the temperature difference between the crop canopy and the air. Dividing the current temperature difference with known upper and lower canopy air temperature values creates a ratio index varying between 0 and 1. The lower-limiting canopy temperature is reached when the crop transpires without water shortage, while the upper-limiting canopy temperature is reached when the plant transpiration is zero due to water shortage [12]. In this study, the *cold* and *hot* anchor LST pixel values can be compared to the equivalent of the lower- and upper-limiting canopy temperatures, respectively, of the CWSI method.

13.4 GOALS

The main objective of this study was to produce ETa estimates using a combination of ET fractions (ETfs) generated from (Moderate Resolution Imaging Spectroradiometer) MODIS thermal imagery and global reference ET data over known irrigated fields in Afghanistan.

13.5 METHODS

13.5.1 STUDY AREA

The study site is located in the Baghlan province of North-central Afghanistan, as shown in Figure 13.1. A polygon was defined around an irrigated area (Figure 13.1) using a combination of Landsat and MODIS data sets. The total area of the polygon is approximately 62,400 hectares and consists of both well-vegetated and sparsely vegetated areas, with some arid/semiarid areas at the periphery of the polygon.

13.5.2 DATA

The primary data sets for this study were derived from the MODIS sensor flown onboard the Terra satellite. MODIS LST data were used to calculate the crucial ETfs explained in Section 13.5.3. Additionally, MODIS NDVI data were used for irrigated area delineation and identifying highly vegetated vs. sparsely vegetated areas within the agricultural zone. The global reference ET data were obtained from the archives of United States Geological Survey (USGS)/FEWS NET operational model outputs. Each data set is further described below.

13.5.2.1 MODIS Land Surface Temperature

Thermal surface measurements were collected from the MODIS eight-day LST/ Emissivity (LST/E) product (MOD11A2). The MODIS instrument provides 36 spectral bands, including 16 in the thermal portion of the spectrum. The LST/E

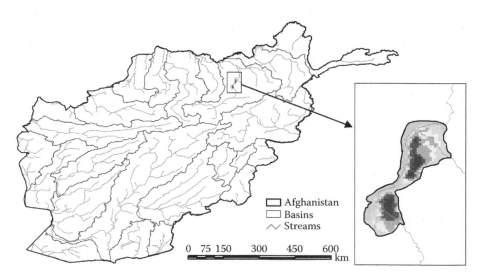

FIGURE 13.1 Study site showing the irrigated fields in the Baghlan province with a network of streams and drainage basins. The streamflow within the basin that includes our study area is northward, originating from the central highlands.

products provide per-pixel temperature and emissivity values at 1-km spatial resolution for eight-day composite products and 5-km resolution for daily products. Temperatures are extracted in degrees Kelvin with a view-angle dependent algorithm applied to direct observations. This study utilized eight-day average daytime LST measurements for eight-day composite periods throughout the growing season. Table 13.1 shows the MODIS composite day representing the first day of the eight-day MODIS LST composite period and corresponding calendar dates. Although MODIS thermal data are available on a daily time step, we used the average daytime LST measurements for eight-day composite periods in order to minimize the effects of cloud contamination on daily observations. We believe that the eight-day composites do not compromise our ability to accurately represent surface temperature conditions that occur before and after irrigation applications during the composite period.

13.5.2.2 MODIS Vegetation Index (VI)

MODIS Vegetation Index (VI) products use reflectance measures in the red (620–670 nm), near infrared (841–876 nm), and blue (459–479 nm) bands to provide spectral measures of vegetation greenness. The MODIS VI products include the standard NDVI and the Enhanced Vegetation Index (EVI). Both indices are available at 250-m, 500-m, and 1-km spatial resolution. The primary difference between the two indices is that EVI uses blue reflectance to provide better sensitivity in high biomass regions. Since this study concentrated on irrigated agriculture in an otherwise dryland environment, we used the NDVI product at 250-m resolution for this analysis.

These data are distributed by the Land Processes Distributed Active Archive Center, located at the U.S. Geological Survey's EROS http://LPDAAC.usgs.gov.

13.5.2.3 Reference ET

The global one-degree reference ET (ETo), based on the six-hourly Global data assimilation systems (GDAS) model output, is calculated daily at the Earth Resources Observation and Science (EROS) Center on an operational basis [13]. The

---

TABLE 13.1

MODIS Composite Day and Corresponding Calendar Dates for each Growing Season

MODIS Composite Day	161	177	193	209	225	241
2000, 2004	9–16 June	25 June to 2 July	11–18 July	27 July to 3 August	12–19 August	28 August to 4 September
2001, 2002, 2003	10–17 June	26 June to 3 July	12–19 July	28 July to 4 August	13–20 August	29 August to 5 September

GDAS ETo uses the standard Penman–Monteith equation, as outlined in the FAO publication for short-grass reference ETo by Allen et al. [10]. The feasibility of using the GDAS ETo for such applications was recommended by Senay et al. [13] after a comparison with station-based daily ETo showed encouraging results with r² values exceeding 0.9. Daily global reference ET values were available for all days between 2001 and 2004. For 2000, the daily reference ET values were not complete. This study used six thermal image dates during the main growing season. Three of the six image dates in 2000 did not have a corresponding reference ET. For the missing time periods, the average reference ET from 2001 to 2004 was used.

13.5.3 DESCRIPTIONS OF METHODS

A set of three hot and three cold pixels were selected for each eight-day composite period for each year of the growing season data. An average of the three pixels was used to represent the *hot* and *cold* values throughout the study area. The pixels were selected using a combination of MODIS 250-m NDVI, MODIS LST, and Landsat ETM+imagery when available. For a given time period, cold pixels, representing well-vegetated and well-watered crops, were selected based on either visual interpretation of the Landsat imagery or high values in the MODIS NDVI. Similarly, hot pixels representing low-density vegetation and relatively dryland were identified either visually or by selecting pixels with very low NDVI values. LST data were used to verify that the selected pixels adequately represented the temperature contrast within the study area for each eight-day composite period.

LST values for each of the six pixels (three hot, three cold) were extracted using ArcGIS software [14]. The resulting database files were imported into an Excel spreadsheet, where average hot and cold pixel values were calculated.

Since we know that hot pixels experience very little ET and cold pixels represent maximum ET throughout the study area, the average temperature of hot and cold pixels could be used to calculate proportional fractions of ET on a per-pixel basis. The ET fraction (ETf) was calculated for each pixel by applying Equation 13.2 to each of the eight-day MODIS LST scenes.

$$ETf = \frac{TH - Tx}{TH - TC},\qquad(13.2)$$

where TH is the average of the three hot pixels selected for a given scene, TC is the average of the three cold pixels selected for that scene, and Tx is the LST value for any given pixel in the composite scene.

The ETf formula was applied to the six eight-day growing season composites for each year, resulting in a series of six images per season. The images contained ETfs for each pixel that were used to estimate ETa throughout the growing season.

The ETf is used in conjunction with a reference ET (ETo) to calculate the per-pixel ETa values in a given scene. The reference ETo is calculated on both daily and dekadal (ten-day) time steps for the globe at one-degree spatial resolution. The dekadal period that corresponded closest to the eight-day MODIS composite period was used to extract average daily ETo values by dividing the dekadal sum into ten daily values. Since our analysis uses every other eight-day period between June and September, each composite period essentially represents the 16 days between observations. Thus, the average daily reference ETo values were multiplied by 16 to provide a summation of ETo for each 16-day period. The calculation of ETa was achieved using the following formula:

$$ETa = ETf * ETo. \tag{13.3}$$

This simplified energy balance approach allowed us to use known reference ET at a coarse spatial resolution of one-degree to derive spatially distributed ET measurements based on LST variability at 1-km resolution. Improvements in the spatial representation of ET distribution during the growing season can provide important insights into the extent of irrigated crop areas and the quality of the growing season.

Actual crop ET for the five-year period, 2000–2004, was used to assess the quality of each growing season in the North-central Afghanistan study area. Using a mask of the irrigated crop area, described in Figure 13.1, we used the actual 1-km gridded ET values to calculate spatially averaged ET for each season. Furthermore, the results from the SSEB model were compared to a watershed-based analysis of an operational FEWS NET irrigation supply and demand model. The irrigation supply and demand model has been using satellite-derived rainfall and GDAS ETo for modeling growing seasons since 1996, and average supply and demand is determined using monthly climatological data defined using a 30-year monthly rainfall and potential ET from 1961 to 1990 (www.cgiar.org/iwmi/WAtlas/atlas.htm). The results and implications of these comparisons are outlined in detail in the following sections.

13.6 RESULTS AND DISCUSSIONS

The results of the analysis are presented in Figures 13.2 through 13.7. Figure 13.2 shows the hot and cold pixel values for the six time-periods used for 2003. Similar temporal patterns were also observed in the other years. For 2003, the hot and cold pixels were separated by an average of approximately 15°C throughout the season between June and August. Furthermore, they appeared to increase or decrease in the same direction by about the same magnitude during the peak portion of the crop-growing season. However, this separation approaches close to 0.0 (data not shown) during the off-seasons, particularly close to the start of the season. The existence of such temporal patterns between the hot and cold pixels is believed to be potentially useful in detecting time of start of season for crop-monitoring activities. Figure 13.2 shows the boundary (extreme) conditions for surface temperature distribution in the study site for each of the growing seasons. By properly selecting the extreme temperature areas representing the hot (dry/bare) and the cold (wet/vegetated) land areas, the remaining pixels in the study area will fall in between these temperature values. The two extreme temperatures also represent extremes in ET values. The range of these values varies from zero ET for the *hot* and dry areas to a high ET, comparable

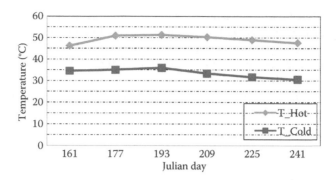

FIGURE 13.2   Temporal variation of the average *hot* and *cold* pixel land surface temperature values during the 2003 crop growing season in °C.

to a reference ET, for the *cold* and wet areas [6]. In this study, we extended this assumption to include the remaining pixels by suggesting that pixels having LST values in between these extremes will experience an ET value in direct proportion to the ETf, as shown in Equation 13.2. One of the challenges of this method is the subjectivity of selecting the hot and cold pixels that truly represent the effect of soil moisture and prevailing climatic conditions of the study area. Although it is harder to find hot pixels with zero evaporation, this problem is less pronounced in irrigated areas than in rain-fed areas where soil moisture can be replenished by rainfall during the eight-day period. We recommend the approach of Allen et al. [6] to determine ETa of the hot pixel using a water balance model for bare soil if rainfall has occurred during the eight-day period.

Figure 13.3 shows the temporal trend of spatially averaged ETfs during the peak-growing season for each of the 5 years used in this study. The ETfs showed both intraseasonal and interseasonal variability of up to 20% from their respective mean values. The major separation between the different years was shown in the middle of the peak season for composite periods beginning on days 193 and 209 (Figure 13.3) when the spatially averaged ETf appears to be lowest. The relatively high ETf in the beginning of the season may be explained by the fact that there is more higher ground cover condition across the study site with the presence of other vegetation that takes advantage of the spring rainfall in addition to crops planted for irrigation. As shown in Figure 13.2, the separation between the *hot* and *cold* pixels on day 161 is the smallest. This suggests that early in the season there are more pixels closer to the *cold* pixel than the *hot* pixel when compared to the remainder of the season. This may be due to the accumulated moisture from the preceding spring, even in nonirrigated areas that would run out of moisture during the remainder of the summer unlike the irrigated areas.

Figure 13.4 shows daily reference ET values that correspond to ETfs from Figure 13.3. For the majority of the points, the daily reference ET varied between 6 and 8 mm per day. Unlike the ETf in Figure 13.3, the reference ET values did not show a marked difference from year to year in any particular time period, with the exception of the 2003 reference ET, which showed distinctly higher ETo values for days 193 and 209. The 2003 data suggest that the region benefited from two important

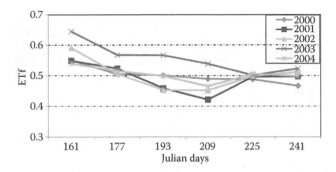

FIGURE 13.3 Temporal patterns of spatially averaged ET fractions for peak-season (June 10 to August 29) 2000–04.

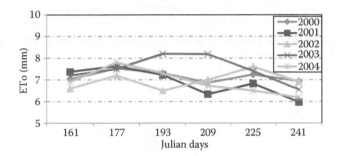

FIGURE 13.4 Temporal patterns of spatially averaged daily reference ET for peak-season (June 10 to August 29) 2000–04.

factors during the 2003 crop-growing season: (1) good water supply as evidenced by the high ETfs; and (2) good energy supply and vapor transport mechanism (clear sky, favorable wind) as evidenced by the high reference ET, which is mainly a measure of the available energy and vapor transport mechanism under optimum water supply conditions.

The actual crop ET is estimated from the product of ETf and the reference ET, as shown by Equation 13.3. Figure 13.5 shows the temporal patterns of the actual crop ET for the five periods. Each data point represents a 16-day ETa estimate that is spatially averaged over the study area. Due to the small size of the study area compared to the spatial resolution of the reference ET, all pixels in the study area have an identical reference ET value for each of the six image dates. Thus, the spatial variation in the ETa is a result of the spatial variation of the ETfs. The spatially averaged seasonal ETa magnitudes are presented in Figure 13.6 to illustrate the year-to-year variability in ETa. Figure 13.6 highlights the fact that 2003 was the best agricultural season during the five-year period, which is corroborated by various field reports that the cycle of three consecutive drought periods, between 2000 and 2002, was alleviated by good precipitation in the 2003 season. These results are comparable to a watershed-based analysis of an operational FEWS NET irrigation supply and demand model output, showing that 2002 and 2004 had below-average supply while the 2003 irrigation water supply met the average demand.

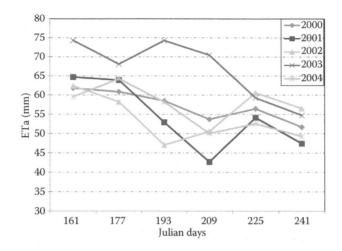

FIGURE 13.5 Temporal patterns of spatially averaged 16-day actual ET for peak-season (June 10 to August 29) 2000–04.

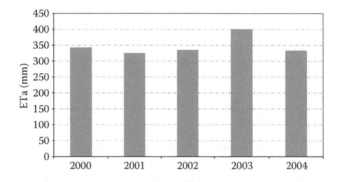

FIGURE 13.6 Spatially averaged peak season (96 days: June 2 to September 6) actual crop ET (mm) from 2000 to 04.

While Figures 13.3 through 13.6 show spatially averaged temporal variation of the ETfs, reference ET, and ETa, Figure 13.7a and b present the spatial variation of the seasonal ETa for 2002 and 2003, respectively. As shown in Figure 13.6, the higher values of the 2003 seasonal ET are illustrated by the expanded extent of higher ETa classes (dark green colors) compared to the 2002 ETa map. A similar pattern of expansion/reduction in greenness for the corresponding years was observed from MODIS seasonal maximum NDVI data. The reduction in ETa values in 2002 has mainly occurred in the downstream irrigated fields (northern fields) where a short tributary joins the main river. The geographic area where lower ETa values were observed seems to suggest downstream irrigators/fields would have access to water only if there was a surplus over the demands of the upstream users. This is more in line with a common practice in regions where water rights are not well established or regulated. Furthermore, Figure 13.7a and b suggest the possibility of using this method of analysis to estimate harvested irrigated areas for a given year based on a threshold of ETa required for successful crop growth. For example,

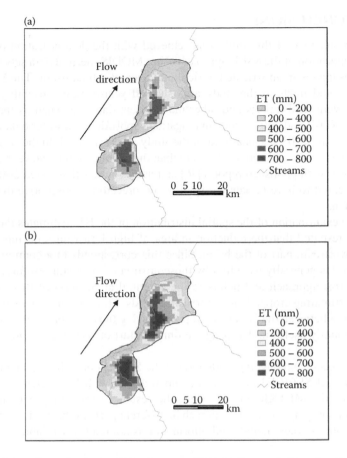

FIGURE 13.7 **(See color insert following page 224.)** (a) Seasonal (June 2 to September 6) actual ET distribution in irrigated fields of the study area in 2002. Arrow indicates the general south-north flow direction of the streams. (b) Seasonal (June 2 to September 6) actual ET distribution in irrigated fields of the study area in 2003. Arrow indicates the general south-north flow direction of the streams.

in the study area where certain fields consumed up to 700 mm in about 3 months, areas that only used half (350 mm) of the water demand by a well-watered crop could be considered as unsuccessful and removed from harvested irrigated areas [1,2,15]. Although the accuracy of the magnitudes of the estimated ETa values requires field validation using other methods or field studies, the methodís relative performance in terms of capturing the year-to-year variability suggests that the method has a potential to characterize irrigated field crop performance in relative terms on reasonably homogeneous flat irrigated fields. Although seasonal ETa values can be a surrogate for crop production, historical crop yield and production data are required to develop a statistical relationship between the two. Without field data, we can only present relative changes in water use that would also indicate directional changes in production. In order to improve the accuracy of ETa magnitude estimates, a method is being developed to downscale the coarser reference ETo to 10 km.

13.7 CONCLUSIONS

The main objective of this study was achieved with the demonstration of the successful application of the SSEB approach using MODIS thermal data sets in producing ETa estimates in an irrigated agricultural area of Afghanistan. The ETa values of the irrigated fields in the study area showed year-to-year variability that was consistent with field reports and other independent data sets, such as the seasonal maximum NDVI and output from an irrigation supply/demand water balance model. Particularly, the 2003 seasonal ETa of the study area was much higher than the rest of the studied years, and about 15% more than the average of the five years. This was in agreement with published reports [16] and news sources that stated 2003 precipitation appeared to have broken the spell of the preceding consecutive dry years in Afghanistan.

A close examination of the spatial distribution of the ETa estimates during 2003 and 2002 revealed that the reduction in area of high ETa values during 2002 was in the downstream part of the basin. Since this corresponds to a common practice where water is generally used first by those upstream, the result reinforces the reliability of this approach and points to the potential application of this method for spatially estimating cropped area in irrigated fields. The existence of a variable temporal pattern between the *hot* and *cold* pixels during the crop season is also believed to be potentially useful in detecting the time of start of season for crop-monitoring activities.

A more recent and ongoing evaluation of the SSEB approach in the United States has shown that SSEB ETa estimates compared well with Lysimeter data and ETa output from the METRIC model. Thus, the results of this study demonstrate the potential of applying the SSEB approach in different parts of the world, especially in remote locations where field-based information is not readily available.

REFERENCES

1. Verdin, J. and Klaver, R., Grid cell based crop water accounting for the famine early warning system, *Hydrological Processes*, 16, 1617–30, 2002.
2. Senay, G.B. and Verdin, J., Characterization of yield reduction in Ethiopia using a GIS-based crop water balance model, *Canadian Journal of Remote Sensing*, 29, 6, 687–92, 2003.
3. Bastiaanssen, W.G.M. et al., A remote sensing surface energy balance algorithm for land (SEBAL): 1) Formulation, *Journal of Hydrology*, 212, 213, 213–29, 1998.
4. Bastiaanssen, W.G.M. et al., SEBAL model with remotely sensed data to improve water resources management under actual field conditions, *Journal of Irrigation and Drainage Engineering*, 131, 1, 85–93, 2005.
5. Bastiaanssen, W.G.M., SEBAL-based sensible and latent heat fluxes in the irrigated Gediz Basin, Turkey, *Journal of Hydrology*, 229, 87–100, 2000.
6. Allen, R.G. et al., A Landsat-based energy balance and evapotranspiration model in Western US Water Rights Regulation and Planning, *Journal of Irrigation and Drainage Systems*, 19, 3–4, 251–268, 2005.
7. Su, H. et al., Modeling evapotranspiration during SMACEX, comparing two approaches for local and regional scale prediction, *Journal of Hydrometeorology*, 6, 6, 910–22, 2005.

8. Senay, G.B. et al., A coupled remote sensing and simplified surface energy balance approach to estimate actual evapotranspiration from irrigated fields, *Sensors*, 7, 979–1000, 2007.

9. Sadler, E.J. et al., Site-specific analysis of a droughted corn crop, water use and stress, *Agronomy Journal*, 92, 403–10, 2000.

10. Allen, R.G. et al., *Crop Evapotranspiration*, Food and Agriculture Organization of the United Nations, Rome, Italy, 1998.

11. Jackson, R.D., Canopy temperature and crop water stress, in *Advances in Irrigation*, D. Hillel (Ed.), Vol. 1., Academic Press, New York, 1982, 43–85.

12. Qiu, G.Y. et al., *Detection of Crop Transpiration and Water Stress by Temperature Related Approach under Field and Greenhouse Conditions*, Department of Land Improvement, National Research Institute of Agricultural Engineering, Tsukuba, Ibraki, Japan, 1999.

13. Senay, G.B. et al., Global reference evapotranspiration modeling and evaluation, *Journal of American Water Resources Association*, 44, 3, 1–11, 2008.

14. ESRI (Environmental Systems Research Institute), *ArcGIS 9.0.*, Environmental Systems Research Institute, Redlands, CA, 2004.

15. Doorenbos, J. and Pruitt, W.O., *Crop Water Requirements*, FAO Irrigation and Drainage Paper No. 24, Food and Agriculture Organization of the United Nations, Rome, Italy, 1977.

16. FEWS NET (Famine Early Warning System Network), *Afghanistan Monthly Food Security Bulletin*, September, 2003 (www.fews.net).

14 Satellite-Based Assessment of Agricultural Water Consumption, Irrigation Performance, and Water Productivity in a Large Irrigation System in Pakistan

Mobin-ud-Din Ahmad
Commonwealth Scientific and Industrial
Research Organization

Hugh Turral
On the Street Productions

Aamir Nazeer and Asghar Hussain
International Water Management Institute

CONTENTS

14.1 INTRODUCTION

Many of the chapters in this book deal with remote-sensing-based techniques for mapping irrigated area, primarily aimed at better estimating production, current water use, and future potential for development. However, such measures are all essentially *indirectly* calculated from estimated area and secondary data on the crop pattern, water supply, or evapotranspiration (ET) calculated by well-known techniques, such as the Penman–Monteith Procedure [1]. It is highly desirable to be able to calculate water consumption and production data directly, and to account for their temporal and spatial variability. This chapter deals with the use of both remote sensing to directly estimate water consumption by irrigated crops, and secondary agricultural statistics to consequently derive irrigation performance and water productivity (WP) indicators.

14.2 ET

Molden [2] notes that, at basin and system scales, it is more relevant to measure the depletion of water supplied to crops, rather than the supply itself, as was the accepted convention in the analysis of irrigation efficiency. Where there is no degradation of water quality, return flows can be reused within irrigation systems and further downstream in a river basin [3–5].

Conventional techniques to estimate ET, based on point meteorological measurements, do not adequately describe spatial variation in ET at broad spatial scales, such as river basins or within large irrigation systems [6], which are increasingly recognized as the management unit for irrigation and other agricultural water use [7].

Energy balance techniques that use meteorological data (such as Penman–Monteith) estimate reference ET and then require knowledge of crop type and growth stage to convert this to a potential value for a well-watered crop, with otherwise perfect growing conditions [1]. In typical field conditions, soil moisture may be limiting and other factors, such as plant disease, may result in actual evapotranspiration (ET_a) being less than the calculated potential. Assessment of irrigation performance has generally been assumed to require knowledge of water supplied as well as an estimate of the amount depleted. Collating information on water supply for a large irrigation system, composed of a large number of primary, secondary, and tertiary channels, is most often a very difficult task. In many countries, such detailed information is often ill-managed, inconsistent, and unreliable. Estimation and recoding of information on actual water depletion are not practiced yet, despite the advances in available methods.

ET_a and water depletion can be estimated from satellite remote sensing [6,8–12]. Such methods provide a powerful means to compute ET_a from an individual pixel to an entire raster image. Emerging developments in the field of remote sensing make it possible to overcome information limitations on soil water status, the actual evaporative depletion, and estimation of net groundwater use for agriculture [13,14].

Gowda et al. [15] give an excellent summary of the development, use, and limitations of energy balance methods (SEBAL, SEBS, SEBIS, METRIC, and so on) for determining ET_a using remote sensing. These techniques appear to be efficient, economical, and standardized means of observing regional ET_a at daily and seasonal time scales. Computation of seasonal ET_a is limited by the frequency of satellite overpasses and corresponding cloud-free conditions.

Daily National Oceanic and Atmospheric Administration/ Advanced Very High Resolution Radiometer (NOAA-AVHRR) and/or Moderate Resolution Imaging Spectroradiometer (MODIS) data are most commonly used for this reason, not least because they are free. However, because the thermal band data from these sensors (used to determine surface temperature, a critical input to energy balance models) have a 1-km pixel resolution, they have limited use for direct irrigation scheduling or for the analysis of small irrigation schemes. Cloud cover can still present problems during monsoonal periods in Asia, and time-based composited surface temperature products cannot be used in the energy balance procedures. Currently, it is not possible to gain more than a snapshot of ET at finer scales (using Landsat TM or ASTER data, with thermal band resolutions of 120 and 90 m, respectively), as the availability and likelihood of repeat pass data to accumulate a seasonal picture are limited.

Among various surface energy balance models, the Surface Energy Balance Algorithm for Land (SEBAL) is one of the most widely used models to assess the ET from cropped lands [16]. SEBAL results have been validated in a number of countries using field instruments, including weighing lysimeter, scintillometer, Bowen ratio, and eddy-correlation towers, indicating that daily ET estimates have errors of about 16% or lower at 90% probability (www.waterwatch.nl). However, there is continuing debate about the absolute accuracy of the technique under different agroclimatic conditions [17], and model estimates are very difficult to validate at irrigation-system or river-basin scale [18,19]. Nevertheless, the spatial patterns and trends derived are accepted as consistent and useful. The technique is most appropriate for large, relatively homogenous, irrigated areas, such as those in the Indus Basin in Pakistan.

14.3   WP

As competition for water increases, both intersectorally and within agriculture itself, WP becomes an increasingly useful indicator of how well water is being used [20]. The spatial dimension to this issue is also increasingly important, as the means to improve WP lies substantially at farm scale in terms of crop management and at system scale in terms of irrigation water distribution and, possibly, allocation [21].

WP as an indicator requires an estimate of production in physical or financial terms (numerator) and an estimate of water supplied or depleted as ET (denominator). Although it is possible to estimate crop production directly from remote sensing, through regression on the Normalized Difference Vegetation Index (NDVI) [22,23], or through more complex estimation of biomass accumulation [24,25], such techniques are either empirical and therefore locally confined, or immature in terms of linking biomass accumulation to actual yield, as the harvest index varies considerably between crops, varieties, and growing conditions [26].

When direct measures of crop production are not available, it is possible to aggregate production data across different cropping systems in monetary terms ($), using available crop statistics. Although these statistics are usually available only at the scale of large aggregated administrative units, such as districts or provinces, these can be transformed to suitable hydrographic levels and combined with remote-sensed estimates of water consumption to provide general disaggregated estimates of WP that have significance at the irrigation-system level. This chapter provides an example for Pakistan.

Pakistan maintains a high population growth rate (about 2%) in an already densely populated region of South Asia, and is rapidly urbanizing. Historically, the vast irrigation systems of the Indus Basin were developed to drought-proof the region, prevent famine, provide tax revenue, and settle a partly nomadic population. Almost all the water available in the system is committed to agriculture, and there has been a revolution in groundwater use over the last 20 years as private pumping displaced public-owned pumped drainage. This has changed the nature of irrigation from being "protective" to being "productive." In the Rechna Doab, 98% of water use is conjunctive (from groundwater and surface water), which is typical of the upper Indus Basin today. In general, productivity is low due to erratic canal supplies and salinity in the groundwater. However, the best farmers in the best conditions are highly productive [27]. High rates of groundwater development and use have occurred, particularly through a drought period from 1999 to 2002.

There are a number of well known and emerging irrigation management objectives and needs in Pakistan. In the Punjab, large quantities of irrigation flows are derived from unaccounted groundwater and there are fears of long-term overabstraction and also of degradation due to salinity mobilization from existing saline areas. The surface system is supplied by snowmelt from the Himalaya, and varies with snowfall and glacier melt behavior, which is not thought of as being severely modified by global warming. In addition to supply side challenges, water distribution in Pakistan is complex and easily subject to manipulation. Although groundwater use is widespread, surface water is highly valued for its good quality, but equity in distribution is well known to be poor, with tail-enders suffering irregular and limited deliveries. Surface water and groundwater interaction/linkages and their quantification at basin scale are not well understood, but they underpin the long-term sustainability of irrigated agriculture in the region. Remote sensing, Geographic Information System (GIS), and geostatistical approaches, along with limited field data, have been used to estimate the distributed water balance and net groundwater use for agriculture [5].

In 2006, the Government of Punjab launched a new program to maintain a computerized database for irrigation releases, to improve irrigation management, reduce rent seeking, increase transparency, and demonstrate to users who is getting how much water (http://irrigation.punjab.gov.pkk). These initiatives are expected to improve data management and availability of surface supplies, yet the information on overall water consumption (surface water and groundwater) at various scales will be essential for judicious and efficient water resources management in Pakistan. There is a need to study the water distribution and consumption patterns, and their impacts on productivity.

It is also very useful for policy makers to be able to link water allocation and irrigation system management performance to productivity in order to address the continuing challenge of feeding the population. Better estimates of crop area and actual water consumption are required, since surface water supplies are not only used directly in the field, but also provide a substantial, but unquantified, portion of groundwater recharge. Remotely sensed estimates of ET_a reflect factors that reduce water consumption below potential, such as salinity, crop condition, and poor irrigation scheduling—all of which are important in Pakistan. These remote-sensing methods also allow spatial and temporal variation to be analyzed and provide better understanding of the effectiveness and equity of system operation (in conjunction with delivery data), and of the role and extent of groundwater use, either conjunctively or on its own.

14.4   OBJECTIVES

The objectives of the work included:

- Application of the SEBAL approach to estimate ET_a in Punjab, with particular focus on the Rechna Doab, an approximately 3 million hectares (Mha) of alluvial plain that lies between the Ravi and Chenab rivers.
- Quantification of agricultural water consumption in different irrigation subdivisions of the Rechna Doab.
- Understanding the seasonal and spatial patterns of ET in the Rechna Doab and their linkage to groundwater use and quality.
- Completion of irrigation and water use performance diagnoses in terms of crop WP and water consumption at nested scales, using secondary agricultural statistics.

14.5   MATERIALS AND METHOD

14.5.1   RESEARCH LOCALE

This study was conducted for the Punjab in Pakistan, with particular focus on the Rechna Doab, a land unit lying between the Ravi and Chenab rivers of Pakistan (Figure 14.1). The gross area of the Doab is about 2.97 Mha, around 80% of which is currently cultivated. It has a maximum length of 403 km, maximum width of 113 km and lies between longitude 71°48′–75°20′ E and latitude 30°31′–32°51′ N. Climatologically, the area is subtropical and designated as semiarid, and is characterized by large seasonal fluctuations in temperature and rainfall. Summers are

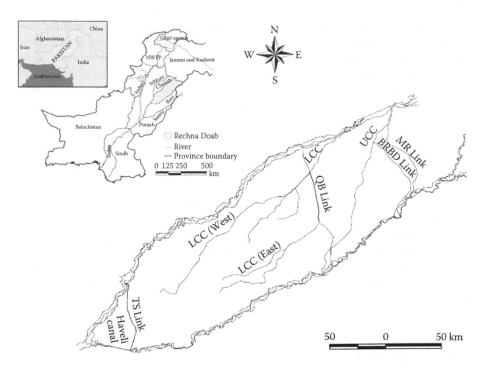

FIGURE 14.1 Location of the Rechna Doab, Pakistan, and configuration of the canal network.

long and hot, lasting from April through September, with temperatures ranging from 21°C to 49°C. The winter season lasts from December through February, with maximum temperatures ranging between 25°C and 27°C. The mean annual precipitation is about 650 mm in the upper Doab, falling to 375 mm in the central and lower areas. Nearly 75% of the annual rainfall occurs during the monsoonal season, from mid-June to mid-September.

The prevailing temperature and rainfall patterns govern two distinct cropping seasons: *kharif* (summer season) and *rabi* (winter season). The Rechna Doab falls under the rice-wheat (upper parts) and sugarcane-wheat agroclimatic zone (the middle to the lower part of the Rechna Doab) of the Punjab Province, with rice, cotton, and forage crops dominating in kharif, whereas wheat and forage are the major crops in rabi. In the central Rechna Doab, sugarcane is the dominant crop, which is a long-season annual crop with a duration of around 11 months. Minor crops include oil seeds, vegetables, and orchards.

As the total crop water requirement is more than double the annual rainfall, it is obvious that irrigation is essential to maintain the current level of agricultural productivity. Irrigation water was originally provided through a network of irrigation canals and then supplemented with unaccounted irrigation supplies from groundwater.

The canal irrigation in the Rechna Doab was introduced in 1892 with the construction of the Lower Chenab Canal (LCC) irrigation system, which draws water from the Chenab River [28]. The original design objective of the irrigation system was to spread the limited water over a large area, supporting a cropping intensity of approximately 65%, to protect against crop failure, prevent famine, and generate employment and revenue. Before the introduction of surface irrigation systems, the groundwater table was about 30 m in the Punjab Province (and about 12–15 m deep in the Sindh Province). The only sources of groundwater recharge were rivers, seasonal floods, and rainfall, and a steady natural hydrological balance was maintained between the rivers and the groundwater table. However, massive and widespread surface water irrigation development in the nineteenth and twentieth centuries altered the natural hydrological balance, due to increased recharge from earthen canals and irrigated fields. Over the years, persistent seepage from this huge gravity flow system has gradually raised the groundwater table [28,29]. By the middle of the last century, at some locations, the groundwater had risen to the surface or very close to the root zone, causing waterlogging and secondary salinity, badly affecting agricultural productivity. While describing these negative impacts of irrigation development in the Indus, the scientific literature has tended to neglect the massive and beneficial freshwater recharge and storage that have occurred in the highly permeable unconfined aquifer of the Indus Basin system. As a result, surface supplies are augmented by groundwater irrigation, initially developed by the government as part of a vertical drainage program (SCARP), starting in the 1960s and greatly increased by private sector investment over the ensuing 25 years [27]. With additional irrigation supplies from groundwater, cropping intensities have increased to 150% in some areas over the last two to three decades, and today groundwater has become a key input in agricultural production.

The Rechna Doab has both nonperennial and perennial irrigation systems. A non-perennial irrigation system is located in the upper Rechna Doab, an area underlain by fresh groundwater. The Marala-Ravi (MR) (internal) and the Upper Chenab Canal (UCC) serve this area; the Bambanwala-Ravi-Bedian-Depalpur (BRBD) (internal) provides water only in the kharif season. The middle and the lower Rechna receive perennial canal supplies from the LCC System and the Haveli Canal. The entire canal irrigation system is divided into 28 irrigation subdivisions, the lowest administrative units of irrigation management (Figure 14.2). About 0.45 Mha of the Rechna Doab region lies outside the canal command area, and most of this area is cultivated through groundwater irrigation, rather than being rain-fed.

14.5.2 METHOD FOR ET COMPUTATION

SEBAL is used for ET_a calculations. SEBAL is an image-processing model that computes a complete radiation and energy balance along with resistances for momentum, heat, and water vapor transport for each pixel [10,30]. The key input data for SEBAL consist of spectral radiance in the visible, near-infrared, and thermal infrared part of the spectrum. In addition to satellite images, the SEBAL model requires the routine weather data parameters (wind speed, humidity, solar radiation, and air temperature). SEBAL is a well-tested and widely used method to compute ET_a [6,10,12,31–33].

From the reflectance and radiance measurements of the bands, first land surface parameters such as surface albedo [34,35], vegetation index, emissivity [36], and surface temperature are estimated [37].

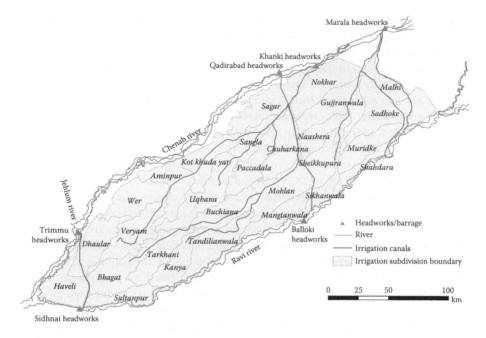

FIGURE 14.2 Canal network and irrigation subdivisions in the Rechna Doab, Pakistan. Note: A subdivision is the lowest administrative unit of irrigation management in Pakistan.

Then, evaporation is calculated from the instantaneous evaporative fraction, Λ, and the daily averaged net radiation, R_{n24}. The evaporative fraction, Λ, is computed from the instantaneous surface energy balance at satellite overpass on a pixel-by-pixel basis:

$$\lambda E = R_n - (G_0 + H), \tag{14.1}$$

where λE (W/m^2) is the latent heat flux (λ is the latent heat of vaporization and E is the actual evaporation), R_n (W/m^2) is the net radiation, G_0 is the soil heat flux (W/m^2), and H (W/m^2) is the sensible heat flux. The latent heat flux describes the amount of energy consumed to maintain a certain ET rate.

The energy balance equation (Equation 14.1) can be decomposed further into its constituent parameters. R_n is computed as the sum of incoming and outgoing short- and long-wave radiant fluxes. G_0 is empirically calculated as a G_0/R_n fraction using vegetation indices, surface temperature, and surface albedo. Sensible heat flux is computed using wind speed observations, estimated surface roughness, and surface-to-air temperature differences that are obtained through a self-calibration between dry ($\lambda E \approx 0$) and wet ($H \approx 0$) pixels. The dry and wet pixels are manually selected, based on vegetation index, surface temperature, albedo, and some basic knowledge of the study area, which make SEBAL somewhat subjective and difficult to automate. SEBAL uses an iterative process to correct for atmospheric instability caused by buoyancy effects of surface heating. More details and recent references/literature on SEBAL are available at (www.waterwatch.nl). Then, instantaneous latent heat flux, λE, is the calculated residual term of the energy budget, and it is then used to compute the instantaneous evaporative fraction $\Lambda(-)$:

$$\Lambda = \frac{\lambda E}{\lambda E + H} = \frac{\lambda E}{R_n - G_0}. \tag{14.2}$$

The instantaneous evaporative fraction Λ expresses the ratio of the actual to the crop evaporative demand when the atmospheric moisture conditions are in equilibrium with the soil moisture conditions. The instantaneous value can be used to calculate the daily value, because the evaporative fraction tends to be constant during daytime hours, although the H and λE fluxes vary considerably. The difference between the instantaneous evaporative fraction at satellite overpass and the evaporative fraction derived from the 24-hour integrated energy balance is often marginal and may, in many cases, be neglected [31,38,39]. For time scales of 1 day, G_0 is relatively small and can be ignored, and net available energy $(R_n - G_0)$ reduces to net radiation (R_n). At daily time scales, ET_a, ET_{24} (mm/day), can be computed as:

$$ET_{24} = \frac{86,400 \times 10^3}{\lambda \rho_w} \Lambda R_{n24}, \tag{14.3}$$

where R_{n24} (W/m²) is the 24-hour averaged net radiation, λ (J/kg) is the latent heat of vaporization, and ρ_w (kg/m³) is the density of water.

For time scales longer than 1 day, ET_a can be estimated using the relation proposed by Bastiaanssen et al. [6]. The main assumption is that Λ, specified in Equation 14.3, remains constant over the entire time interval between the capture of each remote-sensing image so that

$$ET_{int} = \frac{dt \times 86,400 \times 10^3}{\lambda \rho_w} \Lambda R_{n24t}, \tag{14.4}$$

where ET_{int} (mm/interval) is the time integrated ET_a, R_{n24t} (W/m²) is the average R_{n24} for the time interval dt measured in days. R_{n24t} is usually lower than R_{n24} because R_{n24t} also includes cloudy days.

14.5.3 DATA COLLECTION AND PREPROCESSING

Daily meteorological data on temperature, humidity, wind speed, and sunshine hours for Faisalabad and Lahore were collected from the Pakistan Meteorological Department for the period May 2001–May 2002. Rainfall data were collected from eight gauging stations in and around the Rechna Doab and were interpolated to obtain gridded values of seasonal rainfall.

Nineteen cloud-free MODIS scenes (Table 14.1) from April 2001 to May 2002, covering the entire Rechna Doab, were downloaded from Earth Observing System

TABLE 14.1
List of MODIS Images used for Surface Energy Balance Analysis in the Rechna Doab, Pakistan

	Rabi (Winter Season)				Kharif (Summer Season)					Rabi	
	January	February	March	April	May	June	July	August	September	October	November	December
2001		25		21	11, 27	3		2, 20, 29	25	18	22	1, 23
2002	27		4, 16	10, 21	3							

Data Gateway (EOSDG) of NASA (currently, this information is available at NASA Goddard Space Flight Center website: http://ladsweb.nascom.nasa.gov/data/search. html). For a few months during the monsoonal period, especially June and July, the frequency of cloud-free image availability was reduced. This will have some impact on the estimation of overall ET_a, but it was unavoidable. For SEBAL processing, only nine bands (i.e., the first seven bands in the visible and infrared range and two thermal bands 31 and 32) of MODIS were used.

Data describing surface flow diversions at the head of the main irrigation canals were collected for the same time period from the Punjab Irrigation and Power Department for the main canals systems of the Rechna Doab.

14.6 WP – DETERMINATION AND TRANSFORMATION OF SECONDARY CROPPING DATA

WP was analyzed to evaluate the behavior of current water consumption at different scales. WP can be defined in a number of ways as, for example, the physical mass of production or the economic value of production measured against gross inflows, net inflow, depleted water, process depleted water, or available water [2,40]. The expression is most often given in terms of mass of produce, or monetary value, per unit of water. For systems with multiple agricultural enterprises, economic WP is used to aggregate different forms of physical output. As there are complex cropping patterns in the Rechna Doab, land, and WP are discussed in terms of gross value of production (GVP) over ET_a.

For this analysis, data on crop area, production, and output prices were collected from secondary sources, including the Economic Wing of the Ministry of Food, Agriculture and Livestock (MINFAL), the Bureau of Statistics (BoS), the Government of the Punjab, and the Directorate of Economics and Marketing of the Punjab Agriculture Department. Data on cropped area and production were available with MINFAL and BoS for all rabi and kharif crops. The following crops were included in productivity analysis for the computation of GVP at the district and subdivision scales.

Kharif crops: rice, sugarcane, cotton, maize, millet, sorghum, guar, groundnut, sesame, kharif pulses, kharif vegetables, and kharif fruits.

Rabi crops: wheat, gram, barley, castor, linseed, sunflower, potato, onion, tomato, chili, coriander, garlic, peas, *barseem* (a winter fodder very similar to alfalfa), other rabi vegetables, rabi pulses, and rabi fruits.

Data on crop output prices available at the bigger local markets, usually known as *Mandi* (Grain Market), were obtained from the Directorate of Economics and Marketing. These market prices refer to the wholesale prices, which were collected on a daily basis through agricultural market committees at smaller scales and then integrated to the larger market level. The collected data were first adjusted to seasonal/annual values. Thereafter, the district values were calculated with reference to the nearest grain market(s). Price data from different markets revealed that price differentials across the area were very nominal. This is because the major agricultural commodity markets in Pakistan are well integrated with each other and price shocks in one market are instantly observed in other markets [41]. However, the price of sugarcane was not reported at the market level as it is sold directly to sugar mills. Generally, the government announces the price of sugarcane every year. Therefore, the price was taken from the Economic Survey of Pakistan [42] for the corresponding years, represented as the mill gate price and was taken to be the same for all districts during the reported year. Although the prices of some specific crops were not available (usually for the commodities that are not marketed, like fodder and some vegetables), prices were sourced from various studies based on field/primary data. Price data that were not available for the same year were then deflated to the reporting year (2001–02). Finally, these prices were adjusted to constant prices using the wholesale price index (WPI) to avoid the effect of price fluctuations between the years and to conduct our analysis in terms of real values.

The GVP was then computed for all districts in the Rechna Doab. As a first step, the secondary data from different sources at the whole district level were converted to uniform units. These data mainly include land utilization statistics, crop area, production of major crops, and output prices. After aggregating the available secondary data at district level, an overlay analysis was performed using GIS to calculate the fraction of district area falling in a specific irrigation subdivision. This was mainly done to overcome the limitation of district boundaries, which do not match those of the irrigation subdivisions. Finally, this fraction was used to transform secondary data to district boundaries within Rechna and then to the irrigation subdivision scale.

Total GVP and per hectare GVP were computed at the district level for 2001–02 as well as for kharif 2001 and rabi 2001–02. Total GVP for each district was computed by multiplying total production of the individual crop with its price, and reported in US dollars. Per hectare annual and seasonal GVP was calculated on the basis of total cultivated, as well as total cropped, area at district and then at subdivision level.

14.7 SPATIOTEMPORAL VARIATION IN ET

Daily evaporative fraction Λ and ET_a were calculated by Equations 14.1 through 14.3, using cloud-/haze-free MODIS images for May 2001–May 02. Then, daily values were integrated at appropriate intervals (Equation 14.4) to calculate monthly, seasonal, and annual ET. The resultant map showing the annual variation in ET_a in May 2001–02 is presented in Figure 14.3.

The annual ET_a varies from less than 100 mm/year in desert/barren areas to about 1650 mm/year over large water bodies in the processed image covering the Rechna Doab and other parts of the Indus Basin. However, annual ET_a from cropped areas only range between 500 and 1050 mm/year in the Rechna Doab. Brown areas in the Figure 14.3 delineate the desert and barren areas with lowest ET, whereas the dark green-blue line features, showing high ET_a in the image, represent the rivers and the canal network. Due to the heterogeneous cropping pattern, it was difficult to identify pure pixels for particular crops. However, for the Punjab rice-wheat area, the average ET_a is about 970 mm/year, whereas it is generally much lower, i.e., 800 mm/year or less in the lower Rechna due to lower cropping intensity and low delta crops. The average ET_a over the Rechna Doab is about 850 mm/year, which is

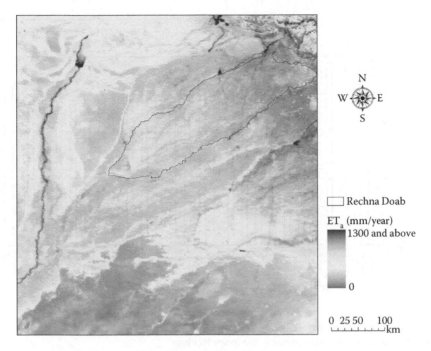

N
W ⊕ E
S

☐ Rechna Doab

ET_a (mm/year)
1300 and above

0

0 25 50   100
‖‖‖‖‖‖‖km

FIGURE 14.3 **(See color insert following page 224.)** Annual ET_a (May 2001–May 2002) in the Rechna Doab and other parts of the upper Indus Basin, Pakistan.

almost double the average rainfall and almost 60% of the reference crop ET. The irrigated areas close to the main canals or the river have higher ET_a due to better access to canals and groundwater for agriculture, and are distinct in Figure 14.3.

Seasonal and annual variations in ET_a in different irrigation subdivisions of the Rechna Doab were computed through an overlay analysis, using GIS coverage of irrigation subdivisions, and seasonal and annual ET_a results. Figure 14.4 illustrates the average seasonal and annual variation of ET_a in irrigation subdivisions of the Rechna Doab. ET_a variations in rabi, dominated by wheat and fodder, between different subdivisions is 64 mm and overall rabi ET_a has a slightly declining trend from subdivisions in the upper Rechna to the middle and lower Rechna Doab. This declining trend of ET_a from the upper Rechna to the lower Rechna is much more profound for the kharif season and governs the annual variations, except for the Sultanpur subdivision. The reason for higher ET_a in the Sultanpur subdivision, which is located in the lower Rechna, is its proximity to the Ravi River and availability of good soil and groundwater quality due to recharge from the river.

14.8 PERFORMANCE OF AGRICULTURAL WATER USE

There is a wide range of irrigation and water use performance indicators [43–45] to assist in achieving efficient and effective use of water by providing relevant feedback to the scheme/river basin management at all levels. Satellite-interpreted raster maps (such as ET_a, evaporative fraction) can be merged with vector maps of the irrigation

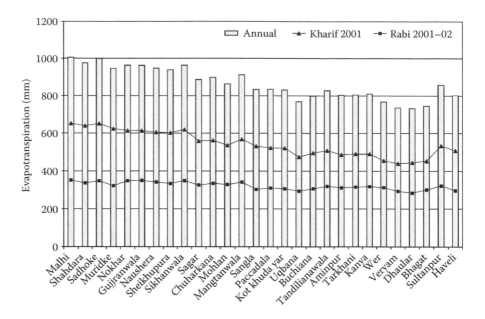

FIGURE 14.4 Seasonal and annual variation in ET_a for irrigation subdivisions in the year 2001–02, arranged from the upper to the lower Rechna.

water delivery systems to understand the real-time performance under actual field conditions. In this study, five indicators representing *equity, adequacy, reliability, sustainability,* and (*land and*) *water productivity* were selected to evaluate the performance of irrigation system in the Rechna Doab.

14.8.1 EQUITY

Traditionally, equity is calculated from the supply side. However, in a water-scarce system as in the Rechna Doab, equity in water consumption is more relevant from the farmer's perspective. ET_a maps were used to calculate the variability in water consumption within different subdivisions (Figure 14.5). The results show that the upper Rechna (Punjab rice-wheat zone) and subdivisions close to rivers have higher ET_a and lower intra-subdivision level variability, which indicate equitable water consumption through surface water and groundwater resources. Higher ET_a in these areas is related to fresh groundwater availability for irrigation [46]. Subdivisions with poor groundwater quality, falling in the middle and lower Rechna Doab, showed the highest value of coefficient of variation, representing high inequity in water consumption. Competition for surface water was greater in irrigation subdivisions of the middle and lower Rechna as compared to the upper Rechna, mainly due to the quality of groundwater, leading to higher inequality in the distribution of surface water. The information summarized in Figure 14.5 is useful to water managers for evaluating options for additional water requirements or reallocation between different parts of the Rechna Doab to achieve equity.

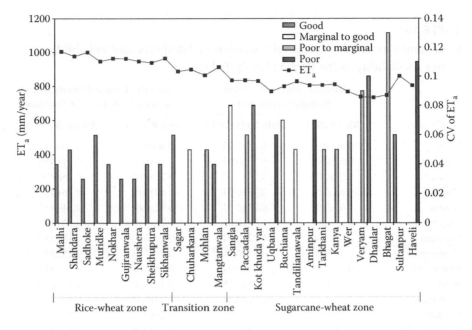

FIGURE 14.5 (See color insert following page 224.) Subdivision-level variation in actual evapotranspiration (ET$_a$), coefficient of variation of ET$_a$ and groundwater quality in the Rechna Doab, Pakistan. Groundwater quality is based on farmers' perception. (*Groundwater Quality Data Source:* IWMI Socioeconomic survey 2004.)

14.8.2 ADEQUACY AND RELIABILITY

Adequacy is the quantitative component, and is defined as the sufficiency of water use in an irrigation system. In contrast, reliability has a time component and is defined as the availability of water supply upon request. Both adequacy and reliability of water supplies to the cropped area can be assessed using the evaporative fraction maps, as they directly reveal the crop supply conditions [44,47].

In this study, adequacy is more specifically defined as the average seasonal evaporative fraction, and reliability as the temporal variability, temporal coefficient of variation, of evaporative fraction in a season. Evaporative fraction values of 0.8 or higher indicate no stress [44], and below 0.8, they reflect increases in moisture shortage to meet crop water requirements as a result of inadequate water supplies. Similarly, the lower values of the coefficient of variation represent the more reliable water supplies throughout the cropping season. This is also evident from the results of a socioeconomic study (unpublished) conducted by the International Water Management Institute (IWMI) in the Rechna Doab during 2003–04, indicating a higher proportion (3.45%) of sample farmers in the upper Rechna reported as receiving adequate canal water compared to 2.97% in the lower and only 0.88% in the middle Rechna. Similarly, canal water reliability (in terms of availability of adequate quantity at the right time) was also reported higher (11% of the sample farmers) in the upper Rechna compared with the middle (4%) and lower (less than 1%) Rechna. Adequacy and reliability were computed for different subdivisions of the Rechna

TABLE 14.2
Inter- and Intrasubdivision Level Variation in Adequacy and Reliability of Water Availability in the Rechna Doab, Pakistan

Subdivision	Adequacy (Average Seasonal Evaporative Fraction)		Reliability (Temporal Coefficient of Variation in Evaporative Fraction)	
	Kharif 2001	Rabi 2001–2002	Kharif 2001	Rabi 2001–2002
Malhi	0.69±0.03	0.75±0.03	0.30±0.06	0.13±0.02
Shahdara	0.67±0.04	0.72±0.04	0.27±0.09	0.14±0.02
Sadhoke	0.68±0.03	0.74±0.03	0.31±0.06	0.15±0.01
Muridke	0.65±0.04	0.70±0.04	0.30±0.07	0.16±0.02
Nokhar	0.64±0.04	0.74±0.04	0.32±0.04	0.14±0.02
Gujjranwala	0.64±0.03	0.75±0.03	0.29±0.05	0.15±0.02
Naushera	0.63±0.04	0.73±0.04	0.30±0.05	0.15±0.02
Sheikhupura	0.63±0.03	0.73±0.03	0.29±0.06	0.15±0.02
Sikhanwala	0.66±0.04	0.74±0.04	0.25±0.07	0.14±0.02
Sagar	0.58±0.05	0.71±0.05	0.35±0.07	0.17±0.03
Chuharkana	0.57±0.04	0.72±0.04	0.32±0.05	0.18±0.02
Mohlan	0.58±0.05	0.72±0.05	0.29±0.07	0.17±0.03
Mangtanwala	0.62±0.03	0.74±0.03	0.29±0.05	0.14±0.02
Sangla	0.56±0.05	0.68±0.05	0.34±0.07	0.22±0.06
Paccadala	0.56±0.04	0.68±0.04	0.29±0.05	0.21±0.03
Kot khuda yar	0.58±0.06	0.69±0.06	0.23±0.07	0.21±0.04
Uqbana	0.52±0.05	0.66±0.05	0.29±0.07	0.23±0.03
Buchiana	0.56±0.04	0.69±0.04	0.24±0.05	0.18±0.03
Tandilianwala	0.57±0.04	0.71±0.04	0.30±0.06	0.15±0.02
Aminpur	0.54±0.05	0.69±0.05	0.24±0.07	0.20±0.03
Tarkhani	0.56±0.04	0.70±0.04	0.29±0.06	0.15±0.03
Kanya	0.56±0.03	0.72±0.03	0.32±0.07	0.13±0.02
Wer	0.50±0.05	0.69±0.05	0.32±0.09	0.19±0.04
Veryam	0.48±0.07	0.66±0.07	0.39±0.13	0.22±0.06
Dhaular	0.50±0.08	0.66±0.08	0.38±0.15	0.22±0.06
Bhagat	0.52±0.10	0.68±0.10	0.36±0.14	0.19±0.07
Sultanpur	0.65±0.04	0.72±0.04	0.21±0.06	0.13±0.02
Haveli	0.61±0.06	0.68±0.06	0.24±0.08	0.18±0.06
Out of CCA	0.68±0.08	0.70±0.08	0.30±0.06	0.14±0.04

Doab and are presented in Table 14.2, for kharif and rabi, respectively. The analysis reveals that crops in the subdivisions of the upper Rechna Doab have relatively reliable and adequate water supplies in both seasons. This better performance can be attributed to both higher canal supplies to subdivisions located close to the canal head and/or access to good-quality groundwater for irrigation (as also evident from Sultanpur subdivision in the lower Rechna). The temporal coefficient of variation of evaporative fraction in kharif is much higher than in rabi, indicating less reliable supplies in the kharif season.

TABLE 14.3
Seasonal and Annual Variation in Rainfall (*P*), Canal Supplies (I_{cw}), Actual Evapotranspiration (ET$_a$), and Net Contributions from Groundwater (I_{ngw}) in the Rechna Doab

	MR and BRBD (Internal)	UCC	LCC	Haveli	Out of Command	Total CCA	The Overall Rechna Doab
Canal Command							
Kharif 2001 (million m³)							
P	1708	2268	5,953	239	2537	10,168	12,705
I_{cw}	684	1695	4,369	NA	0	NA	NA
ET$_a$	2068	2828	7,785	503	2803	13,184	15,987
I_{ngw}	–323	–1135	–2536	NA	266	NA	NA
Rabi 2001–2002 (million m³)							
P	128	148	420	23	240	719	959
I_{cw}	12	293	2,147	NA	0	NA	NA
ET$_a$	1094	1612	4,857	294	1475	7,857	9,332
I_{ngw}	954	1171	2,290	NA	1235	NA	NA
Annual (million m³)							
P	1835	2416	6,373	263	2777	10,887	13,664
I_{cw}	696	1988	6,516	NA	0	NA	NA
ET$_a$	3163	4440	12,642	797	4278	21,042	25,319
I_{ngw}	631	35	–246	NA	1501	NA	NA

Note: Canal supplies for command area served by the Haveli Canal in the Rechna Doab are not readily available. Canal water represents the water diversion for irrigation at the head of main canal system. MR and BRBD (internal): Malhi, Shahdara, Shadhoke Muridke subdivisions; UCC: Nokhar, Gujjranwala, Naushera, Sheikupura, Mangtanwala, Sikhanwala; Haveli: Haveli; remaining subdivisions of the Rechna Doab are part of the LCC system.

14.8.3 SUSTAINABILITY

The impact of water use patterns on sustainability was assessed in terms of the extent and quality of groundwater use and its likely impact on secondary salinization in different parts of the basin. As illustrated in Table 14.3, annual evaporation from the Rechna Doab is more than 25,000 million m³ of water. It is interesting to note that the evaporation outside the command areas (areas outside the canal irrigation network) is about 4000 million m³ of water. A significant proportion of ET$_a$ from out of the command area (mostly in the upper Rechna Doab) is due to unaccounted groundwater irrigation in a nominally rain-fed area. As discussed earlier, even in canal command areas, a large proportion of irrigation supplies comes from groundwater pumpage, but despite this, we still see that the evaporative fraction is typically less than 0.8 in all subdivisions. Satellite-based ET$_a$ results were compared with canal supplies and rainfall, to calculate the net groundwater contribution in different seasons in the main canal commands of the Rechna Doab, as subdivision level canal flow data for the period of study were not easily accessible (Table 14.3). The analysis showed that in kharif, due to higher rainfall and canal supplies, there was generally a net groundwater recharge (indicated by negative values), except outside the command area, where net groundwater contribution to ET_a was about 10%. It is important to remember that canal head flow data are used to compute net groundwater contribution at the canal command level, and this is different from irrigation from groundwater. In reality, as much as 50–60% of canal head water flow percolates from the earthen conveyance system and irrigated fields, and a large fraction is pumped back for irrigation by farmers through private tube wells [5].

The sustainability of the system is evaluated by comparing the location of net-recharge or net-groundwater use in irrigated areas in consideration with groundwater quality; however, the interaction of rivers and link canals with groundwater is not considered in this study. The highest net-groundwater use was found in rabi, with 47% (LCC) to 87% (MR and BRBD internal) net contribution to ET_a coming from groundwater (Table 14.3). High groundwater reliance in (dry) rabi compared to (wet) kharif can be explained by the annual canal closure period, low flows, and (for MR and BRBD internal) seasonally operated canals. It appears that, in kharif, most of the recharge is occurring in the LCC system, which is largely underlain by marginal to highly saline groundwater aquifer. As a result, recharge from good-quality canal water, after mixing with brackish groundwater, is becoming marginally fit to unfit for irrigation. The use of such brackish groundwater in these areas of the Rechna Doab has already caused secondary salinization and sodicity [48,49]. Considering the availability of detailed digital canal flow data for recent years from the Punjab Monitoring and Implementation Unit (PMIU), we strongly suggest that these analyses be conducted at the irrigation subdivision or distributary command area scales, for systematic tracking of vulnerable areas and for exploring management options to ensure the sustainability of irrigated agriculture in the Rechna Doab.

14.8.4 LAND AND WP

Land and WP values were calculated using the subdivision-level GVP (transformed from the district-level secondary agricultural statistics) and ET_a for the kharif, rabi, and annual time step (Figure 14.6). The analysis reveals that WP in rabi is high and relatively less variable than kharif values across the Rechna Doab. This is mainly due to the higher percent cropped area in rabi than in kharif [50]. This is directly related to low evaporative demand in rabi (winter season) and efficient use of limited canal supplies in marginal to poor groundwater quality areas. High rabi WP is also partly related to government policies, especially the support price of wheat, which is the major rabi crop. Lower WP in kharif is offset by higher gross returns from rice and sugarcane per unit of land or farm, and reflects the incentives to farmers to grow high water-using but high income-generating crops [51].

The annual values of land and WP are presented in Figure 14.7. High annual WP is found in subdivisions with good groundwater quality areas (mostly in the upper Rechna) and with adequate and reliable water supplies. The highest WP was found

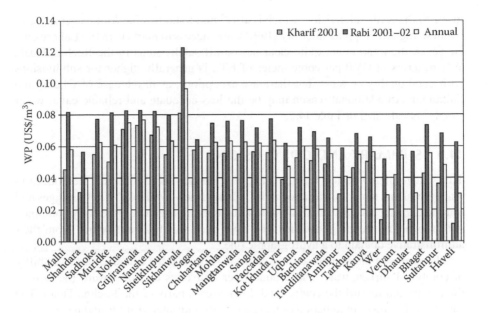

FIGURE 14.6 Seasonal and annual variation in subdivision-level water productivity in terms of GVP per unit of ET_a in the Rechna Doab (US$1.00=PRs.60.55).

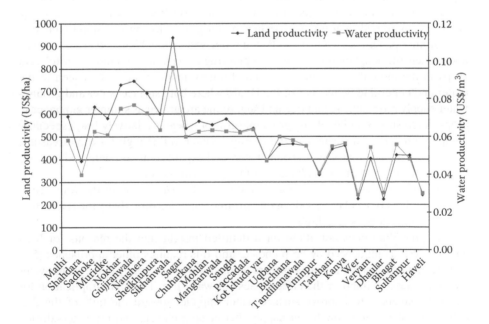

FIGURE 14.7 Spatial variation in subdivision-level annual land and water productivity in the Rechna Doab (US$1.00=PRs.60.55).

in the Sikhanwala subdivision, which is attributed to high-value fruit, vegetable cultivation, highest cropping intensity (199%), and access to markets in the Lahore city. However, the trend is not quite clear across all subdivisions in the Rechna Doab. WP, in terms of GVP per cubic meter of ET_a, is generally higher for subdivisions with good quality of water, but there are exceptions, as in the case of Dhular and Sultanpur. An additional reason may be the less adequate and reliable canal water supply, as indicated in Table 14.2.

14.9 SUMMARY AND CONCLUSIONS

This chapter demonstrates the application of surface energy balance techniques to map spatial and temporal variation in ET_a, using freely available MODIS images and routine climatic data. The analysis shows that these data can be effectively combined with secondary cropping statistics that have been suitably transformed from their administrative domain to a hydrographic one.

The results of the remote-sensing analysis show that the adequacy and reliability of combined surface water and groundwater deliveries decline toward the tails of the canals and toward the central and downstream parts of the Rechna Doab. The causes of this are a combination of increasing groundwater salinity, and more erratic surface water supplies.

The analysis reveals that sustainability of groundwater irrigation faces contrasting challenges. In the upper Rechna Doab, groundwater use in irrigated areas is higher than recharge, indicating the possible decline in groundwater stocks (if the trends continue). In contrast, the challenge in the lower Rechna, where the recharge is more than groundwater use, is to reduce recharge to saline groundwater (sinks).

A similar pattern is seen in the WP values derived in both kharif and rabi seasons, and reflects the adaptation by farmers of planting lower-value and more salt-tolerant crops in the more downstream areas. The highest annual water use is seen in the upper and upper-middle reaches of the Doab, where both surface water and groundwater supply are plentiful, and rice and sugarcane are grown as summer/annual cash crops. Interestingly, although the gross margin for wheat is relatively low, especially compared to rice and sugarcane, the average winter WP is high for three reasons: low water use, judicious use of groundwater, and expanded areas compared to areas in kharif, resulting from a combination of the first two factors.

The one anomalous result is for a subdistrict that runs close to the Ravi River (Sagar subdivision) and has good-quality groundwater, fed directly by the river. Here both WP and water use are high.

The procedure does not allow for a detailed insight into the reasons for high and low WP as influenced by crop choice, and this requires other approaches [52]. However, it does show the bigger picture, and shows where policy makers and water managers need to improve the effectiveness of water use. The main implication for water managers is to improve surface water supplies toward the tails of the distribution systems, to allow better supply, better crop choice, and mitigate salinity. However, a more detailed analysis is needed in order to understand (1) how far the soil and salinity limitations of the downstream areas can be overcome by simply improving the adequacy, reliability, and quality of water delivered to the farmer; and

(2) the longer-term implications of increasing the use of fresh groundwater in the head reaches; the long-term sustainability of a changed allocation policy would be compromised if the quality of current fresh groundwater degrades due to the development of internal water table gradients that drive mixing from the saline areas. From this analysis, we realize that managing WP to improve the overall average value through changed allocation has long-term temporal perspectives as well as spatial ones.

Although there is some natural energy price and water quality control on groundwater abstraction in Pakistan, this analysis is also useful in identifying both where groundwater use is not accounted for and how much water is being used. This information can also be put into more sophisticated analyses in combination with data on surface water supply within the command, for instance using seasonal and annual surface water supply to each subdivision. Similarly, analyses could be undertaken below subdivision level, as a second level of enquiry: this would require nested water balances to be calculated on an annual and seasonal basis, within the over-riding water balance of the whole Doab.

A more specific analysis of the performance and the WP of different cropping systems requires better crop mapping and identification. This could be done by higher-resolution remote sensing, but would then require better spatial disaggregation of ET_a. It remains to be seen if the patterns observed in individual snapshots (using Landsat or ASTER) are reflected in the seasonal water allocations, which can be investigated in the future. However, the loss of Landsat ETM+ functionality in 2003 due to the failure of the scan line corrector (SLC), and the current low likelihood of a replacement satellite having a thermal band, raise questions for high-resolution applications in the future. Researchers are currently looking for alternatives that do not require thermal band information, but it remains to be seen how robust and accurate such techniques will be.

In a country such as Pakistan, with dwindling water availability, and further threats from climatic change, set against a continually rising population and food demand, this information is vital for irrigation policy makers and managers. Both policy analysts and managers work in conditions in which it is difficult to obtain reliable data with field measurements, but techniques such as those described in this chapter offer an interim step to improving long-term management of water resources and to making the required improvements in average WP.

ACKNOWLEDGMENTS

The authors thank the Punjab Irrigation Department, Pakistan Meteorological Department, Directorate of Economics and Marketing of Provincial Agriculture Department for their cooperation in sharing the valued data sets to conduct this study. Satellite images used in this study were downloaded, free of cost, from Earth Observation System Data Gateway (EOS Data Gateway of NASA). For this, the authors specially thank NASA and USGS. This research was conducted as a part of the IWMI unrestricted project scaling-up water productivity in Pakistan and Challenge Program on Water and Food (CPWF) Indo-Gangetic Basin Focal Project, for which we thank the IWMI and CPWF donors.

REFERENCES

1. Allen, R. et al., *Crop Evapotranspiration* (guidelines for computing crop water requirements), FAO Irrigation and Drainage Paper No. 56, FAO, Rome, Italy, 1998.
2. Molden, D., *Accounting for Water Use and Productivity*, SWIM Paper 1, International Irrigation Management Institute, Colombo, Sri Lanka, 1997.
3. Seckler, D., *New Era of Water Resources Management: From "Dry" to "Wet" Water Savings*, IIMI Research Report 1, International Irrigation Management Institute, Colombo, Sri Lanka, 1996.
4. Keller, J., Keller, A., and Davids, G., River basin development phases and implications of closure, *Journal of Applied Irrigation Science*, 33, 2, 145–64, 1998.
5. Ahmad, M.D., Bastiaanssen, W.G.M., and Feddes, R.A., Sustainable use of groundwater for irrigation: A numerical analysis of the subsoil water fluxes, *Irrigation and Drainage*, 51, 3, 227–41, 2002.
6. Bastiaanssen, W.G.M., Ahmad, M.D., and Chemin, Y., Satellite surveillance of evaporative depletion across the Indus, *Water Resources Research*, 38, 12, 1273–82, 2002.
7. Molden, D.J., Sakthivadivel, R., and Habib, Z., *Basin Level Use and Productivity of Water: Examples from South Asia*, Research Report 49, International Water Management Institute, Colombo, 2000.
8. Engman, E.T. and Gurney, R.J., *Remote Sensing in Hydrology*, Chapman and Hall, London, 1991.
9. Kustas, W.P. and Norman, J.M., Use of remote sensing for evapotranspiration monitoring over land surfaces, *Hydrological Sciences Journal*, 41, 4, 495–516, 1996.
10. Bastiaanssen, W.G.M. et al., A remote sensing surface energy balance algorithm for land (SEBAL), Part 1: Formulation, *Journal of Hydrology*, 212–13, 198–212, 1998.
11. Su, Z., The surface energy balance system (SEBS) for estimation of turbulent heat fluxes, *Hydrology and Earth System Science*, 6, 1, 85–99, 2002.
12. Kustas, W.P., Diak, G.R., and Moran, M.S., *Evapotranspiration, Remote Sensing of Encyclopedia of Water Science*, Marcel Dekker, New York, 2003, 267–74.
13. Scott, C.A., Bastiaanssen, W.G.M., and Ahmad, M.D., Mapping root zone soil moisture using remotely sensed optical imagery, *Journal of Irrigation and Drainage Engineering*, ASCE, 129, 5, 326–35, 2003.
14. Ahmad, M.D., Bastiaanssen, W.G.M., and Feddes, R.A., A new technique to estimate net groundwater use across large irrigated areas by combining remote sensing and water balance approaches, Rechna Doab, Pakistan, *Hydrogeology Journal*, 13, 653–64, 2005.
15. Gowda, P.H. et al., Remote sensing based energy balance algorithms for mapping Et: Current status and future challenges, Invited Paper, *ASABE*, 2006.
16. Bastiaanssen, W.G.M. et al., SEBAL model with remotely sensed data to improve water-resources management under actual field conditions, *ASCE Journal of Irrigation and Drainage Engineering*, 131, 1, 85–93, 2005.
17. Timmermans, W.J. et al., An intercomparison of the surface energy balance algorithm for land (SEBAL) and the two-source energy balance (TSEB) modeling schemes, *Remote Sensing of Environment*, 108, 369–84, 2007.
18. Allen, R.G., Tasumi, M., and Trezza, R., Satellite-based energy balance for mapping evapotranspiration with internalized calibration (METRIC)-model, *ASCE Journal of Irrigation and Drainage Engineering*, 133, 4, 395–406, 2007.
19. Chávez, J.L. et al., Comparing aircraft-based remotely sensed energy balance fluxes with eddy covariance tower data using heat flux source area functions, *Journal of Hydrometeorology, AMS*, 6, 923–40, 2005.
20. Kijne, J.W., Molden, D., and Barker, E., Eds., *Water Productivity in Agriculture: Limits and Opportunities for Improvement*, Comprehensive Assessment of Water Management in Agriculture Series, No. 1, CAB International, Wallingford, 2003.

21. Biggs, T.W. et al., *Closing of the Krishna Basin: Summary of Research, Hydronomic Zones and Water Accounting*, Research Report 111, International Water Management Institute, Colombo, Sri Lanka, 2007.
22. Sakthivadivel, R. et al., *Performance Evaluation of the Bhakra Irrigation System, India, Using Remote Sensing and GIS Techniques*, IWMI Research Report 28, International Water Management Institute, Colombo, Sri Lanka, 1998.
23. McVicar, T. et al., Monitoring regional agricultural water use efficiency for Hebei Province on the North China Plain, *Australian Journal of Agricultural Research*, 53, 55–76, 2002.
24. Lobell, D.B. et al., Remote sensing for regional crop production in the Yaqui Valley, Mexico: Estimates and uncertainties, *Agriculture, Ecosystems and Environment*, 94, 2, 205–20, 2003.
25. Bastiaanssen, W.G.M. and Ali, S., A new crop yield forecasting model based on satellite measurements applied across the Indus Basin, Pakistan, *Agricultural, Ecosystems and Environment*, 94, 321–40, 2003.
26. Nangia, V., Turral, H., and Molden, D.J., Increasing water productivity with improved N fertilizer management, *Irrigation and Drainage Systems*, 22, 193–207, 2008.
27. Ahmad, M.D. et al., *Water Savings Technologies and Water Savings: Myths and Realties Revealed in Pakistan's Rice-Wheat Systems*, Research Report 108, International Water Management Institute, Colombo, Sri Lanka, 2007.
28. Ahmad, N. and Chaudhry, G.R., *Irrigated Agriculture of Pakistan*, Shahzad Nazir, Lahore, 1988.
29. Wolters, W. and Bhutta, M.N., Need for integrated irrigation and drainage management, example of Pakistan, *Proceedings of the ILRI Symposium, Towards Integrated Irrigation and Drainage Management*, ILRI, Wageningen, the Netherlands, 1997.
30. Bastiaanssen, W.G.M., SEBAL-based sensible and latent heat fluxes in the irrigated Gediz Basin, Turkey, *Journal of Hydrology*, 229, 87–100, 2000.
31. Farah, H.O., Estimation of regional evaporation using a detailed agro-hydrological model, *Journal of Hydrology*, 229, 1–2, 50–58, 2001.
32. Tasumi, M. et al., *U.S. Validation Tests on the SEBAL Model for Evapotranspiration via Satellite,* ICID Workshop on Remote Sensing of ET for Large Regions, ICID, Montpellier, France, 17 September, 2003.
33. Ahmad, M.D. et al., Application of SEBAL approach and MODIS time-series to map vegetation water use patterns in the data scarce Krishna river basin of India, *IWA Journal of Water Science and Technology*, 53, 10, 83–90, 2006.
34. Liang, S., Narrowband to broadband conversions of land surface albedo: I Algorithms, *Remote Sensing of Environment*, 76, 213–38, 2000.
35. Liang, S. et al., Narrowband to broadband conversions of land surface albedo: II Validation, *Remote Sensing of Environment*, 84, 25–41, 2002.
36. van de Griend, A.A. and Owe, M., On the relationship between thermal emissivity and the Normalized Difference Vegetation Index for natural surfaces, *International Journal of Remote Sensing*, 14, 6, 1119–31, 1993.
37. Tasumi, M., Progress in operational estimation of regional evapotranspiration using satellite imagery, PhD dissertation, University of Idaho, 2003.
38. Brutsaert, W. and Sugita, M., Application of self-preservation in the diurnal evolution of the surface energy budget to determine daily evaporation, *Journal of Geophysical Research*, 97, D17, 18, 322–77, 1992.
39. Crago, R.D., Conservation and variability of the evaporative fraction during the day time, *Journal of Hydrology*, 180, 173–94, 1996.
40. Molden, D. and Sakthivadivel, R., Water accounting to assess use and productivity of water, *Water Resources Development*, 15, 55–71, 1999.
41. Tahir, Z. and Riaz, K., Integration of agricultural commodity markets in Punjab, *The Pakistan Development Review*, 36, 3, 1997.

42. GoP (Government of Pakistan), *Economic Survey, 2005–2006*, Ministry of Finance, Islamabad, 2006.
43. Rao, P.S., *Review of Selected Literature on Indicators of Irrigation Performance*, International Irrigation Management Institute, Colombo, Sri Lanka, 1993.
44. Bastiaanssen, W.G.M. and Bos, M.G., Irrigation performance indicators based on remotely sensed data: A review of literature, *Irrigation and Drainage Systems*, 13, 291–311, 1999.
45. Bos, M.G., Burton, M.A., and Molden, D.J., *Irrigation and Drainage Performance Assessment: Practical Guidelines*, CAB International, Wallingford, 2005, 158.
46. Khan, S. et al., *Investigating Conjunctive Water Management Options Using a Dynamic Surface-Groundwater Modelling Approach: A Case Study of Rechna Doab*, Technical Report 35/03, CSIRO Land and Water, Australia, 2003.
47. Alexandridis, T., Asif, S., and Ali, S., *Water Performance Indicators Using Satellite Imagery for the Fordwah Eastern Sadiqia (South) Irrigation and Drainage Project*, Pakistan Research Report 87, International Water Management Institute, Lahore, Pakistan, 1999.
48. Sarwar, A., A transient model approach to improve on-farm irrigation and drainage in semi-arid zones, PhD dissertation, Wageningen University and Research Center, 2000.
49. Kelleners, T.J., Effluent salinity of pipe drains and tube-wells: A case study from the Indus plain, PhD dissertation, Wageningen University and Research Center, 2001.
50. Gamage, M.S.D.N., Ahmad, M.D., and Turral, H. Semi-supervised technique to retrieve irrigated crops from Landsat ETM+ imagery for small fields and mixed cropping systems of South Asia, *International Journal of Geoinformatics*, 3, 1, 45–53, 2007.
51. Jehangir, W.A. et al., *Sustaining Crop Water Productivity in Rice-Wheat Systems of South Asia: A Case Study from the Punjab*, Pakistan Working Paper 115, International Water Management Institute, Colombo, Sri Lanka, 2007.
52. Ahmed, S. et al., *Limits and Opportunities to Improve Farm Level Land and Water Productivity of Major Crops in Rechna Doab*, Pakistan, Unpublished Report, International Water Management Institute, Colombo, Srilanka, 2007.

Section V

Rain-Fed Cropland Areas of the World

15 Global Map of Rainfed Cropland Areas (GMRCA) and Statistics Using Remote Sensing

Chandrashekhar M. Biradar
University of Oklahoma

Prasad S. Thenkabail
U.S. Geological Survey

Praveen Noojipady and Y.J. Li
University of Maryland

Venkateswarlu Dheeravath
United Nations Joint Logistics Centre

Manohar Velpuri
South Dakota State University

Hugh Turral
On the Street Productions

Xueliang L. Cai and MuraliKrishna Gumma
International Water Management Institute

Obi Reddy P. Gangalakunta
National Bureau of Soil Survey and Land Use Planning (ICAR)

Mitchell A. Schull
Boston University

Ranjith D. Alankara and Sarath Gunasinghe
International Water Management Institute

Xiangming Xiao
University of Oklahoma

CONTENTS

15.1 INTRODUCTION

One of the key challenges now is to ensure sufficient food production and global food security. The capability of individual nations to meet food demand is directly related to water resources and management. Worldwide, about 60% of food production comes from the rain-fed croplands and these constitute the backbone of the marginal or subsistence farmers. The improved rain-fed cropping system could be a better alternative to irrigated agriculture to meet global food production due to its environmental friendliness and sustainability. Approximately 80%

of the global agricultural land is under rain-fed systems, with generally low-yield levels and high on-farm water losses [1]. The substantially increased food demand is putting enormous pressure on freshwater resources. The challenge of feeding tomorrow's world population is, to a large extent, dependent on the improved crop and water productivity within present land use, specifically in rain-fed crop-land areas (CA) [2]. This clearly indicates a significant window of opportunity for improvement.

The areal extent and spatial distribution of agricultural croplands, including rain-fed and irrigated areas, are important for the study of global food and environmental security. However, rain-fed and irrigated cropland maps at the global scale were not available. The overarching goal of this study was to produce a global map of rain-fed cropland areas (GMRCA) and calculate country-by-country rain-fed area statistics using remote sensing data.

The speculative estimation of the potential CAs based on climatic and edaphic conditions is 3.29 billion hectares (Bha) [3] to 4.15 Bha [4]. However, productivity of a large proportion of rain-fed croplands is limited due to poor farm practices and soil fertility, insufficient rainfall, and poor pest and disease management. To increase the rain-fed production by converting the national vegetation and protected areas would be environmentally costly [5] and ecologically unacceptable. Therefore, any increases in grain production have to be met by increasing the intensification of rain-fed croplands and the use of improved varieties and better agronomic practices.

To our knowledge, based on the existing literature reviews, the world's cropland estimates vary from 1.11–3.62 Bha [6–10]. It is well known that a number of factors determine the outcome, which leads to the fact that no two global data sets match [11] as a result of differences in definitions, methods, data sources, data types, data calibration, and data acquisition modes. The increase in CA has been only modest, from 1.36 Bha in 1960 to 1.47 Bha in 1990 [12,13]. Various estimates show that it has remained at around 1.5–1.8 Bha at the end of the millennium [12,14], of which about 16–18% is irrigated [15–17]. However, the above efforts do not differentiate between the rain-fed and irrigated croplands.

The available global land use and land cover (LULC) maps do not distinctly separate rain-fed croplands. The commonly used global LULC data sets are primarily non-remote-sensing-based, but are produced using data from various maps at 100-km grid by Matthews [18], and 50-km grid by Olson and Watts [19], Olson [20], and Wilson and Henderson-Sellers [21]. More recently, Advanced Very High Resolution Radiometer (AVHRR) and Moderate Resolution Imaging Spectroradiometer (MODIS) sensor data have been widely used to produce global LULC. The 1992–1993 AVHRR 1-km data were used by the United States Geological Survey (USGS) [9,22] and the University of Maryland [23,24] to produce global LULC data sets. The most recent LULC products are from Boston University using MODIS [25,26].

Even though we may have access to the national- and regional-level cropland statistics, the spatial pattern is not known. Also, the spatial distributions of CAs keep on changing due to soil degradation, land conversion, urbanization, desertification, and global warming. Between the early 1960s and the late 1990s, the world's CA increased by only ~11%, while the population grew almost doubled, according to the 1960 estimate. As a result, cropland per person reduced by 40%, from 0.43 to 0.26 ha [27]. In the future, most of the crop production in developing countries will have to come from agricultural intensification: higher yield varieties, increased multiple cropping, and shorter fallow periods. At the same time, CAs are increasing in certain parts of the world, such as the African and Amazonian rain forests, where expansion of slash-and-burn agriculture with decreasing fallow periods and expansion of biofuel cultivation are major factors. In addition, changes occur in the cropping pattern (e.g., from grain crop to vegetables) and intensification (e.g., from single crop to double crop).

Therefore, there is a clear need to determine the precise extent of rain-fed CAs and their spatial distribution. Given this importance, the present research has been carried out to produce the first satellite-sensor-based GMRCA. The first objective of the present research was to develop methods to map rain-fed CAs at the global scale using remote sensing and secondary data. The second objective was to calculate rain-fed CAs for every country in the world and quantify accuracies and errors of such an estimate.

15.2   DATA SETS

15.2.1   Satellite and Ancillary Data

The satellite sensor and ancillary data, with global coverage, used in this research consisted of: (a) National Oceanic and Atmospheric Administration (NOAA) AVHRR at ~10-km monthly time series for 1997–1999 with four spectral wavebands (red, near-infrared, and two thermal infrared bands). The primary data have gone through several stages of calibrations and corrections [28–31], making it a high-quality scientific data set. The primary data came from NASA GSFC DAAC. The original data were available from (a) NASA DAAC (http://daac.gsfc.nasa.gov), (b) Satellite Pour l'Observation de la Terre Vegetation (SPOT VGT) 1-km monthly time series for 1999 (http://free.vgt.vito.be/), (c) the Global 30 arc second elevation data (GTOPO30) digital elevation data from the USGS EROS Data Center (http://edcdaac.usgs.gov/gtopo30/gtopo30.html), and (d) the University of East Anglia Climate Research Unit's (CRU's) 50-km precipitation monthly data for 1961–2001 (http://www.cru.uea.ac.uk/~mikeh/datasets/global/).

Mapping and monitoring agricultural seasonality at the global level require information at high temporal frequency. Time series help differentiate the dynamics of agriculture to delineate rain-fed croplands from other LULC types. The above-mentioned time-series satellite remote sensing and secondary data sets are stacked together to produce a single megafile of 159 layers with a massive volume of 100 GB (Table 15.1). This single megafile was resampled and saved as a 1-km grid file whose volume was 110 GB. This consisted of 144 AVHRR layers from three years (12 layers from one band per year×four bands, including a Normalized Difference Vegetation Index (NDVI) band×three years), 12 SPOT layers from one year, and

TABLE 15.1

Megafile Data Characteristics

Band Number or Primary Source (No.)	Wavelength Range (μm)	Duration (Year)	No. of Bands and Radiometry (No.; One per Month)	Data Final Format Z-Scale (Percent: for Reflectance)	Range
		Satellite Sensor Data			
AVHRR 10-km					
Band 1 (B1)	0.58–0.68	1997–1999	36	Reflectance at ground, 8-bit	0–100
Band 2 (B2)	0.73–1.1	1997–1999	36	Reflectance at ground, 8-bit	0–100
Band 4 (B4) (Top-of-atmosphere)	10.3–11.3	1997–1999	36	Brightness temperature	160–340
NDVI	(B2–B1)/(B2+B1)	1982–2000	36	Unitless, 8-bit scaled NDVI	–1 to+1
SPOT VGT 1-km					
NDVI	(B2–B1)/(B2+B1)	2000	1	Unitless, 8-bit scaled NDVI	–1 to+1
		Secondary Data			
GTOPO30 1-km					
One-band	DCW, DTM, and others	1 time	1	Meters, 16-bit	–1 to+1
Rainfall 1-km					
One-band	Mean of monthly 40-years	1961–2001	1	Millimeters, 16-bit	0–65,536
Forest cover 1-km					
One-band	None	1992–1993	1	Class names, 8-bit	0–256

Note: A single global megafile of 159 bands was composed by layer stacking of NOAA AVHRR (~10-km), SPOT VGT (1-km), and a suite of secondary data.

single layers of elevation, mean rainfall for 40 years, and forest cover (Table 15.1). Table 15.1 describes the data set used and its characteristics.

15.2.2 Groundtruth Data

Collection of groundtruth data is important to understand the real situation on the ground for land cover class identification, naming, and accuracy assessment. All the groundtruth data sets collected for this project are pooled together in a standard Geographic Information System (GIS) format (Arcview shapefile) with unique attributes and class characteristics. These data sets are made available to the public through the website http://www.iwmidsp.org. The three distinct groundtruth data are described below.

15.2.2.1 GMRCA Project Groundtruth Data

During various field campaigns in India, China, Central Asia, West Africa, the Middle East, and southern Africa, 1861 groundtruth points were collected. The data consist of precise location, LULC, irrigated croplands, rain-fed croplands, canopy cover percentage, and digital photos.

15.2.2.2 Google Earth Groundtruth Data

The Google Earth high-resolution images over 11,000 randomly selected full zoom-in views (at full scale) are used as "groundtruth points." These areas of interest (AOI) provide visual information on various landscape features and classes, such as agricultural, forest, and barren and on the existence of irrigation structures (e.g., canals, tanks, reservoirs). The AOI have the advantage over point-based groundtruth data in that they provide information on much larger areas and, therefore, constitute a more representative view than is normally sampled directly on the ground.

15.2.2.3 Degree Confluence Project Data

The Degree confluence project (DCP) [32] is a voluntary effort of organized sampling of the entire world at every single degree latitude and longitude intersection (http://www.confluence.org/). We picked these groundtruth data points and identified individual classes of LULC categories through detailed descriptions provided by the visitor who collected that particular point. Unfortunately, every groundtruth point does not provide a detailed description of the type of land use, agriculture, irrigated, or rain-fed, etc. Therefore, we dropped many points whose description was insufficient to label LULC classes. Finally, we were left with only 3,982 confluence points and have LULC type, precise latitude, longitude, and a digital photo.

15.3 METHODS

The GMRCA is derived using a method similar to that used in mapping irrigated areas [16]. The suite of techniques and methods is explained in Chapter 2 of this volume and in a recently published article by Thenkabail et al. [16]. In the concepts and methods, we have used the same approaches used to derive and synthesize global irrigated area with additional calculations and attributes to distinguish rain-fed

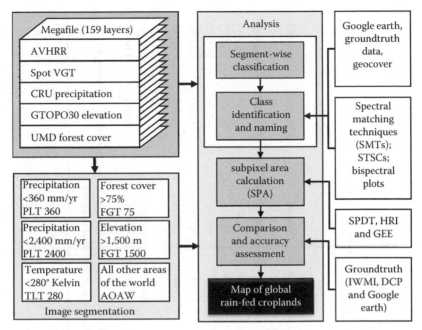

UMD- University of Maryland; CRU-Climate Research Unit; PLT-Precipitation less
than; FGT-Forest cover greater than; TLT-Temperature less than; AOAW-All other areas
of the world; STSC-Space-time spiral-curves; HRI-High Resolution Images; SPDT-Sub-
pixel decomposition techniques; GEE-Google earth estimate; IWMI-International Water
Management Institute; and DCP-Degree Confluence Project.

FIGURE 15.1   Flow chart for the global map of rain-fed cropland areas (GMRCA).

croplands from the irrigated and other land cover classes. In the following sections,
we will briefly describe the steps used in these approaches.

We have derived the global rain-fed CA map by integrating satellite sensor data,
groundtruth data, and secondary/ancillary information. The distinct class types have
been identified by segmenting the global megafile data sets (Figure 15.1) into characteristic regions that are easier to analyze, performing an unsupervised classification
on each segment, grouping similar classes through spectral matching techniques
(SMTs) (Table 15.2), setting up a class identification and labeling process (Figures 15.2
through 15.5), resolving the mixed classes (Figure 15.6), and calculating the subpixel
areas (SPAs) (Figure 15.7). As far as possible, class naming is standardized with earlier global land cover classifications (Table 15.2). The specific methods are discussed
in the following sections.

15.3.1 Class Aggregation and Identification

At the global scale, it is expected to identify and label an enormous number of the
classes. This is quite a tedious and time-consuming process if we employ the conventional approach of using groundtruth data and class signatures. With the advantage
of the new technique, the SMT [33–36] has been used to group classes. Time series

TABLE 15.2
Spectral-Matching Technique (SMT) to Group Similar Classes

Class No.	Class 83	Class 84	Class 85	Class 86	Class 87	Class 88	Class 89		Class 102	Class 103
Class 83	1.0000	-0.5056	-0.5709	-0.5683	-0.5649	-0.5755	-0.5375	...	-0.3969	-0.4792
Class 84	-0.5056	1.0000	0.7806	0.7678	0.7671	0.7701	0.6105	...	0.4025	0.4536
Class 85	-0.5709	0.7806	1.0000	0.9920	0.9923	0.9944	0.9325	...	0.1228	0.1649
Class 86	-0.5683	0.7678	0.9920	1.0000	0.9999	0.9988	0.9542	...	0.0463	0.1072
Class 87	-0.5649	0.7671	0.9923	0.9999	1.0000	0.9989	0.9530	...	0.0436	0.1044
Class 88	-0.5755	0.7701	0.9944	0.9988	0.9989	1.0000	0.9495	...	0.0605	0.1260
Class 89	-0.5375	0.6105	0.9325	0.9542	0.9530	0.9495	1.0000	...	-0.0705	-0.0038
..								..		
Class 102	-0.3969	0.4025	0.1228	0.0463	0.0436	0.0605	-0.0705	...	1.0000	0.9360
Class 103	-0.4792	0.4536	0.1649	0.1072	0.1044	0.1260	-0.0038	...	0.9360	1.0000

Note: Spectral correlation similarity between R^2-values (SCS R^2-values) and a shape measure of any one class with other classes. The higher the SCS R^2-value, the greater the shape similarity between two classes. This will help rapidly identify similar classes and analyze them as a group.

of NDVI (e.g., sample illustration in Figure 15.2a and b) of the SMT help group classes with similar spectral characteristics (e.g., Figure 15.2b and c). The spectral correlation similarity (SCS) R^2 values, which identify and match classes with similar shaped time-series spectra (e.g., NDVI, band reflectivity), were used to group classes (e.g., Figure 15.2b and c) whose spatial distribution is illustrated in Figure 15.2a.

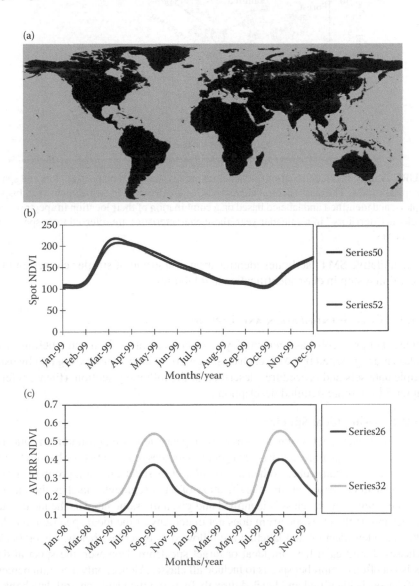

FIGURE 15.2 Class identification and labeling process based on time-series spectral curves. The classes generated by classifying the multiple-sensor megafile data are identified and labeled based on several approaches described in this chapter. One such method is illustrated here, (a) shows the spatial distribution where classes are identified and labeled based on time-series characteristics of classes in (b) SPOT VGT and (c) AVHRR.

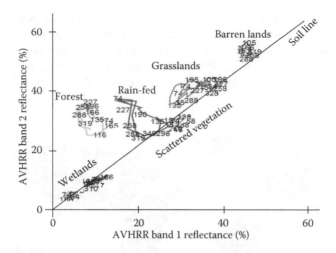

FIGURE 15.3 Class identification and labeling process based on bispectral plots and space-time spiral curves. The class spectral characteristics are depicted in two-dimensional bispectral plots, and identified and labeled based on a combination of their location in spectral space and distinct "territory" in a calendar year; these characteristics are referred to as space-time spiral curves.

The quantitative SMT facilitates identification of a group of similar classes and is a powerful first step in class identification and labeling.

15.3.2 Class Identification and Labeling

A wide set of protocols has been employed for class identification and labeling. Once the classes are grouped by SMT, each class in a group is further investigated by using multiple data sets and procedures described in the following section. (Please refer to Chapter 2 for a more detailed description.)

15.3.2.1 Using Ideal Spectra

The ideal or target spectra signatures have been generated using groundtruth data. The time-series class signature consists of vegetation indices (e.g., NDVI) or band reflectivity using megafile data-cubes. When a particular class spectrum (or a group of similar classes) generated, based on the classification of each segment, matches the ideal or target spectrum well, the class is identified and given the same label as the ideal or target spectrum. If the class spectrum does not match any of the ideal spectra in the data bank, class identification and labeling are followed as described below. Theoretically, all classes should each have an ideal or target spectrum. However, the spectral data bank is not often comprehensive as to include all classes. Hence, only a certain percentage of classes is identified and labeled directly by using the ideal spectral data bank.

15.3.2.2 Using Spectral Properties

The inherent properties of the spectral information from remote-sensing data have been used in discriminating croplands from other LULC classes. The spectral

Legend

- Rain-fed croplands (corn, soybean, and wheat-dominant)
- Rain-fed croplands (wheat, cotton, and barley-dominant)
- Rain-fed croplands and grasslands
- Rain-fed croplands and shrublands
- Rain-fed croplands and woodlands
- Grassland-dominant with rain-fed croplands
- Shrublands-dominant with rain-fed croplands
- Woodland-dominant with rain-fed croplands
- Forest-dominant with rain-fed croplands
- Other areas and ocean

Scale

1,000   0   1,000 km

FIGURE 15.4 (See color insert following page 224.) Global map of rain-fed cropland areas (GMRCA). This is a 9-class GMRCA map at nominal 1-km resolution produced using a fusion of 1-km SPOT VGT, 10-km NOAA AVHRR, and numerous secondary data. The first five classes are dominated by rain-fed classes, the next three are dominated by savannas with significant rain-fed croplands, and class 9 is dominated by forests with significant rain-fed cropland fragments.

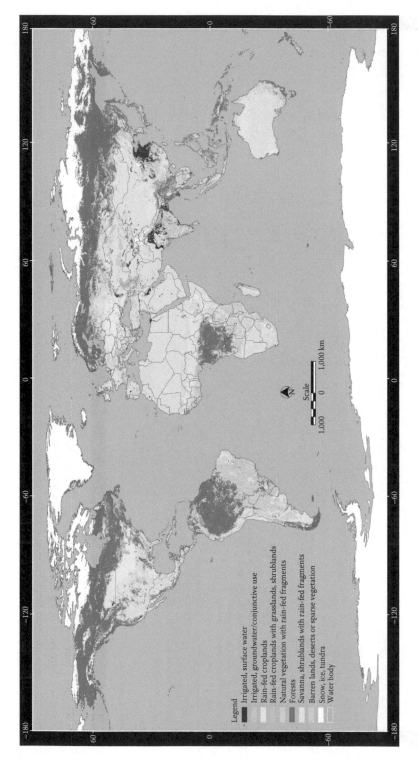

FIGURE 15.5 (See color insert following page 224.) Global rain-fed croplands along with irrigated crops and other land use/land cover. Classes 3 and 4 are rain-fed croplands, class 5 is natural vegetation with significant rain-fed fragments, classes 1 and 2 are irrigated croplands, and the rest of the classes are noncroplands.

Legend
Irrigated, surface water
Irrigated, groundwater/conjunctive use
Rain-fed croplands
Rain-fed croplands with grasslands, shrublands
Natural vegetation with rain-fed fragments
Forests
Savanna, shrublands with rain-fed fragments
Barren lands, deserts or sparse vegetation
Snow, ice, tundra
Water body

Scale
1,000   0   1,000 km

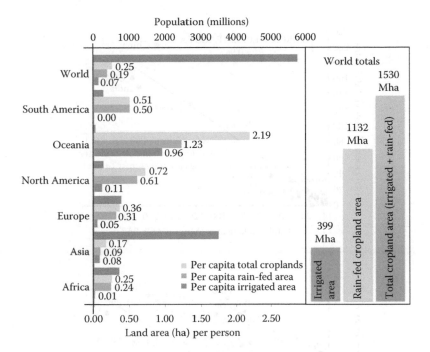

FIGURE 15.6   World irrigated, rain-fed, and total cropland area and per capita land area in each category.

characteristics of the classes that need to be identified are plotted in the bispectral plots. The specific location of a class in a two-dimensional feature-space is indicative of the class type, which is further verified using groundtruth and other knowledge bases. Also, when classes cluster in a specific location of the feature space, all the clustered classes are likely to be similar classes.

The spectral curves plotted using time-step data depict time-series characteristics of a class in a given feature space for the set of months. Each class has a territory in which it moves around every year (Figure 15.3). They show distinct territories and paths during certain periods of the year. The unique time-component properties of the spectral signatures are an added advantage to track class movement in a feature space continuously over a season or a year.

15.3.3   RESOLVING UNRESOLVED PIXELS

As explained in Chapter 2, in spite of the rigorous methods and availability of a large volume of groundtruth data, a number of classes could not be identified and labeled due to intermixing of the nearest groups of the classes. We have reclassified those patches and regions while masking out the resolved portion of the images. Along with reclassifications, we have also employed decision-tree algorithms [24] and spatial modeling [15].

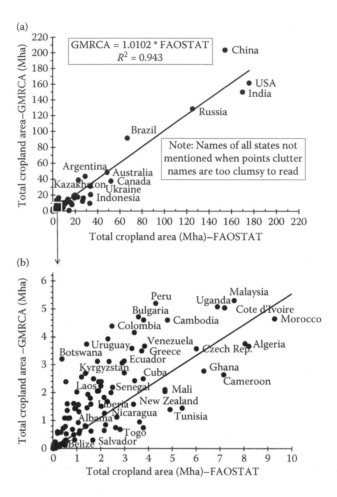

FIGURE 15.7 Comparison of the remote-sensing-based total cropland areas with FAO Aquastat-gathered national-level statistics.

15.3.4 RAIN-FED AREA CALCULATIONS

It is now well known that accurate estimates of areas from coarse-resolution imagery can only be achieved based on SPA calculation methods [37,38]. A series of well-established methods and approaches has been used to calculate the precise area as much as possible. The subpixel rain-fed CA is calculated by multiplying full pixel areas of that pixel with rain-fed area fractions, and the latter are determined by the three methods mentioned in Thenkabail et al. [37]: (a) Google Earth Estimates (GEE), (b) high-resolution imagery (HRI), and (c) subpixel Decomposition Techniques (SPDT). Of these three methods, SPDT is found to be the best option, automated and relatively fast. The SPDT involves plotting the reflectivity of red vs. near-infrared [39] for every pixel in a class in a feature space. Depending on the pixel location in the feature space plot, the percent area cultivated is determined, providing us with the rain-fed area fraction.

15.4   RESULTS AND DISCUSSIONS

As per our knowledge at this moment, this is the first rain-fed cropland map at the global scale. Initially, we had rain-fed CA maps with a large number of disaggregated classes. After considering the advice of the global modeling group and policy makers, we grouped a large number of disaggregated cases into the major rain-fed class to produce the simplest aggregated 9-class rain-fed map at the global scale (Figure 15.4). There is also a disaggregated GMRCA map with 66 classes, not presented in this chapter, which is made available for downloading from the web portal http://www.iwmigiam.org/. This map will be specifically useful for those interested in creating more refined class maps for their AOI. The results and discussion of this chapter are limited to the 9-class rain-fed CA with some extent of total CA and irrigated area map.

The final 9-class global rain-fed CA map with the most succinct class name description is listed here: (i) Rain-fed cropland (with dominant crops of corn, soybean, and wheat in the order of preference): this is the pure rain-fed class with the three most dominant crops mentioned parenthetically, and has little or no other classes mixing with them; (ii) rain-fed croplands (wheat, cotton and barley dominant): this is another pure rain-fed class with the three most dominant crops mentioned parenthetically; (iii) rain-fed cropland with grasslands: this class has significant rain-fed fractions with a mix of grassland fragments. Similarly, classes (iv) and (v) have significant rain-fed cropland fractions with mixes of shrublands and woodlands, respectively. The nonsignificant fractions of rain-fed croplands are classes (vi) to (ix); these classes have dominant fragments of grasslands, shrublands, and woodlands, with moderate or small amounts of rain-fed croplands. And class (ix) is forestland-dominant with significant fragments of rain-fed croplands (Figure 15.4, Table 15.3). All classes have mixed crops due to the coarse spatial resolution (10-km) of the primary data sets.

The crop dominance for the first two classes is determined based on groundtruth data (Table 15.3) and is also compared with the geographic distribution of the major crops of the world, produced by Leff et al. [40].

The fractions of the rain-fed croplands are used to estimate the SPAs. Classes 1 and 2 have over 50% of rain-fed area fractions, classes 3 to 5 have between 34 and 40% of fractions, and classes 6–9 have between 12 and 25% of fractions (Table 15.3). Each rain-fed cropland class (or each pixel within a class) has its own characteristics in terms of its vegetation dynamics, seasonality, topographic location, and biophysical properties—the dynamic properties of an individual class from which one can study variables such as cropping calendars, crop growth stages, biomass levels, and rain-fed fractions.

The total area of all nine classes combined is estimated to be a staggering 1.13 Bha of rain-fed CA. The first two major rain-fed classes account for ~36.5% of rain-fed CA. These two pure rain-fed classes are predominantly rain-fed croplands with very little other land use. This is followed by rain-fed croplands with grasslands (class 3), accounting for about 18.6% of rain-fed CA. The grassland dominant with rain-fed croplands (class 6) is the second biggest class among the nine classes, with a total contribution of about 20% of rain-fed CA. Rain-fed croplands with shrublands and woodlands account for 7.4% of the total rain-fed croplands of the world. Other

TABLE 15.3
Areas of Global Map of Rain-Fed Cropland Area (GMRCA) Classes

Class No.	Class Name	Full Pixel Area (Area in ha)	Rain-Fed Area Fractions (RAF)	Subpixel Area (SPA) (Area in ha)	Percent of Total Rain-Fed Area
1	Rain-fed croplands (corn, soybeans, and wheat dominant)	423,204,269	0.55	232,762,348	20.57
2	Rain-fed croplands (wheat, cotton, and barley dominant)	348,389,483	0.52	181,162,531	16.01
3	Rain-fed croplands and grasslands	569,862,800	0.37	210,849,236	18.63
4	Rain-fed croplands and shrublands	69,362,335	0.34	23,583,194	2.08
5	Rain-fed croplands and woodlands	150,842,832	0.40	60,337,133	5.33
6	Grassland-dominant with rain-fed croplands	900,008,806	0.25	225,002,201	19.88
7	Shrublands-dominant with rain-fed croplands	416,508,118	0.13	54,146,055	4.79
8	Woodland-dominant with rain-fed croplands	359,324,051	0.17	61,085,089	5.40
9	Forest-dominant with rain-fed croplands	671,021,180	0.12	82,624,536	7.30

Note: The areas are computed by taking the Full Pixel Area (FPA) and multiplying it with the rain-fed area fraction (RAF). The SPAs provide the actual areas.

natural vegetation (shrublands, woodlands, and forest) with fragmented rain-fed croplands together contribute about 17.5% of rain-fed CA (classes 7–9). These three classes geographically distributed at larger extent after the first three main rain-fed classes (classes 1–3), however their (classes 7–9) minimal rain-fed fractions lead to smaller subpixel rain-fed areas.

The analysis of the rain-fed croplands along latitude gradients indicates that the most intensive rain-fed area falls in the temperate latitudes of both hemispheres. The rain-fed croplands are prevalent in the plains of the United States, central European Union, and southwestern Russian plains (Figure 15.4) in the Northern Hemisphere. The large extent of the rain-fed CAs in the Southern Hemisphere occurs mainly in the plains of Brazil, Argentina, Uruguay, and Paraguay. However, most of the areas in the Southern Hemisphere are found to be fragmented croplands with natural vegetation mixtures. The highly fragmented rain-fed croplands with other natural vegetation also spread across South and southeastern Africa. The following section further highlights the finding of the rain-fed CA mapping with irrigated and total CAs at continental to country level, with key statistical estimates.

15.4.1 GLOBAL AND REGIONAL PROSPECTIVE

The global distribution of the total rain-fed CA is estimated at 1.13 Bha (Table 15.3, Figure 15.4). The total CA is estimated at 1.53 Bha (Table 15.4, Figure 15.5) by combining rain-fed CAs and the irrigated CAs. Despite some disagreements in a few country-level estimates, our results agree with the total CAs at the global scale estimated by FAOSTAT [7], which is based on the country statistics (Table 15.5, Figure 15.5), and Tillman et al. [41] estimate a total CA of 1.54 Bha, which is based on a temporal extrapolation from the available data around the nominal year 1998. A literature review shows that the world's total croplands (rain-fed plus irrigated) around the end of the last millennium (1999) was anywhere between 1.11 and 3.62 Bha. The total CAs estimated by different studies vary significantly (Table 15.4) due to

TABLE 15.4
Cropland Areas of the Continents and the World

Sl. No.	Continents (Name)	Irrigated Area (Mha)	Rain-Fed Area (Mha)	Total Cropland Area (Mha)	Population 2000 (Million)
1	Africa	8.69	189.05	197.74	794
2	Asia	290.64	327.29	617.93	3672
3	Europe	33.94	227.89	261.83	727
4	North America	35.43	190.67	226.1	314
5	Oceania	29.71	38.22	67.93	31
6	South America	0.13	158.44	158.57	314
	World	399	1132	1530	6057

Note: Rain-fed, irrigated, and total cropland areas for the continents and the world. For irrigated areas, refer to Chapter 2, Table 2.2 (this volume).

TABLE 15.5

Cropland Areas of the World Estimated from Various Sources

Total Cropland area (Bha)	Year	Source
1.11	2000	Hansen et al. [6]
1.34	1994	Warnanat et al. [42]
1.36	1999	Houghton [43]
1.39	1999	Loveland et al. [9]
1.40	1991	FAO [44,45]
1.48	2001	Klein Goldewijkm [10]
1.48	1998	Amthor and MoEW Group [46]
1.50	1991	Lal and Pierce [47]
1.50	2003	WRI [48]
1.51	1998	FAOSTAT [7]
1.53	1999	Our estimation (Table 15.4)
1.54	2000	Tilman et al. [41]
1.60	1998	WBGU [49]
1.79	1998	Ramankutty and Foley [12]
+2.79	2000	WRI [8]
3.62	2000	Wood et al. [50] WRI [8]

Note: The cropland areas (rain-fed+irrigated) of the world estimated from different sources for the end of the last millennium show huge differences from one source to another.

different approaches, method, and definitions adopted. Most global digital maps [18,19,22] overestimate agricultural areas as a result of the full-pixel-based area calculations; and other studies reported estimates of potential CAs [3,4]. A pixel when classified as agriculture is automatically taken to have 100% croplands in digital global maps. In reality, only a certain percentage of a pixel is in croplands and that percentage can vary substantially. As a result, the total agricultural lands estimated in various digital maps were 2.7 Bha by Olson and Watts [19] using a 50-km grid, 3.2 Bha by Matthews [18] using a 100-km grid, and 2.8 Bha by IGBP and USGS using a 1-km grid [22].

FAOSTAT [7] estimated 1.51 Bha for 1998 and Tilman et al. [41] estimated 1.54 Bha for the nominal year 2000. Our estimate for 1999 (1.53 Bha) falls between these two estimates. It shows there is an increase of about 0.03 Bha of CA globally. Some differences in these three estimates may also be due to differences in methods and approaches employed in these three data sets. Grubler [51] estimated that an increase of 1 Bha of arable lands would be needed for an additional world population of five billion in this century. The overall trend at a global perspective clearly indicates that the increased human population had put tremendous pressure on land resources from the mid-1970s to the 1980s, which led to clearing of further land areas for food production. In 1960, per capita cropland land availability was about 0.5 ha when the world population was only three billion [52]. However, by the end of the last millennium, per capita CA dropped by half (0.25 ha) while the human population doubled and reached more than 6 billion (Figure 15.7).

Figure 15.5 and Table 15.4 summarize both the global and continental level CAs, including irrigated, rain-fed, and per capita availability of these three categories. Asia ranks first in all three categories of croplands (irrigated, rain-fed, and total croplands).

Of the total 1.13 Bha of rain-fed CA in the world (Table 15.4), Asia ranks first with 29%, followed by Europe (20%), North America (17%), Africa (17%), South America (14%), and Oceania (3%). Asia also has the largest proportion (73%) of irrigated CAs in the world, followed by Europe (9%), North America (9%), and Africa (7%). Out of the total CA in the world, Asia accounts for 40%, followed by Europe (17%), North America (15%), and Africa (13%) (Table 15.4 and Figure 15.5).

The total CA of the world has not changed as significantly as the changes in irrigated CAs. Grain production has increased with the tremendous expansion of the irrigation systems and the use of synthetic fertilizer-pesticide inputs and high-yielding varieties, which have peaked in developed countries. Developing countries are importing food from the developed nations whose human population is less but with a high crop production yield. In the future, approximately 80% of increased crop production in developing countries will have to come from intensification: higher yields, increased multiple cropping, and shorter fallows [27].

15.4.2 NATIONAL SCENARIOS

The CA statistics are estimated for the 182 countries. Table 15.6 provides detailed country-level irrigated, rain-fed, and total CAs organized following the ranking of the rain-fed CAs. It is obvious that some countries like India and China have relatively low rain-fed areas, but an overwhelming proportion of irrigated areas (Table 15.6). The rain-fed croplands (Figure 15.4) are dominant in the United States (11.8% of 1.13 Bha), followed by Russia (10.1%), China (8.1%), Brazil (7.72%), India (4.31%), Australia (3.25%), Canada (3.09%), and Argentina (3.03%). These eight countries account for 51.4% of all the global rain-fed CAs. There are 12 countries that have a 1.1–2.8% proportion of the global rain-fed CAs. The rest of the countries have less than 1% of the total global rain-fed CAs. The first 20 countries account for 70% of the global rain-fed croplands and the first 40 countries account for 84.2% of the global rain-fed croplands.

We have also compared the total CAs with the FAO statistics [7]. We aggregated the remote-sensing-based irrigated and rain-fed CAs to produce the total CA, which enables a direct comparison between the remote-sensing-based data set and the agricultural census data set. Hereafter, we call the remote-sensing-based total cropland data set as the RSD-Crop data set, and the national agricultural census-based total CA data set as the ACD-Crop data set. According to the RSD-Crop data set, China has the largest CA, approximately 204 Mha, followed by the United States (162 Mha), India, (150 Mha), Russia (129 Mha), and Brazil (92 Mha). These five nations have a

TABLE 15.6
Rain-Fed Cropland Area Ranking by Country

Rank	Countries	Rain-Fed Cropland Area (ha)	Irrigated Cropland Area* (ha)	Total Cropland Area (ha)
1	USA	133,571,602	24,309,188	161,617,081
2	Russia	114,788,560	11,203,530	128,675,415
3	China	91,635,702	151,802,086	203,624,473
4	Brazil	87,408,556	4,085,844	91,603,674
5	India	48,824,269	132,253,854	150,059,162
6	Australia	36,758,302	5,373,409	48,623,546
7	Canada	34,944,402	2,874,252	37,602,699
8	Argentina	34,318,900	8,766,412	43,623,158
9	Kazakhstan	31,722,986	6,469,685	38,950,704
10	Ukraine	28,290,153	2,381,799	31,285,731
11	France	17,648,821	2,687,153	20,048,339
12	Indonesia	17,573,608	3,322,443	20,746,487
13	Zambia	16,677,106	536	16,677,885
14	Tanzania	16,410,652	46,998	16,457,674
15	Congo	15,815,336	20,375	15,837,169
16	Spain	15,392,046	3,025,823	18,813,770
17	Poland	14,424,037	454,111	14,775,550
18	Mozambique	13,726,544	60,742	13,782,960
19	Angola	13,454,118	34,158	13,477,434
20	Mexico	12,497,923	3,608,730	16,352,595
21	Belarus	10,968,114	60,926	11,052,202
22	Turkey	10,603,366	1,577,313	12,356,748
23	Ethiopia	10,564,343	162,808	10,748,582
24	South Africa	10,097,803	828,491	10,918,843
25	Thailand	9,931,747	7,397,368	16,542,332
26	Nigeria	9,572,789	216,154	9,770,698
27	Germany	8,998,878	3,001,674	11,196,575
28	Bolivia	8,803,829	163,036	9,017,920
29	Zimbabwe	8,781,932	3,533	8,786,677
30	Sudan	7,816,063	1,930,592	9,553,181
31	Romania	7,563,254	2,049,888	9,938,493
32	Philippines	7,479,645	1,789,108	9,022,274
33	Italy	6,436,452	2,644,140	9,265,975
34	Myanmar	6,257,996	6,306,671	10,710,993
35	Vietnam	5,967,528	4,949,533	10,351,550
36	Kenya	5,944,333	104,527	6,029,734
37	Paraguay	5,538,996	25,029	5,567,578
38	Iran	5,509,694	2,488,558	8,133,031
39	Colombia	5,359,287	592,495	5,905,473
40	Madagascar	5,345,476	75,156	5,417,835
41	Malaysia	5,042,468	274,565	5,301,234

(Continued)

TABLE 15.6 (Continued)

Rank	Countries	Rain-Fed Cropland Area (ha)	Irrigated Cropland Area* (ha)	Total Cropland Area (ha)
42	UK	5,014,629	1,060,204	5,985,362
43	Uganda	5,012,869	30,586	5,042,886
44	Cote d'Ivoire	4,986,024	101,890	5,081,162
45	Peru	4,846,774	374,954	5,202,729
46	Hungary	4,358,475	186,221	4,600,188
47	Cambodia	3,868,166	938,441	4,604,483
48	Pakistan	3,642,557	15,959,342	17,678,708
49	Morocco	3,603,724	1,153,817	4,648,843
50	Algeria	3,520,819	136,946	3,665,169
51	Japan	3,428,667	2,468,596	5,953,762
52	Bulgaria	3,416,518	1,012,064	4,718,322
53	Uruguay	3,354,348	360,055	3,735,751
54	Venezuela	3,256,971	807,078	4,151,851
55	Botswana	3,198,620	4,278	3,204,037
56	Nepal	3,131,060	1,477,303	4,383,047
57	Czech Republic	3,068,209	701,727	3,586,245
58	Serbia	2,947,604	234,348	3,119,543
59	Ecuador	2,844,430	281,166	3,133,011
60	Uzbekistan	2,821,987	5,295,515	6,423,474
61	Greece	2,757,498	766,678	3,665,237
62	Portugal	2,756,177	313,908	3,115,042
63	Afghanistan	2,748,082	923,490	3,756,220
64	Ghana	2,716,307	71,764	2,776,954
65	Lithuania	2,651,512	41,591	2,708,784
66	Cameroon	2,591,767	52,128	2,644,461
67	Bangladesh	2,536,292	7,166,028	7,771,342
68	Chile	2,412,213	1,445,230	3,927,135
69	CA Republic	2,393,214	1,086	2,394,369
70	Congo	2,386,480	0	2,386,480
71	Guinea	2,190,800	320,350	2,493,433
72	Mongolia	2,136,984	376,378	2,559,316
73	Mali	2,051,073	65,879	2,107,428
74	Latvia	2,040,565	7,325	2,053,248
75	Burkina Faso	2,025,961	14,660	2,041,623
76	Cuba	2,007,424	637,159	2,494,322
77	Malawi	1,996,142	2,794	1,999,435
78	Kyrgyzstan	1,990,967	770,274	2,691,843
79	Senegal	1,980,242	290,572	2,191,659
80	Korea, Rep.	1,928,760	1,313,755	3,121,229
81	Laos	1,917,269	107,734	2,022,854
82	Austria	1,822,194	98,551	1,938,650
83	Slovakia	1,690,815	75,488	1,800,719

(Continued)

TABLE 15.6   (Continued)

Rank	Countries	Rain-Fed Cropland Area (ha)	Irrigated Cropland Area* (ha)	Total Cropland Area (ha)
84	Papua New Guinea	1,607,752	0	1,607,752
85	Liberia	1,598,806	300	1,599,043
86	Korea, DR.	1,598,207	2,053,625	3,065,469
87	Croatia	1,551,680	44,630	1,586,883
88	Moldova	1,533,012	229,433	1,827,082
89	New Zealand	1,459,699	141,686	1,585,089
90	Guatemala	1,440,867	91,313	1,510,240
91	Sri Lanka	1,439,246	809,579	2,387,275
92	Azerbaijan	1,398,784	821,980	2,234,411
93	Honduras	1,384,346	77,729	1,454,930
94	Iraq	1,356,711	2,626,564	3,576,735
95	Sierra Leone	1,336,205	29,037	1,358,012
96	Bosnia/Herzegovina	1,303,620	14,203	1,314,387
97	Tunisia	1,284,882	100,647	1,394,025
98	Nicaragua	1,241,957	22,720	1,258,396
99	Tajikistan	1,190,392	449,153	1,573,635
100	Somalia	1,189,487	403,574	1,561,962
101	Taiwan	1,111,947	677,877	1,610,990
102	Belgium	1,101,425	507,430	1,426,221
103	Gabon	1,084,861	0	1,084,861
104	Estonia	1,052,562	14,476	1,077,199
105	Sweden	1,040,821	71,108	1,124,739
106	Panama	1,037,572	45,048	1,086,641
107	Chad	925,287	27,698	950,520
108	Syria	879,249	596,263	1,446,239
109	Albania	864,549	225,864	1,088,326
110	Burundi	798,743	8,490	810,536
111	Georgia	796,878	146,141	925,416
112	Costa Rica	772,096	15,791	784,724
113	Togo	725,130	23,843	746,856
114	Finland	721,148	71,961	846,455
115	Rwanda	710,557	64,806	790,624
116	Niger	703,697	4,317	707,826
117	Macedonia	695,920	131,620	865,762
118	Switzerland	690,849	36,976	720,372
119	Namibia	672,697	9,303	683,224
120	Benin	668,742	15,415	683,915
121	Ireland	630,766	0	630,766
122	Lesotho	571,627	3,681	577,303
123	Netherlands	564,102	1,011,340	1,434,345
124	Dominican R.	550,415	79,648	621,291

(Continued)

TABLE 15.6   (Continued)

Rank	Countries	Rain-Fed Cropland Area (ha)	Irrigated Cropland Area* (ha)	Total Cropland Area (ha)
125	Turkmenistan	542,979	1,999,984	2,065,350
126	Slovenia	542,220	510	542,659
127	Denmark	517,119	979,539	1,681,824
128	Haiti	486,161	53,903	537,009
129	Norway	485,336	1,453	487,408
130	Swaziland	446,942	97,004	596,216
131	Armenia	415,591	118,324	522,285
132	Libya	412,158	210,022	642,814
133	Montenegro	364,360	13,908	374,691
134	El Salvador	294,667	10,401	306,258
135	Egypt	281,590	3,292,726	2,425,690
136	Eritrea	232,850	13,776	249,868
137	Belize	228,077	3,510	231,964
138	East Timor	222,063	4,061	225,863
139	Yemen	206,310	79,188	297,998
140	The Gambia	197,119	63,415	236,991
141	Guinea-Bissau	191,959	155,389	300,001
142	Guyana	184,027	102,930	280,303
143	Bhutan	154,068	1,396	155,065
144	Cyprus	148,942	4,863	156,041
145	Puerto Rico	138,701	11,253	150,664
146	Lebanon	126,708	25,268	151,455
147	Luxembourg	111,041	0	111,041
148	Israel	101,665	104,542	201,470
149	Mauritania	100,469	20,036	115,592
150	Suriname	88,593	20,774	108,439
151	Brunei	86,509	1,002	87,308
152	Jordan	75,548	52,541	148,266
153	Oman	66,449	30,145	84,302
154	West Bank	64,136	1,542	65,748
155	Saudi Arabia	63,518	551,066	742,196
156	Jamaica	61,139	4,556	66,020
157	Eq. Guinea	57,260	2,644	60,072
158	French Guyana	18,869	2,822	21,729
159	Singapore	8,941	0	8,941
160	Trinidad and Tobago	8,590	1,720	10,449
161	Andorra	5,776	0	5,776
162	Liechtenstein	4,839	0	4,839
163	San Marino	2,060	797	3,162
164	Gaza Strip	693	6,790	6,601
165	Monaco	132	53	206

(Continued)

TABLE 15.6   (Continued)

Rank	Countries	Rain-Fed Cropland Area (ha)	Irrigated Cropland Area* (ha)	Total Cropland Area (ha)
166	United Arab Emirates	0	70,603	93,810
167	Qatar	0	27,596	38,509
168	Kuwait	0	26,753	37,333
169	Mauritius	0	3,910	5,312
170	Antigua and Barbuda	0	2,468	2,270
171	Guadeloupe	0	2,022	1,894
172	St. Kitts and Nevis	0	1,445	1,650
173	Djibouti	0	587	905
174	Virgin Islands	0	1,015	827
175	Reunion	0	846	651
176	Anguilla	0	404	489
177	Comoros	0	417	241
178	Turks and Caicos Islands	0	170	214
179	St. Pierre and Miquelon	0	59	70
180	Montserrat	0	115	69
181	Seychelles	0	44	66
182	Cayman Islands	0	55	66
		1,131,552,270	466,757,680	1,530,079,274

Note: Ranking of the countries based on rain-fed cropland areas. For irrigated area estimations, Thenkabail, P.S. et al., A global irrigated area map (GIAM) using time-series satellite sensor, secondary, Google Earth, and groundtruth data, *International Journal of Remote Sensing*, 2008 (in press).

total area of 736 Mha of CAs, accounting for 52% of the total CA in the world (Table 15.6). Figure 15.7 shows the comparison in total CA at the national level between the RSD-Crop data set and the ACD-Crop data set in 1999/2000. The root mean square error (RMSE) between these two data sets at the national level is ~6 Mha. To some degree, there is a similarity in the spatial distribution of total cropping areas. Figure 15.7 shows the linear relationship in the national-level statistics between the RSD-Crop and ACD-Crop data sets ($R^2 = 0.94$; $n = 182$).

Approximately one-third of the world's population lives in two countries (China and India); however, these two most populous countries in the world have only 23% of the global CA. The per capita agricultural production in these two countries has been below demand. As a result, their agricultural imports tend to be more than their exports.

The geographical distribution of the total rain-fed CA estimated in this study is closely correlated with that of the human population. The high percent of total rain-fed croplands occurs mostly in countries with high populations. The United States, Russia, China, Brazil, and India account for the highest rain-fed CAs in the world.

However, per capita rain-fed area in India and China is quite low compared to the rest of the top five rain-fed nations.

This study reviews the major changes in global distribution of croplands at the end of the last millennium. The total global CA base changed with great magnitude from approximately 0.75 ha per person in 1900 [53] to approximately 0.25 ha per person in 1999 (this study). The per capita availability of the total CA has not changed uniformly throughout the world. More than half the world's populations reside in the developing countries with less than two-thirds of the per capita CA.

The world population has increased from ~1500 million in 1900 to more than 6000 million by 1999, which is about a 1.3% increase per year, adding a population of roughly 78 million to our planet every year [53,54]. Figure 15.8 shows the comparison of the population map for 2000 with CAs for 1999; we found that, in general, countries with high populations have larger CAs (Figure 15.8). Countries like India, China, the United States, Russia, and Brazil have the highest population density with the largest CAs. The world's two most populated nations, India and China, mostly depend on irrigated CAs rather than on rain-fed croplands.

15.4.3 SPATIAL DISTRIBUTION OF RAIN-FED CROPLAND AREAS

The study of cropland distribution in this research project estimates that 1.5 Bha of the earth's land surface has been put under agricultural crop production (irrigated and rain-fed area). This estimate suggests that roughly one-third of the total potential area available for crop production is being utilized. There is still a large surface area available for further expansion of agricultural croplands, especially in Latin American and African regions. However, expansion of croplands in these regions depends on crop management practices, where soil degradation will determine the success of crop production. Highly populated nations, such as China and India, have little room for further expansion or intensification of CAs.

The spatial distribution of globally rain-fed croplands relative to globally irrigated croplands and other global LULC classes (http://www.iwmigiam.org) is shown in Figure 15.5. China and India with one-sixth of the world population (~2.6 billion) mostly depend on irrigation and, often, on double-cropping, to feed their populations. In contrast, North America and Europe, with a combined population of about 1.3 billion, depend on rain-fed agriculture with only one crop per year. They also export large quantities of food grains to other continents. Globally, on average, only 0.25 ha per capita of cropland (including irrigated and rain-fed) is available for food production. The per capita availability of the world's total CA has decreased due to population growth, soil degradation, and salanization [55–57]. In contrast, North America and Europe have 0.55 ha per capita CA to support the diverse food systems. Over 60% of the world population lives in Asia with support from 40% of the total CA. In others words, Europe and North America account for 16% of the global CA with only 9% of the total population of the world. Asia accounts for only 0.17 ha of total CA available per capita, which is lower than 0.26 ha of the global average. North America has the highest per capita CA of 0.74 ha, followed by Europe (0.36), South America (0.31), and Africa (0.26).

Asia is hosting over 60% of the world population with 40% of CAs. Most of the food production comes from 73% of the irrigated and from 29% of the rain-fed cropping

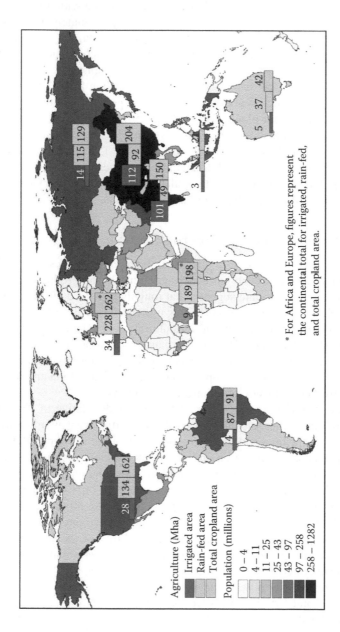

FIGURE 15.8 (See color insert following page 224.) Human population in the twentieth century compared with total cropland area. The most populated countries are in dark blue shades, showing the histogram of irrigated, rain-fed, and total cropland area.

system. In fact, increased food demand in Asia, in general, is met by a large increase in water and land productivity. On the other hand, developed countries such as the United States, Russia, and European nations with roughly one-tenth of the world population, contain nearly one-third of the global CA. Africa is the second largest continent on earth, occupying 20% of the earth's land mass, but it shares only 13% of the CA of the world, which is mostly rain-fed (17%) and just 2% of irrigated areas. Both Americas (North and South) have 5% of the world population, but they share 17% and 14% rain-fed croplands, respectively. The Oceania (including Australia) has just 1% of the world's population with 3% of the total rain-fed CAs, resulting in the highest per capita rain-fed cropland (1.13 ha) as well total CA (2.19 ha).

15.4.4 ACCURACY ASSESSMENTS

In this study, we produced the first GMRCA, and also faced a big challenge in carrying out a comprehensive quality assessment, as no comparable data sets were available for comparison. Limited efforts have been made to characterize the spatial distribution of the irrigated areas using national statistics [58,59] and remote-sensing techniques [16,37]. Other efforts include the global cropland mapping by Ramankutty and Foley [12] and Wood et al. [50].

The map accuracies and errors of our rain-fed cropland mapping are assessed based on pooling two unique and independent data sets that are not used during the class identification and labeling process. Altogether, 1924 groundtruth data points were used in the class-accuracy assessment. First, the 915 groundtruth data points reserved for accuracy assessment from GMRCA field campaigns were pooled and the accuracy was assessed. Second, the 1009 randomly generated groundtruth data points were also used separately. Finally, all the points from groundtruth and Google Earth were pooled and an accuracy assessment was conducted using 1924 points. Table 15.7 shows the accuracy of the rain-fed croplands with errors of omission and commission. The accuracy of the rain-fed croplands varied between 91% and 98%. However, the errors of omission were 2–8% and errors of commission were 19–36%. The pooled data from the two sources provided a rain-fed cropland accuracy of 94% with errors of omission of 5% and errors of commission of 27% (Table 15.7).

The large error of commission indicates that a certain fragmented proportion of the other LULC areas was also mapped as rain-fed croplands. However, there are a number of fundamental issues related to accuracy assessments at such large scales as 1-km or 10-km resolution pixel size, as explained by Thenkabail et al. [15]. First, because of the lack of a complete groundtruth database at the global scale, there are considerable difficulties in groundtruthing and establishing the exact percent of rain-fed fraction at the pixel size of ~10,000 ha (AVHRR at ~10×10 km). This will lead to the pixel being labeled as rain-fed in groundtruth data in one corner of the ~10,000-ha pixel, whereas in reality it has some percent of other LULC fragments. Satellite sensors capture the average reflectivity from the pixel and hence are influenced by both the rain-fed and the non-rain-fed crops within the pixel, leading to average spectra for the pixel. This can lead to somewhat higher errors of omission and commission. The phenomenon is acute when dealing with pixels of other LULC

TABLE 15.7

Accuracy of the Rain-Fed Cropland Areas of the World

Level of Accuracy Assessment	Total Groundtruth Sample Size of Rain-Fed and Noncultivated Areas (No.)	Total Groundtruth Sample Size of Rain-Fed Areas (No.)	Accuracy of Rain-Fed Area Classes (Rain-Fed Gt Points Falling on Rain-Fed Areas) (%)	Errors of Omission (Rain-Fed GT points Falling on Non-Rain-Fed Areas) (%)	Errors of Commission (Non-Rain-Fed GT Points Falling on Rain-Fed Areas) (%)
(a) IWMI GT data	915[a]	549	92	8	36
(b) GE GT data	1009[b]	647	98	2	19
(c) IWMI GT+GE GT	1924[c]	1196	95	5	27

Note:　The rain-fed classes are assessed for accuracies using data from three sources: (a) IWMI project groundtruth, (b) Google Earth groundtruth data, and (c) pooled data of the IWMI project groundtruth and Google Earth groundtruth data.

[a]　Groundtruth (GT) data points of the world for rain-fed and non-rain-fed areas collected by IWMI. The same data were also used for identification of the disaggregated classes. These are nonindependent data sets.

[b]　Google Earth groundtruth (GEGT) data points of the world for rain-fed and noncultivated areas collected by IWMI. The GEGT points were randomly generated and are completely independent. These 1,009 points were not used in the class identification and labeling process.

[c]　The combined IWMI GT data and GE GT data. Most of the points used in accuracy assessment were not used in the class identification and labeling process.

areas (e.g., grasslands, shrublands, etc.) with fragmented rain-fed croplands that lead to higher errors of commission.

15.5 CONCLUSIONS

Rain-fed agriculture plays an important role in global food production and the economy. Despite the increased crop production, irrigated area and rain-fed croplands account for 60% of the total area under food crops. For maintaining food security even at the current nutritional levels, approximately 1.5 billion tons of food grains need to be produced additionally by 2030. Therefore, tracking changes in spatial distribution and changing patterns of rain-fed croplands are essential for understanding and planning the food and nutritional demands of expanding populations of the world. In this context, this research outcome provides a benchmark measure of rain-fed CAs of the world at the end of the last millennium. The SPA of rain-fed croplands provides realistic estimates of the actual area cultivated.

A GMRCA is produced using the multiple-resolution, multiple-platform remote sensing at nominal resolution of 10-km and secondary data for 1999–2000. SPA statistics on rain-fed croplands and total CA are determined for the 182 countries, using a suite of geospatial techniques. Then, methods for mapping rain-fed CAs are developed, investigated, and described. Key areas of the present research findings include (a) climatic- and topographic-based global image partition; (b) use of spectral characteristics, spectral signatures, and SMTs for grouping similar classes; (c) techniques for resolving mixed classes by masking and reclassifying decision-tree algorithms and spatial modeling; and (d) subpixel rain-fed and total CA estimation. The global rain-fed CA estimate from the GMRCA 9-class map is 1.13 Bha.

Due to inadequate access to water and concerns about the sustainable environmental flows, the scope for further expansion of irrigated areas is limited in many parts of the world. Except in Asia, the bulk of the world's food production comes from the rain-fed CAs. Of the 1.53 Bha of croplands worldwide, only 399 Mha (26%) constitute irrigated land; the remaining 1.13 Bha (74%) constituting rain-fed land. Even though almost all agroclimatic zones of the world practice rain-fed croplands, the importance and intensity of rain-fed agriculture vary from region to region. Rain-fed agriculture in Africa, Europe, South America, and Australia contributes more than 80% of the total cereal production of these regions. However, in many countries, productivity remains low due to less than optimal rainfall, unfavorable land conditions, and lack of proper management of these resources.

This study estimated the global rain-fed CAs as 1.13 Bha at the end of the last millennium. This is three times the net global irrigated area estimated at 399 Mha by Thenkabail et al. [16]. A country-by-country comparison between total CAs estimated in this study with the agricultural census data set shows the linear relationship $R^2 = 0.94$; $n = 182$. The estimated mean square error (EMSE) between these two data sets at the national level is ~6 Mha. There is some degree of similarity in spatial distribution of CAs. The accuracies and errors of GMRCA are assessed using groundtruth and Google Earth data points. The accuracy varies between 92% and 98% with errors of omission of 2–8% and errors of commission of 19–36%.

The total global CA (rain-fed plus irrigated) is 1.53 Bha, a value that closely agrees with two other estimates based on national statistic sources. The FAOSTAT [7] estimated 1.51 Bha for 1998 and Tilman et al. [41] estimated 1.54 Bha for the nominal year 2000. We estimated 1.53 Bha for 1999, which falls between the above two national agricultural census-based estimates. The above three estimates, 1.51 Bha (1998), 1.53 Bha (1999), and 1.54 Bha (2000) show there is an increase of about 0.03 Bha of CA globally from 1998 to 2000. However, we expect some differences in these three estimates due to differences in the methods and approaches employed in the three data sets. Nevertheless, there could be minute differences in these estimates, as the total cropped area in the world remained static around 1.53 Bha. So, these areas do not serve as a source of increased output. Thus, the increased yield must come from improved land productivity.

This study reveals the major changes in global distribution of croplands at the end of the last millennium. The total global CA changed with great magnitude from approximately 0.75 ha per person in 1900 to 0.25 ha per person in 1999. When we compared the population map for 2000 with CAs for 1999 we found that, in general, countries with high populations have larger CAs and/or higher cropping intensities. Countries like India, China, the United States, Russia, and Brazil have the highest population density with the largest CAs. The world's two most populated nations, India and China, mostly depend on irrigated CAs to feed their one-sixth of the world population from barely 8% (India 2% and China 6%) of their total land mass of the world. These countries have to meet most of the food demand from the high intensification and irrigated and as well rain-fed CAs. This could even apply to many parts of the world, such as sub-Saharan Africa, Central Asia, and South and Southeast Asia. Globally, on an average, only 0.25 ha per capita of cropland (including irrigated and rain-fed) is available for food production; however, the demand is twice the present per capita estimation. The world population needs at least 0.5 ha per capita of CA to fill the gap between supply and demand and to meet the food requirement of the steadily increasing human populations with little scope for further expansion of the irrigated and rain-fed croplands. Ultimately, the challenge lies in increasing food production by improving land productivity as well as water productivity across the potential cropland zones of the world. Can we expect another green revolution in Asia and Latin America and the beginning of one in Africa?

ACKNOWLEDGMENTS

This study was supported by the IWMI core funding. We acknowledge the vision and support shown by Professor Frank Rijsberman, Director of Water and Climate Adaptation Initiatives, Google.org (formerly Director General of IWMI) and initial funding support from the Comprehensive Assessment of Water Management of Agriculture led by Dr. David Molden. We also acknowledge the effort of numerous people whose direct or indirect contributions led to the successful completion of the project.

REFERENCES

1. Rockstrom, J., Barron, J., and Fox, P. Water productivity in rain-fed agriculture: Challenges and opportunity for smallholder farmers in drought-prone tropical agroecosystems, in *Water Productivity in Agriculture: Limits and Opportunities for Improvement*, J.W. Kijne, R. Barker, and D. Molden, Eds., CAB International, Wallingford, 2003, 145–62.
2. CA (Comprehensive Assessment), Comprehensive assessment (CA) of water management in agriculture, in *Water for Food and Water for Life: A Comprehensive Assessment of Water Management in Agriculture*, M. David, Ed., Earthscan, London, and International Water Management Institute, Colombo, 2007.
3. Xiao, X. et al., *Transient Climate Change and Potential Croplands of the World in the 21st Century*, Joint Program on the Science and Policy of Global Change, Report no. 18, Massachusetts Institute of Technology, Cambridge, MA, 1997.
4. Cramer, W.P. and Soloman, A.M., Climatic classification and future global redistribution of agricultural land, *Climate Research*, 3, 97–110, 1993.
5. Richards, J.F., Land transformation, in *The Earth as Transformed by Human Action*, B.L. Turner, Ed., Cambridge with Clark University, New York, 1990, 163–78.
6. Hansen, M.C. et al., Global land cover classification at 1km spatial resolution using a classification tree approach, *International Journal of Remote Sensing*, 21, 1331–64, 2000.
7. FAOSTAT, *On-line Statistical Database of the Food and Agriculture Organization of the United Nations, 2000*, http://apps.fao.org (accessed February 9, 2004).
8. WRI (World Resources Institute), *World Resources 2000–2001: People and Ecosystems*, World Resources Institute, Washington, DC, 2000.
9. Loveland, T.R. et al., Development of a global land cover characteristics database and IGBP DISCover from 1 km AVHRR data, *International Journal of Remote Sensing*, 21, 1303–30, 2000.
10. Goldewijk, K.K., Estimating global land use change over the past 300 years: The HYDE database, *Global Biogeochemical Cycles*, 15, 417–34, 2001.
11. Defries, R.S. and Townshend, J.R.G., Global land cover: Comparison of ground based data sets to classifications with AVHRR data, in *Environmental Remote Sensing from Regional to Global Scales,* G.M. Foody and P.J. Curran, Eds., Wiley, New York, 84–110,t 1994.
12. Ramankutty, N. and Foley, J.A., Characterizing patterns of global land use: An analysis of global croplands data, *Global Biogeochemical Cycles*, 12, 667–85, 1998.
13. Ramankutty, N. and Foley, J.A., Estimating historical changes in global land cover: Croplands from 1700 to 1992, *Global Biogeochemical Cycles*, 13, 997–1027, 1999.
14. World Resources, *World Resources 1992–1993: A Guide to the Global Environment*, Oxford University Press, New York, 1992.
15. Thenkabail, P.S. et al., *An Irrigated Area Map of the World (1999) Derived from Remote Sensing*, Research Report 105, International Water Management Institute, Colombo, 2006.
16. Thenkabail, P.S. et al., A global irrigated area map (GIAM) using time-series satellite sensor, secondary, Google Earth, and groundtruth data, *International Journal of Remote Sensing*, 2008 (in press).
17. Siebert, S., Hoogeveen, J., and Frenken, K., *Irrigation in Africa, Europe, and Latin America, Update of the Digital Global Map of Irrigation Areas to Version 4*, Frankfurt Hydrology Paper 05, University of Frankfurt (Main), Germany, and FAO, Rome, Italy, 2006.
18. Matthews, E., Global vegetation and land use: New high resolution databases for climate studies, *Journal of Climate and Applied Meteorology*, 22, 474–87, 1983.

19. Olson, J.S. and Watts, J.A., *Major World Ecosystem Complex Map,* Oak Ridge National Laboratory, Oak Ridge, TN, 1982.
20. Olson, J.S., *Global Ecosystem Framework: Definitions,* USGS EROS Data Center Internal Report, Sioux Falls, SD, 1994.
21. Wilson, M.F. and Henderson-Sellers, A., A global archive of land cover and soils data or use in general circulation models, *Journal of Climatology,* 5, 119–43, 1985.
22. Loveland, T.R. et al., An analysis of the IGBP global land-cover characterization process, *Photogrammetric Engineering and Remote Sensing,* 65, 1021–32, 1999.
23. Defries, R. et al., Global discrimination of land cover types from metrics derived from AVHRR Pathfinder data, *Remote Sensing of Environment,* 54, 209–22, 1995.
24. Defries, R. et al., Global land cover classifications at 8 km resolution: The use of training data derived from Landsat imagery in decision tree classifiers, *International Journal of Remote Sensing,* 19, 3141–68, 1998.
25. Friedl, M.A. et al., Global land cover mapping from MODIS: Algorithms and early results, *Remote Sensing of Environment,* 83, 287–302, 2002.
26. Zhan, X. et al., The 250 m global land cover change product from the Moderate Resolution Imaging Spectroradiometer of NASA's Earth Observing System, *International Journal of Remote Sensing,* 21, 1433–60, 2000.
27. FAO (Food and Agriculture Organization of the United Nations), *World Agriculture Towards 2015/30,* Summary Report, Food and Agriculture Organization of the United Nations, Rome, 2002.
28. Smith, P.M. et al., The NOAA/NASA Pathfinder AVHRR 8-km land data set, *Photogrammetric Engineering and Remote Sensing,* 63, 12–31, 1997.
29. Rao, C.R.N., *Nonlinearity Corrections for the Thermal Infrared Channels of the Advanced Very High Resolution Radiometer: Assessment and Recommendations,* NOAA Technical Report NESDIS-69, NOAA/NESDIS, Washington, DC, 1993.
30. Rao, C.R.N., *Degradation of the Visible and Near-Infrared Channels of the Advanced Very High Resolution Radiometer on the NOAAP9 Spacecraft: Assessment and Recommendations for Corrections,* NOAA Technical Report NESDIS-70, NOAA/NESDIS, Washington, DC, 1993.
31. Kidwell, K., *NOAA Polar Orbiter Data User's Guide,* NCDC/SDSD, National Climatic Data Center, Washington, DC, 1991.
32. DCP (Degree Confluence Project), *Degree Confluence Project, 2003 m,* Degree Confluence Project (DCP), Northampton, MA, 2005, http://www.confluence.org/ (accessed July 25, 2005).
33. Granahan, J.C. and Sweet, J.N., An evaluation of atmospheric correction techniques using the spectral similarity scale, *Proceedings of the IEEE 2001 International Geoscience and Remote Sensing Symposium,* Vol. 5, 2022–24, 2001.
34. Homayouni, S. and Roux, M., Material mapping from hyperspectral images using spectral matching in urban area, in *IEEE Workshop on Advances in Techniques for Analysis of Remotely Sensed Data,* Greenbelt, MD, October 2003.
35. Thenkabail, P.S. et al., Spectral matching techniques to determine historical land use/ land cover (LULC) and irrigated areas using time series AVHRR Pathfinder data sets in the Krishna River Basin, India, *Photogrammetric Engineering and Remote Sensing,* 73, 1029–40, 2007.
36. Bing, Z. et al., Study on the classification of hyperspectral data in urban area, *SPIE,* 3502, 169–72, 1998.
37. Thenkabail, P.S. et al., Sub-pixel irrigated area calculation methods, *Sensors Journal,* 7, 2519–38, 2007.
38. Biggs, T. et al., Vegetation phenology and irrigated area mapping using combined MODIS time-series, ground surveys, and agricultural census data in Krishna River Basin, India, *International Journal of Remote Sensing,* 27, 4245–66, 2006.

39. Kauth, R.J. and Thomas, G.S., The tasseled cap – A graphic description of the spectral-temporal development of agricultural crops as seen by Landsat, in *Proceedings of the Symposium on Machine Processing of Remotely Sensed Data*, Purdue University, West Lafayette, IN, 4B-41–4B-50, 1976.
40. Leff, B., Ramankutty, N., and Foley, J.A., Geographic distribution of major crops across the world, *Global Biogeochemical Cycles*, 18, 1–27, 2004.
41. Tilman, D. et al., Forecasting agriculturally driven global environmental change. *Science*, 292, 281–84, 2001.
42. Warnant, P. et al., Caraib – A global model of terrestrial biological productivity, *Global Biogeochemical Cycles*, 8, 255–70, 1994.
43. Houghton, R.A., The annual net flux of carbon to the atmosphere from changes in land use 1850–1990, *Tellus,* 51B, 298–313, 1999.
44. FAO (Food and Agriculture Organization of the United Nations), *1989 Production Yearbook*, UNFAO, Rome, 1990.
45. FAO, *1990 Production Yearbook*, UNFAO, Rome, 1991.
46. Amthor, J.S. and M.o.E.W. Group, *Terrestrial Ecosystem Responses to Global Change: A Research Strategy*, Oak Ridge National Laboratory, Oak Ridge, TN, 1998.
47. Lal, R. and Pierce, F.G., *Soil Management for Sustainability*, World Association of Soil and Water Conservation and the Soil Science Society of America, Ankeny, IA, 1991.
48. WRI (World Resources Institute), *Earth Trends: The Environmental Information Portal*, World Resources Institute, Washington, DC, 2003.
49. WBGU (Wissenschaftlicher Beirat der Bundesregierung Globale Umweltveränderungen – German Advisory Council on Global Change), *Die Anrechnung Biologischer Quellen und Senken im Kyoto-Protokoll: Fortschritt oder Ruckschlag fur den globalen Umweltschutz*, Sondergutachten, Bremerhaven, Germany, 1998.
50. Wood, S., Sebastian, K., and Scherr, S., *Pilot Analysis of Global Ecosystems: Agroecosystems*, Technical Report, World Resources Institute and International Food Policy Research Institute, Washington, DC, 2000.
51. Grubler, A., Technology, *in Changes in Land Use and Land Cover: A Global Perspective*, W.B. Meyer and B.L. Turner II, Eds., Cambridge University Press, New York, 1994, 287–328.
52. Pimental D. and Sparks D.L., Soil as an endangered ecosystem, *American Institute of Biological Sciences*, 50, 11, 947, 2000.
53. Ramankutty, N., Foley, J.A., and Olejniczak, N.J., People on the land: Changes in population and global croplands during the 20th century, *Ambio*, 31, 3, 251–57, 2002.
54. UN (United Nations), Charting the Progress of Populations, United Nations Population Division, http://www.undp.org/popin/wdtrends/chart/contents.htm, 2000.
55. Worldwatch Institute, *Worldwatch Institute, Vital Signs*, W.W. Norton, New York, 2001.
56. Preiser, R.F., Living within our environmental means: Natural resources and an optimum human population, http://dieoff.org/page50.htm (accessed December 20, 2007), 2005.
57. Thomas, D.S.G. and Middleton, N.J., Salinization: New perspectives on a major desertification issue, *Journal of Arid Environments*, 24, 95–105, 1993.
58. Siebert, S. et al., *Global Map of Irrigated Areas Version 3.0*, Goethe University, Frankfurt am Main, Germany, and Food and Agriculture Organization of the United Nations, Rome, Italy, 2005.
59. Siebert, S. et al., *Global Map of Irrigation Areas Version 4.0.1*, Johann Wolfgang von Goethe University, Frankfurt am Main, Germany, and Food and Agriculture Organization of the United Nations, Rome, Italy, 2007.

Section VI

Methods of Mapping Croplands Using Remote Sensing

16 Multiangle Spectral Measurements: A Way to Distinguish Cropping Areas

Francis Canisius
Canada Center for Remote Sensing

CONTENTS

16.1 INTRODUCTION

Satellite remote sensing can be used to update information on cropped area and crop patterns, in order to evaluate the status and trends in food production. As competition for water increases, it becomes increasingly important to monitor irrigated cropping. With multiple-season cropping and enterprise diversification, remote-sensing analysis requires multiple images collected at different stages through individual crop seasons through the year. However, multitemporal analysis of high-resolution imagery, such as Système Pour l'Observation de la Terre (SPOT) and Landsat, is expensive for large-scale studies and encounters difficulties in obtaining sufficient coverage due to cloud contamination. Therefore, considerable attention has been paid to the analysis of free public domain time-series imagery, such as National Oceanic and Atmospheric Administration/Advanced Very High Resolution Radiometer

393

(NOAA/AVHRR), Moderate Resolution Imaging Spectroradiometer (MODIS), and SPOT Variable Geometry Turbocharger (VGT) data [1–3].

Canisius et al. [1] noted that, in tropical regions, a number of natural land covers appear green all year-round, making the classification of cropped area more difficult, especially if relatively coarse-scale imagery is used. Although croplands sometimes have greening patterns similar to natural vegetation, they are homogenous and the distribution of crops is spatially arranged and, therefore, not random compared to natural vegetation. Also, natural vegetation has different structures: for example, forests show highly organized structures with shoots, branches, tree crowns, and tree groups, while crops have none of these attributes. This indicates the potential for further methodological development to better distinguish cropped areas from natural vegetation.

Multiangle remote sensing provides more complete land surface information in terms of spectral, directional, spatial, and temporal characteristics [4–8]. In many research studies, the directional reflectance of the land surface has been demonstrated using multiangle remote-sensing data [9,10]. The difference in directional reflectance can be regarded either as noise that should be removed from wide-swath remote-sensing data or retained as a useful item of information on the structural features of the land surface [11,12]. The structural information about the land cover, such as vegetation canopy structure, can be obtained by using the bidirectional reflectance distribution function (BRDF) parameters from the directional anisotropic properties of multiangle images [13,14].

A number of studies have been conducted to extract the structural features of land covers from directional observations of different sensors. Sandmeier and Deering [15] used multiangle data acquired with the Advanced Solid-State Array Spectroradiometer (ASAR) over the Canadian boreal forest to analyze dominant species based on their structural parameters. Zhang et al. [16,17] used consistency between cover-type definitions, uniqueness of their signatures, and physics of the Polarization and Directionality of the Earth's Reflectances (POLDER) data to classify biomes. This study analyzes the use of Multiangle Imaging Spectroradiometer (MISR) data to separate different land cover types in order to distinguish cropped area from other land cover types.

16.2 THEORETICAL BACKGROUND

Objects on the earth's surface transmit, absorb, reflect, or emit solar radiation in a specific way. Thus, as the amount of energy absorbed by the target increases, the reflected portion decreases in accordance with the law of conservation of energy. The degree of radiation of an object in the electromagnetic spectrum is influenced by the object's optical properties and structure, whether it falls in direct sunlight or shadow, and by the prevailing atmospheric and weather conditions [9].

The irradiance intensity at an object changes with the sun's position, and the measured radiance changes with sensor position. The positive view zenith angle (back to the sun) and the negative view zenith angle (facing the sun) are the two contributions to the reflectance of each azimuth plane. As illustrated in Figure 16.1, Φ_i is the

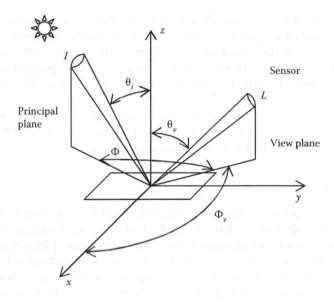

FIGURE 16.1   The viewing geometry of a senso*r*: I and L (solid angle elements), θ (zenith angle), Φ (azimuth angle), i (illumination direction), and v (viewing direction).

principal plane position (position of the sun) and Φ_v is the view plane position. The azimuth relative angle, $\Phi=\Phi_i-\Phi_v$, specifies the view plane position relating to the principal plane. Hence, the viewing geometry of a sensor is determined by the view zenith angle and the view plane position.

16.2.1   BRDF

The BRDF is the ratio of the measured radiance L of a specific wavelength reflected from the surface towards a sensor (positioned at θ_v, Φ_v) to incident irradiance I, illuminating the surface from a direction of the sun's position (θ_i, Φ_i) [18,19].

$$\text{BRDF}\left(\theta_v,\phi_v,\theta_i,\phi_i\right)=\frac{L\left(\theta_v,\phi_v,\lambda\right)}{I\left(\theta_i,\phi_i,\lambda\right)}. \tag{16.1}$$

The amount of radiance reflected from the vegetation cover in a particular view direction for a given illumination is a function of its geometrical and optical properties. Surface orientation with respect to the sun, structure of vegetation, and background properties are some of the principal factors affecting BRDF. The orientation of the surface with respect to the sun directly affects the amount of shadow and its distribution [7,20]. Based on the structural properties of the vegetation surface, the combination of incident ray reflected only once (singly scattered) from the surface to the sensor, and other rays that are reflected multiple times (multiply scattered) from other particles into the field-of-view of the sensor, varies [15]. Absorption and transmittance at the surface are influenced by the background of a surface, which varies depending on the probability of it being seen from the viewpoint [4]. A qualitative analysis of the BRDF of a land cover is not sufficient for satellite image interpretation or classification. Therefore, it is necessary to make quantitative estimates of the deviations of the BRDF behavior of a land cover from an ideal Lambertian surface.

Researchers have come up with various indices, e.g., Bidirectional Reflectance Factor (BRF), Anisotropy Factor (ANIF), and Anisotropy Index (ANIX), in an attempt to quantify the BRDF properties of a land cover.

16.2.2 BRF

The ratio of the radiance reflected from the land cover in a direction L_v, to the radiance L_{ref} reflected from a reference Lambertian reflectance surface in the same direction measured under identical illumination conditions is known as the BRF [18]. An ideal Lambertian surface reflects all the energy, incident upon it (L_{ref}) in all directions equally. The measured radiance depends on the incident illumination I, the direction of the sensor (θ_v, Φ_v), with respect to the surface and the wavelength (λ), and thus the BRF is a function of all these quantities:

$$\mathrm{BRF}\left(\theta_v, \phi_v, \theta_i, \phi_i, \lambda\right) = \frac{L_v\left(\theta_v, \phi_v, \theta_i, \phi_i, \lambda\right)}{L_{ref}\left(\theta_v, \phi_v, \theta_i, \phi_i, \lambda\right)}. \tag{16.2}$$

16.2.3 ANIF

As an ideal Lambertian (isotropic) surface radiates equally in all directions, the observed reflectance in any direction will be identical. Most surfaces in nature are non-Lambertian (anisotropic), i.e., they do not radiate equally in all directions. Different surfaces exhibit different degrees of anisotropy. For example, concrete, which exhibits diffuse reflectance, has a lower anisotropy compared to water, which has a higher anisotropy due to specular reflectance. To measure the degree of anisotropy of a surface, generally the nadir reflectance is used as a reference quantity.

The ANIX of a surface in a particular view direction (θ_v, Φ_v) is defined as the ratio of the reflectance in that direction to the nadir reflectance (θ_0, Φ_0) [21].

$$\mathrm{ANIF}\left(\theta_v, \phi_v, \theta_i, \phi_i, \lambda\right) = \frac{\mathrm{BRF}\left(\theta_v, \phi_v, \theta_i, \phi_i, \lambda\right)}{\mathrm{BRF}\left(0°_v, 0°_v, \theta_i, \phi_i, \lambda\right)}. \tag{16.3}$$

16.2.4 ANIX

The ratio of the maximum and the minimum reflectance factors observed in a given azimuth plane for a given spectral band is known as the Anisotropy Index [15].

$$\text{ANIX}\left(\phi_{v},\theta_{i},\phi_{t},\lambda\right)=\frac{\text{BRF}_{\max}\left(\theta_{v1},\phi_{v},\theta_{i},\phi_{t},\lambda\right)}{\text{BRF}_{\min}\left(\theta_{v2},\phi_{v},\theta_{i},\phi_{t},\lambda\right)}. \tag{16.4}$$

In this equation, θ_{v1} and θ_{v2} are the zenith angles in the azimuth plane Φ_{v}, where the maximum and the minimum reflectance factors are observed. It describes the variation of the reflectance in any given azimuth plane. Generally, the ANIX values in the solar principal plane are used to understand the anisotropy behavior of a land cover.

16.3 METHODOLOGY

The study area lies between 24° and 31° N and 75° to 80° W and covers a significant area of the Indo-Gangetic Basin. The Indo-Gangetic Plain is a fertile, irrigated, agricultural area located in northern India, which makes a major contribution to food production in the country [2]. A research team from the International Water Management Institute (IWMI) collected the extensive groundtruth information in October 2003. Data obtained from the part of the Indo-Gangetic Basin (Figure 16.2) include those from Punjab, Haryana and part of Uttar Pradesh, Madhya Pradesh, and Rajasthan regions, and are used for this study. Cropped land, mainly irrigated agriculture, is the dominant vegetation type (40% of total area) among other land cover types, such as mountain forest, secondary forest shrub, grass, barren land, settlements, and water. Delhi and Agra are the two big cities in this area.

Data from the MISR instrument, obtained on 6 February 2004 over the study area, were used for the analysis. The MISR instrument on board the Terra satellite acquires reflectance data on any earth target in four spectral bands (blue [425–467 nm], green [543–572 nm], red [661–683 nm], and near-infrared (NIR) [846–886 nm]), from nine different directions at a multiple spatial resolution of 275 and 1100 m [22]. The nine along-track angles (Figure 16.3) are 0 (nadir) for the An camera, and 26.1, 45.6, 60.0, and 70.5 forward and backward of nadir for the Af/Aa, Bf/Ba, Cf/Ca, and Df/Da cameras, respectively. The resolution of all four bands in the nadir view and the red band of all the nine angles are in 275 m and other bands are in 1,100 m. One dimension of MISR data fully exploits the multispectral information, while the other dimension capitalizes to assess the anisotropic behavior of terrestrial surfaces with respect to solar radiation.

Calibrated MISR imagery from multispectral and multiangle cameras, after correction of radiometric and geometric errors, is available as a Level 1B2 Georectified Radiance product [23]. The MISR Level 2 product is resized at 1,100 m resolution and is screened for contamination from sources such as clouds, cloud shadows, sun glitter over water, topographically complex terrain, and topographically shadowed regions. Level 2 product contains the hemispherical directional and BRF values with BRF model parameters, bihemispherical and directional-hemispherical reflectance (albedo), leaf-area index, and biome. The MISR Level 1B2 and Level 2 products are mapped to the Space Oblique Mercator (SOM)

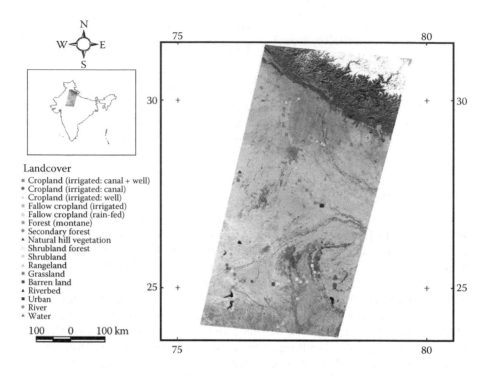

FIGURE 16.2   Study area in northern India.

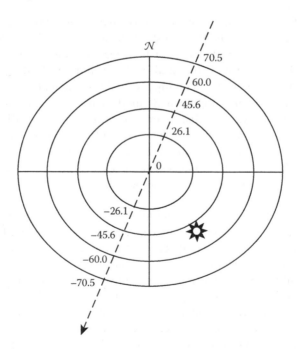

FIGURE 16.3   Explanation of MISR view angles.

projection into an equal-sized blocks of ellipsoidal surface, defined by the World Geodetic System 1984 (WGS84).

The MISR BRF values, representative of several land cover types, were selected using the field data, including irrigated and rain-fed cropland, fallow irrigated and rain-fed croplands, mountain and secondary forest, natural hill vegetation, shrubland, rangeland, grassland, barren land, the riverbeds, urban areas, water, and snow. The irrigated cropland (mostly under winter wheat during the image acquisition period) is the dominant land cover type in the study area. Each land cover type was represented by a number of pixels from different locations (Figure 16.2). The analysis is based on scatterplots, a common technique used in remote sensing to analyze the separability between land cover types.

16.4   ANALYSIS AND RESULTS

16.4.1   Spectral Variation of Land Cover

The extracted spectral features of the land cover classes from MISR Level 2 data (6 February 2004) are shown in two graphs (Figure 16.4a and b). The first graph (Figure 16.4a) is a spectral plot that shows the spectral behavior of the land cover classes. The second graph (Figure 16.4b) is a scatterplot of the pixel values of the red and NIR bands for the different classes. The data points for the nonvegetated classes generally lie on a straight line passing through the origin, called the "soil line." The vegetated land cover classes lie above the soil line due to their higher reflectance in the NIR region relative to the visible region. Pixels falling away from the soil line indicate the densest vegetation or canopy, while pixels near the soil line indicate little to no vegetation. All areas between the dense vegetation and soil line indicate a combination of soil and vegetation. Here (Figure 16.4b) there is a clear spectral separability between very dense vegetation (irrigated cropland, forest) and other land cover types (rangeland, grassland, and barren land). It is noteworthy that the separability between forest and irrigated cropland is not significant.

16.4.2   Anisotropic Variation of Land Cover

The MISR observation is made on an oblique plane about 45° from the perpendicular plane (Figure 16.3), and the contribution of the land cover to the observed total reflectance within the nine view angles may change a great deal based on the vegetation structure. Figure 16.5a and b show the bidirectional reflectance of different land cover classes on an oblique plane observed by MISR in red and NIR wavelengths, respectively. In order to facilitate the interpretation, the ANIF of each land cover class was derived selecting nadir reflectance as the reference quantity. The derived ANIFs for land cover classes on an oblique plane in red and NIR wavelengths are shown in Figure 16.5c and d, respectively.

As shown in Figure 16.5c and d, most of the land cover classes are non-Lambertian (anisotropic) because they do not radiate equally in all observed directions. Different surfaces exhibit different degrees of anisotropy. As noted in Section

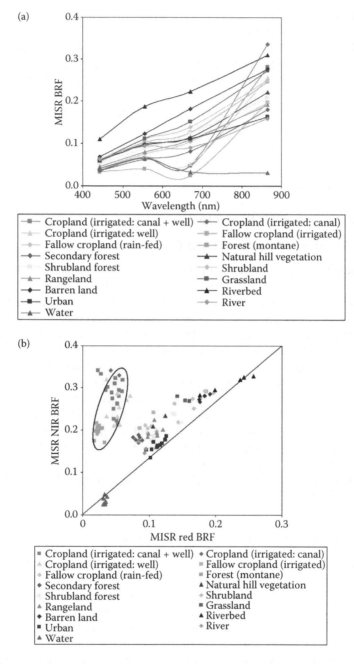

FIGURE 16.4 (a) Spectral behavior of the land cover classes. (b) Scatterplot of the pixel values of the red and NIR bands.

16.2, the riverbeds, which exhibit diffuse reflectance, have the lowest anisotropy, whereas water has the highest anisotropy in red and NIR wavelengths.

The angular patterns of the reflectance curves show some influence of the "hot spot" and the "dark spot" in forest [24], which are most pronounced on the

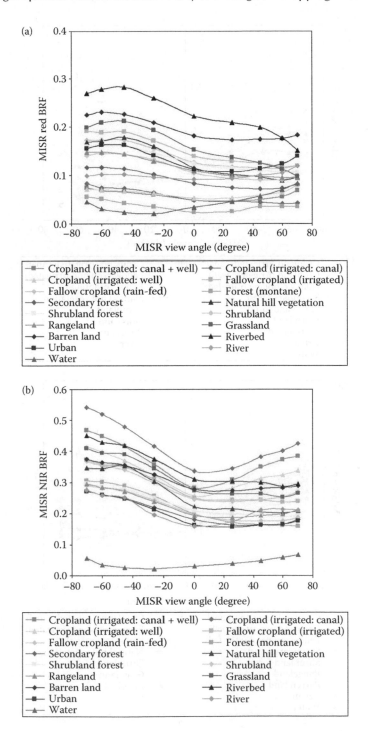

FIGURE 16.5 (a) Bidirectional red reflectance of land covers observed by MISR. (b) Bidirectional NIR reflectance of land covers observed by MISR. (c) Anisotropy factor of land covers in red wavelengths. (d) Anisotropy factor of land covers in NIR wavelengths.

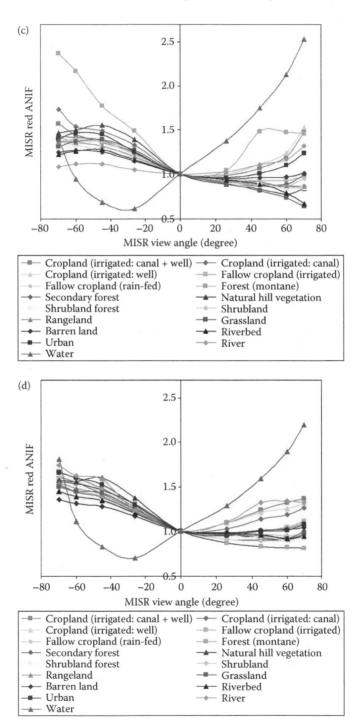

FIGURE 16.5   (Continued)

principal plane, but in the case presented, was less pronounced, as the azimuthal difference between the principal plane and the observation plane was about 45°. When irrigated croplands are compared with other land cover types, they show higher values in the red band in the forward scattering side ("dark spot" side) and much higher values in the NIR wavelength. These differences may be due to their homogeneous surface and their scattering effect compared with other natural vegetation that has specific crown structures causing stronger shadows. The actual situations are, of course, very complex because the reflectance of each pixel viewed from different angles depends on the intrinsic optical properties, background properties, and shadows.

16.4.3 ANGULAR INDEX

The ANIX is defined as the ratio of maximum and minimum reflectance factors observed in a given azimuth plane for a spectral band [15]. Sandmeier and Deering [15] suggest that the reflectance values exhibit maximum contrast in the principal plane and, consequently, show high ANIX and the contrast decreases as the sensor moves to the oblique plane. This is problematic because the observed reflectance is highly sensitive to small angle changes near the "hot spot" [25]. Figure 16.6a and b show that ANIX values of both red and NIR bands associated with reflectance values of the pixels representing each land cover class, are very scattered, which supports Sandmeier's statement. Therefore, MISR-ANIX$_{V60}$ was introduced in this study and was found to effectively increase the separability of the land cover classes while maintaining the homogeneity of the pixels within the classes (Figure 16.6c and d).

$$\text{MISR-ANIX}_{V60}\left(\phi_v,\theta_i,\phi_i,\lambda\right) = \frac{\text{BRF}_F\left(\theta_{-60v},\phi_v,\theta_i,\phi_i,\lambda\right)}{\text{BRF}_B\left(\theta_{+60v},\phi_v,\theta_i,\phi_i,\lambda\right)}. \tag{16.5}$$

In this equation, θ_{-60v} and θ_{+60v} are the forward and backward view zenith angles in the azimuth plane Φ_v, where the reflectance factors are observed.

16.4.4 VEGETATION AND ANGULAR INDEX

In most vegetation indices, visible red and NIR wavelengths are used. These are chosen because visible red corresponds to chlorophyll light absorption, and the NIR is a good indicator of vegetative biomass. The Normalized Difference Vegetation Index (NDVI) is a commonly used vegetation index that brings the multispectral data in a format that is easier to show the amount of vegetation and certainly removes confounding factors, which affect both bands equally. Because of the effect of red and NIR wavelengths (Figure 16.4b), the soils and vegetation in the NDVI derived from MISR should be easily distinguishable.

Further, MISR-ANIX$_{V60}$ can be perceived as an additional source of information to NDVI, since the anisotropy exhibited by surface BRF is based on radiation

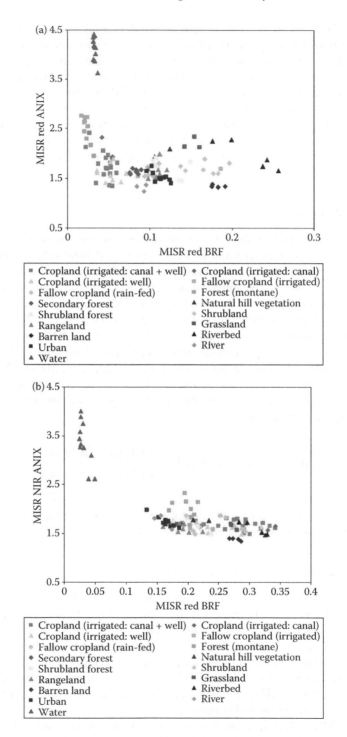

FIGURE 16.6   (a) Anisotropy Index of land covers in red wavelengths. (b) Anisotropy Index of land covers in NIR wavelengths. (c) MISR-ANIX$_{V60}$ of land covers in red wavelengths. (d) MISR-ANIX$_{V60}$ of land covers in NIR wavelengths.

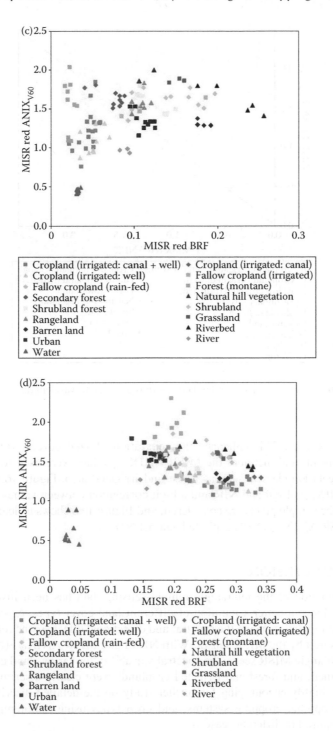

FIGURE 16.6   (Continued)

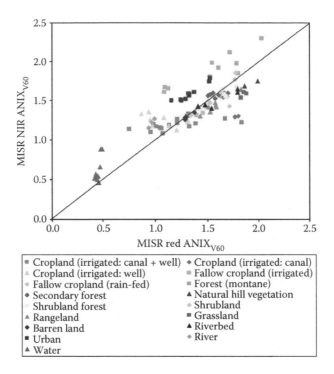

FIGURE 16.7 MISR-ANIX$_{V60}$ of land covers in red and NIR wavelengths.

transfer processes. This enhances the separation of land covers with high vegetation content, and therefore the MISR-ANIX$_{V60}$ values were used to distinguish denser vegetation classes, such as forest and irrigated area. Figure 16.7 shows the MISR-ANIX$_{V60}$ of red and NIR and a high correlation between values is observed when all the sample pixels are considered, and Figure 16.8 shows a clear distinction when MISR-ANIX$_{V60}$ (NIR) is plotted against NDVI.

16.5 CONCLUSION

The contribution of the multiangle remote-sensing data has been investigated in order to achieve a better separation of different land cover types, especially those with dense vegetation, using the spectral and directional effects in the reflected radiance measured by the MISR sensor. With NDVI and MISR-ANIX$_{V60}$ (NIR) from a single multiangle MISR scene, the spectral variability for various land cover classes was identified, and forest and irrigated croplands were clearly discriminated. The results are highly encouraging for further study on the detailed discrimination of rain-fed cropping, cropping systems, and crop types using multitemporal MISR images acquired in different seasons.

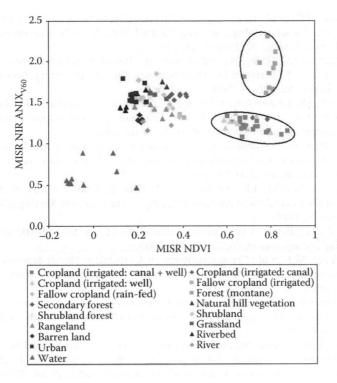

FIGURE 16.8   NDVI and NIR MISR-ANIX$_{V60}$ of land covers.

REFERENCES

1. Canisius, F., Turral, H., and Molden, D., Fourier analysis of historical NOAA time series data to estimate bimodal agriculture, *International Journal of Remote Sensing*, 28, 24, 5503–22, 2007.
2. Thenkabail, P., Schull, M., and Turral, H., Ganges and Indus river basin land use/land cover (LULC) and irrigated area mapping using continuous streams of MODIS data, *Remote Sensing of Environment*, 95, 317–41, 2005.
3. Kamthonkiat, D. et al., Discrimination of irrigated and rainfed rice in a tropical agricultural system using SPOT VEGETATION NDVI and rainfall data, *International Journal of Remote Sensing*, 26, 2527–47, 2005.
4. Canisius, F. and Chen, J.M., Retrieving forest background reflectance in a boreal region from Multi-Angle Imaging SpectroRadiometer (MISR) data, *Remote Sensing of Environment*, 107, 312–21, 2007.
5. Diner, D.J. et al., New directions in earth observing: Scientific applications of multiangle remote sensing, *Bulletin of the American Meteorological Society*, 80, 11, 2209–28, 1999.
6. Hese, S. et al., Global biomass mapping for an improved understanding of the CO_2 balance – the earth observation mission carbon-3D, *Remote Sensing of Environment*, 94, 94–104, 2005.

7. Leblanc, S.G. et al., Investigation of directional reflectance in boreal forests with an improved 4-scale model and airborne POLDER data, *IEEE Transaction on Geoscience and Remote Sensing*, 37, 1396–1414, 1999.
8. Verstraete, M.M., Pinty, B., and Myneni, R., Potential and limitation of information extraction on the terrestrial biosphere from satellite remote sensing, *Remote Sensing of Environment*, 58, 201–14, 1996.
9. Liang, S., *Quantitative Remote Sensing of Land Surface*, John Wiley, Hoboken, NJ, 2005.
10. Sobrino, J.A. (Ed.), *Second Recent Advances in Quantitative Remote Sensing*, Publicacions de la Universitat de Valencia, Spain, 2006.
11. Pinty, B. et al., Uniqueness of multi-angular measurements – part I: An indicator of subpixel surface heterogeneity from MISR, *IEEE Transactions on Geoscience and Remote Sensing*, 40, 1560–73, 2002.
12. Deering, D.W., Eck, T.F., and Banerjee, B., Characterization of the reflectance anisotropy of three boreal forest canopies in spring-summer, *Remote Sensing of Environment*, 67, 205–29, 1999.
13. Gao, F. et al., Detecting vegetation structure using a kernel based BRDF model, *Remote Sensing of Environment*, 86, 2, 198–205, 2003.
14. Chopping, M.J. et al., Canopy attributes of Chihuahuan Desert grassland and transition communities derived from multi-angular airborne imagery, *Remote Sensing of Environment*, 85, 3, 339–54, 2003.
15. Sandmeier, S.R. and Deering, D.W., Structure analysis and classification of boreal forests using airborne hyperspectral BRDF data from ASAS, *Remote Sensing of Environment*, 69, 281–95, 1999.
16. Zhang, Y. et al., Assessing the information content of multiangle satellite data for mapping biomes – II, theory, *Remote Sensing of Environment*, 80, 3, 435–46, 2002.
17. Zhang, Y. et al., Assessing the information content of multiangle satellite data for mapping biomes I, statistical analysis, *Remote Sensing of Environment*, 80, 418–34, 2002.
18. Sandmeier, S.R. et al., Sensitivity analysis and quality assessment of laboratory BRDF data, *Remote Sensing of Environment*, 64, 2, 176–91, 1998.
19. Kukko, A., Hyypp, J., and Kuittinen, R., Use of HRSC-A for sampling bidirectional reflectance, *Journal of Photogrammetry and Remote Sensing*, 59, 6, 323–41, 2005.
20. Chopping, M. et al., Large area mapping of southwestern forest crown cover, canopy height, and biomass using the NASA Multiangle Imaging Spectro-Radiometer, *Remote Sensing of Environment*, 112, 5, 2051–63, 2008.
21. Sandmeier, S. and Itten, K., A Field Goniometer System (FIGOS) for acquisition of hyperspectral BRDF data, *IEEE Transactions on Geoscience and Remote Sensing*, 37, 978–86, 1999.
22. Diner, D.J. et al., Performance of the MISR instrument during its first 20 months in earth orbit, *IEEE Transactions on Geoscience and Remote Sensing*, 40, 7, 1449–66, 2002.
23. Bothwell, G.W. et al., The Multi-Angle Imaging SpectroRadiometer science data system, its products, tools, and performance, *IEEE Transactions on Geoscience and Remote Sensing*, 40, 7, 1467–76, 2002.
24. Chen, J.M. and Leblanc, S.G., A four-scale bidirectional reflectance model based on canopy architecture, *IEEE Transactions on Geoscience and Remote Sensing*, 35, 1316–37, 1997.
25. Chen, J.M., Menges, C.H., and Leblanc, S.G., Global mapping of foliage clumping index using multiangular satellite data, *Remote Sensing of Environment*, 97, 447–57, 2005.

17 Applying Pattern Recognition to Satellite Data for Detecting Irrigated Lands: A Case Study for Georgia, United States

Vijendra K. Boken
University of Nebraska at Kearney

Gerrit Hoogenboom
The University of Georgia

Gregory L. Easson
The University of Mississippi

CONTENTS

17.1 INTRODUCTION

Irrigated lands can withstand the impacts of drought and ensure an increase in agricultural productivity and profits. Therefore, whenever and wherever economically feasible, farmers prefer to irrigate their crops. With increasing demands for surface water and groundwater, regional disputes arise over how to apportion water from the rivers that flow through two or more regions. Such disputes are often prolonged and are not resolved due to various factors, including the lack of reliable data on irrigated area within the territories of the disputed regions. This study presents a methodology for detecting irrigated lands using a pattern recognition (PR) technique on the Advanced Very High Resolution Radiometer (AVHRR) satellite data. As an example, the ongoing dispute among the states of Alabama, Florida, and Georgia in the United States is discussed.

As shown in Figure 17.1, Alabama, Florida, and Georgia share water from several major rivers: Alabama-Coosa-Tallapoosa (ACT) and Apalachicola-Flint-Chattahoochee (ACF). The states of Florida and Alabama often claim that Georgia

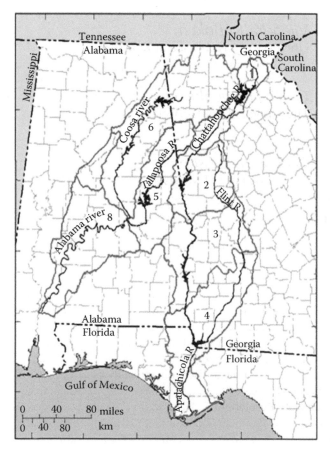

FIGURE 17.1 Major rivers that flow through Alabama, Florida, and Georgia. The polygons representing individual river basins have been numbered.

withdraws more water from these rivers than its fair share; hence the dispute. The dispute has been referred to federal courts, but has remained unresolved for more than 15 years. One of the critical parameters being disputed is the estimate of the total amount of water used (i.e., water usage) in each state by different water-consuming sectors (e.g., agriculture, municipal, industrial, recreational, and hydropower generation).

Water is usually metered in every sector except the agriculture sector. Thus, the proportion of agricultural water usage largely determines the degree of reliability of the data on water usage. The agriculture sector consumes almost 90% of the global water usage [1]. Errors do exist in the estimates of irrigated water usage for the entire world [2]. For example, the United States Department of Agriculture (USDA) collects data on irrigated area simply by sending questionnaires to farmers. In Georgia, the agricultural water usage accounts for approximately 60% of the total water usage in the state. The actual number of irrigated farms in Georgia has not been recorded in state documents, because only the farmers who have installed irrigation devices with pumping capacity equal to or more than 100,000 gallons/day are required to have permits [3]. The number of farmers who operate their irrigation systems without permits is unknown. Moreover, some farmers keep their irrigation data confidential and do not disclose them to the USDA. There is also the likelihood of mistakes in reporting data due to human error. As a result, the data on irrigation currently available with the USDA are subject to errors. Hence, there is a significant need to generate more reliable data on irrigated area that, in turn, may contribute to resolving the water dispute while improving the water planning and agricultural production forecasting.

17.2   OBJECTIVE

The objective of this study was to apply a PR technique on the AVHRR data to detect irrigated lands and improve estimates of the total irrigated area in Georgia.

17.3   PR

PR is a process used to classify an object by analyzing the numerical data that characterize the object. Various academic disciplines, such as image processing, medical engineering, criminology, speech recognition, and signature identification have applied PR to classify objects of interest [4]. Boken et al. [5] applied PR to predict drought in Canada's prairie region. In this study, irrigated land was considered an object to be classified, and the Normalized Difference Vegetation Index (NDVI) derived from the AVHRR data was used as the numerical data to characterize the irrigated land.

17.4   METHODS

17.4.1   STUDY AREA

The study area included 87 counties that comprise approximately 90% of Georgia's agricultural area. These counties also form three main water zones: Flint, Central,

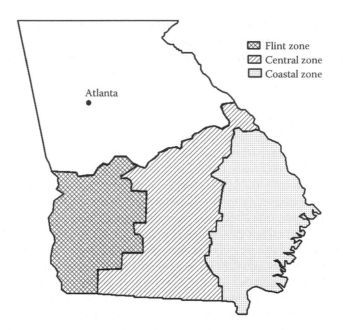

FIGURE 17.2   Study area: Three main water zones of Georgia.

and Coastal, as shown in Figure 17.2. The central pivot system is a common practice used for irrigating the main crops grown in the study area. These crops include cotton, peanut, maize, soybean, wheat, fruits, and vegetables. The combined irrigated area for cotton, peanut, and maize accounted for almost three-fourths of the total irrigated area in Georgia [6].

17.4.2   DATA

We collected NDVI data derived from the AVHRR satellite data for 2000. The AVHRR data are collected in five spectral bands: (a) 0.58–0.68 μm (band 1), (b) 0.725–1.100 μm (band 2), (c) 3.55–3.93 μm (band 3), (d) 10.30–11.30 μm (band 4), and (e) 11.50–12.50 μm (band 5). Various indices have been developed using the AVHRR data for monitoring vegetation conditions over large areas [7–12], but the NDVI, as defined by Equation 17.1, has been widely used for monitoring vegetation [13–15].

$$NDVI = (IR-R)/(IR+R),\qquad(17.1)$$

where IR and R are the reflectance in infrared band and red band, respectively.

The NDVI data were collected from the EROS Data Center in Sioux Falls, SD. The 14-day (biweekly) NDVI composites, resulting in a total of 26 images for a year, were generated using a compositing procedure to minimize cloud coverage. Out of the 26 images, three were found corrupted and could not be used. The original NDVI values, ranging from −1 to+1, were scaled between 0 (for −1) and 200 (for+1).

17.4.3 DESCRIPTION OF METHODS

The goal of this study was to detect irrigated lands using AVHRR-derived NDVI data. The NDVI data are known to detect healthy vegetation. This capability of the NDVI data can also be used to detect irrigated lands, because these lands produce healthier (i.e., greener and spatially more homogeneous) vegetation, while nonirrigated lands may produce moisture-stressed and relatively less homogeneous vegetation.

17.4.3.1 Subsetting

From the US image, a subimage that encompassed the study area was extracted using a Geographic Information System (GIS). The extracted image was then geo-referenced (Figure 17.3). In total, 23 biweekly images were generated for 2000. The next task was to identify polygons representing irrigated and nonirrigated lands in the georeferenced images in order to compute the average NDVI and begin the PR process.

We utilized information available from three sources to identify irrigated and nonirrigated lands on the NDVI images. These three sources included land use-land cover maps, digital orthophoto quarter quads (DOQQ) collected from the Georgia GIS Clearing House (https://www.gis1.state.ga.us), and the locations of the water withdrawal sites used for the Agricultural Water Pumping project (www. AgWaterPumping.net; [16]; Figure 17.4). Subsequently, 23 polygons—17 representing irrigated lands and six representing nonirrigated lands—were delineated. For these polygons, the average NDVI values were computed using a spatial analysis module of a GIS.

FIGURE 17.3 A georeferenced NDVI image of the study area that included the Flint, Central, and Coastal water zones of Georgia.

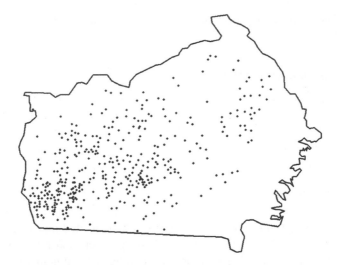

FIGURE 17.4 The water withdrawal sites selected for the Agricultural Water Pumping project of the University of Georgia.

17.4.3.2 Linear Discriminant Analysis

There are various PR techniques available in the literature [14]. We selected a two-category stepwise linear discriminant analysis for the present study. The two categories included irrigated and nonirrigated land. In this type of discriminant analysis, a linear combination of predictor variables is achieved to separate one category from another by maximizing the between-variable variation in relation to the within-category variation.

Statistical software was employed to perform the stepwise linear discriminant analysis to differentiate irrigated land from nonirrigated land, using NDVI values for each. To perform this analysis, the sample of 23 polygons was divided into two groups: a training group and a testing group. The training group consisted of 14 irrigated polygons and four nonirrigated polygons, while the testing group consisted of three irrigated polygons and two nonirrigated polygons.

17.5 RESULTS AND DISCUSSION

Table 17.1 and Figure 17.5 present the variation in the NDVI values for the polygons representing irrigated and nonirrigated lands in Georgia. Following the stepwise discriminant analysis using these NDVI values, four NDVI images (biweek Nos. 1, 9, 20, and 21) were found to be significant enough to distinguish between irrigated lands and nonirrigated lands. These biweekly periods overlapped with the pre- and post-cropping seasons in Georgia. Using the NDVI data for these four images, a linear discriminant model was developed. Table 17.2 presents the coefficients of this model. The accuracy of the model was also determined for both the training group and the testing group. The accuracy in the case of the training group was found to be 100%; in the case of the testing group, the accuracy was only 50%, as shown in Table 17.3.

TABLE 17.1
The NDVI Values for Samples Of Irrigated and Nonirrigated Lands Delineated on the Biweekly NDVI Composites Derived from the AVHRR Data

Biweek No.[b]

Category[a]	1	2	3	4	5	6	8	9	10	11	12	14	15	16	17	19	20	21	22	23
1	150	143	146	137	147	148	144	146	145	160	161	157	154	156	158	165	171	158	143	141
1	157	154	154	153	158	152	155	145	143	140	143	155	161	161	160	162	159	153	143	148
1	150	144	142	135	138	138	145	144	150	150	151	160	155	159	160	162	166	157	145	145
1	168	163	161	158	154	157	169	155	162	167	166	161	142	165	165	166	167	165	147	155
1	148	150	150	144	143	143	141	143	144	143	146	155	163	164	167	167	169	156	147	143
1	149	141	145	141	135	138	150	150	151	150	153	165	167	169	167	168	164	155	149	144
1	145	147	149	150	148	155	160	155	150	149	146	146	158	164	161	168	173	165	143	153
1	142	144	142	143	145	150	146	138	145	143	148	153	158	166	163	167	163	150	148	149
1	150	150	146	143	143	146	146	138	147	152	157	157	155	165	163	156	157	148	142	151
1	147	142	141	135	139	144	147	144	144	147	149	162	166	163	170	167	164	155	150	137
1	154	158	155	151	152	156	168	160	165	167	168	160	169	166	164	169	171	166	151	160
1	148	147	152	144	143	148	142	144	146	142	145	161	161	162	168	164	165	153	143	136
1	161	166	160	160	158	154	154	148	154	154	162	163	155	168	165	175	171	164	148	145
1	149	149	148	144	141	146	146	141	143	147	150	159	165	165	167	169	164	155	146	144
1	154	144	146	142	144	144	142	143	138	142	147	164	163	167	167	174	161	154	141	150
1	148	142	142	138	139	143	154	148	151	149	154	163	166	168	168	171	171	160	142	141
1	152	156	154	151	153	156	163	156	156	155	155	151	161	162	163	164	167	165	139	151
Average_1	151	150	149	145	145	148	151	147	149	150	153	158	160	164	164	166	166	157	145	146

(Continued)

TABLE 17.1 (Continued)

Category[a]	Biweek No.[b]																			
	1	2	3	4	5	6	8	9	10	11	12	14	15	16	17	19	20	21	22	23
2	166	163	160	155	156	158	170	164	169	165	169	156	162	162	164	174	175	164	156	166
2	161	160	158	156	157	157	162	163	157	163	161	158	157	160	156	165	169	159	158	162
2	160	165	158	156	156	157	169	168	169	173	173	160	168	154	162	172	175	163	142	165
2	152	156	148	150	152	155	165	160	163	161	164	152	165	155	160	168	173	161	136	155
2	160	163	160	158	153	156	161	162	162	165	166	156	159	156	157	163	170	163	159	163
2	161	162	154	156	157	159	164	159	168	163	164	166	163	161	152	169	169	168	160	165
Average_2	160	162	156	155	155	157	165	162	165	165	166	158	163	158	159	169	172	163	152	163

[a] 1. Irrigated; 2. Nonirrigated.
[b] The NDVI images for biweek numbers 7, 13, and 18 were found to be corrupt and could not be used.

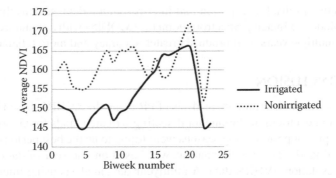

FIGURE 17.5 Biweekly variation in the NDVI values averaged for the irrigated and nonirrigated lands in the Flint, Central, and Coastal water zones of Georgia.

TABLE 17.2

The coefficients for the Linear Discriminant Function to Distinguish Between Irrigated and Nonirrigated Lands in Georgia

Variable	Irrigated	Nonirrigated
Constant	−1323	−1584
Biweek 1	8.777	9.835
Biweek 9	4.944	7.314
Biweek 20	24.686	28.048
Biweek 21	−22.263	−27.538

TABLE 17.3

The Classification Accuracy Achieved by the Pattern Recognition Models for Detecting Irrigated and Nonirrigated Lands in Georgia

	Classification Accuracy for Linear Methods							
	Training Group				Testing Group			
Category	IR	NIR	Total	Percent	IR	NIR	Total	Percent
Irrigated (IR)	14	0	14	100	3	0	3	100
Nonirrigated (NIR)	0	4	4	100	1	1	2	50

17.6 SATELLITE DATA RESOLUTION

In this study, the AVHRR data had a coarse resolution of about 1 km. The AVHRR-derived NDVI data were readily available and prompted the authors to make use of these data. Because of the coarser resolution, it is possible that some of the NDVI pixels may not represent irrigated or nonirrigated lands in their entirety, which may have influenced the results. It is expected that finer-resolution data, such as those derived from the Moderate Imaging Spectroradiometer (MODIS) with a spatial resolution of 250 m for multiple years, will produce greater accuracy and more reliable results.

17.7 CONCLUSION

The stepwise linear discriminant analysis of PR using the AVHRR-based NDVI data was found to be a useful technique for detecting irrigated lands for Georgia, United States. The pre- or postcropping seasons were found to be the best periods for detecting irrigated lands. The present analysis was based on one year of data pertaining to coarse-resolution AVHRR data. A multiyear data analysis using finer-resolution data, such as MODIS, may improve the reliability of results for detecting irrigated lands.

ACKNOWLEDGMENTS

The National Oceanic and Atmospheric Administration/National Environmental Satellite Data and Information Services, Camp Springs, MD, funded this project through Grant No. NA16EC2373. The AVHRR-based NDVI composites were provided by James Rowland of EROS Data Center, Sioux Fall, SD.

REFERENCES

1. Doll, P. and Siebert, S., Global modeling of irrigation water requirement, *Water Resources Research*, 38, 4, 8.1–8.10, 2002.
2. Droogers, P., *Global Irrigated Area Mapping: Overview and Recommendation*, Working Paper 36, International Water Management Institute, Colombo, Sri Lanka, 2002.
3. Houser, J.B. et al., Using a small sub-sample to project state-wide agricultural water use in 2000, in *Proceedings of the 2001 Georgia Water Resources Conference*, K.J. Hatcher, Ed., Institute of Ecology, The University of Georgia, Athens, GA, 110–13, 2001.
4. Duda, R.O., Hart, P.E., and Stork, D.G., *Pattern Recognition and Scene Analysis*, John Wiley, New York, 2001.
5. Boken, V.K., Haque, C.E., and Hoogenboom, G., Predicting drought using pattern recognition, *Annals of Arid Zone*, 46, 2, 133–44, 2007.
6. USDA (United States Department of Agriculture), *The 1997 Census of Agriculture*, Vol. 1, Geographic Area Series 1A, 1B, 1C, CD-ROM set, National Agricultural Statistics Service, Washington, DC, 1999.
7. Tucker, C.J., Gatlin, J.A., and Schnieder, S.R., Monitoring vegetation in the Nile Delta with NOAA-6 and NOAA-7 AVHRR imagery, *Photogrammetric Engineering and Remote Sensing*, 50, 1, 53–61, 1984.
8. Gallo, K.P. and Flesch, T.K., Large-area crop monitoring with NOAA AVHRR: Estimating the silking stage of corn development, *Remote Sensing of Environment*, 27, 73–80, 1989.
9. Gutman, G.G., Vegetation indices from AVHRR – an update and future prospects, *Remote Sensing of Environment*, 35, 121–36, 1991.
10. Laguette, S., Combined use of NOAA/AVHRR indices and agrometeorological models for yield forecasting, in *Agrometeorological Model: Theory and Applications in the MARS Project*, J.F. Dallemand and P. Vossen, Eds., 197–208, 1995.

11. Leprieur, C. and Kerr, Y.H., Critical assessment of vegetation indices from AVHRR in a semi-arid environment, *International Journal of Remote Sensing*, 17, 13, 2549–63, 1996.

12. Thenkabail, P.S., Smith, R.B., and Pauw, E.D., Hyperspectral vegetation indices and their relationships with agricultural crop characteristics, *Remote Sensing of Environment*, 71, 158–82, 2000.

13. Plessis, W.P., Linear regression between NDVI, vegetation and rainfall in Etosha National Park, Namibia, *Journal of Arid Environment*, 42, 235–60, 1999.

14. Boken, V.K. and Shaykewich, C.F., Improving an operational wheat yield model for the Canadian prairies using phenological-stage-based normalized difference vegetation index, *International Journal of Remote Sensing*, 23, 20, 4157–70, 2002.

15. Anyamba, A. et al., Monitoring drought using coarse-resolution polar-orbiting satellite data, in *Monitoring and Predicting Agricultural Drought: A Global Study*, V.K. Boken, A.P. Cracknell, and R.L. Heathcote, Eds., Oxford University Press, New York, 2005, 57–78.

16. Thomas, D.L. et al., Status of ag. water pumping: A program to determine agricultural water use in Georgia, in *Proceedings of the 2001 Georgia Water Resources Conference*, K.J. Hatcher, Ed., Institute of Ecology, The University of Georgia, Athens, GA, 2001, 101–04.

18 Mapping Irrigated Crops from Landsat ETM+ Imagery for Heterogeneous Cropping Systems in Pakistan

M.S.D.N. Gamage
International Water Management Institute

Mobin-ud-Din Ahmad
Commonwealth Scientific and Industrial
Research Organisation

Hugh Turral
On the Street Productions

CONTENTS

18.1   INTRODUCTION

Irrigated areas have played a vital role in meeting the global food and fiber demands of the rapidly growing world population over the last five decades. Irrigated areas have expanded from the early 1940s with 95 million hectares (Mha) to 280 Mha at the beginning of the twenty-first century [1], following the "green revolution" in the 1960s. Irrigation remains a key plank in global and regional food-security strategies. Options to expand or enhance agricultural production require information on the extent, location, and type of irrigated crops in the world. Traditionally, this information is collected through laborious field surveys. The accuracy of the resulting maps largely depends on the sampling design and proportion of vegetated area surveyed. Furthermore, due to different levels of accuracy, often the results of these surveys cannot be used to compare spatial and temporal land use changes. With the emergence of satellite remote sensing, several attempts have been made to develop a consistent and rapid approach to map land cover and generate crop statistics [2] at a variety of scales.

Remote-sensing data can be obtained from both active and passive sensors and can be used for crop classification [3,4]. Bastiaanssen [5] categorized the various fields that could use remote-sensing techniques for irrigation, planning and water allocation, productivity measures, catchment hydrology, and water flow relevant to irrigation, and the environment and health of the watershed and irrigation. Brisco and Brown [6] evaluated single-band Synthetic Aperture Radar (SAR) and Thematic Mapper (TM) visible and near-infrared (VNIR) data to classify crops frequently grown in western Canada. They concluded that VNIR data are superior to the SAR data for single date classifications due to the multispectral information content. Multidate classification with SAR data can improve the classification results (up to 74%), but it has a lower accuracy than multidate VNIR (90%). One of the main advantages of active sensors (such as SAR) is that they can operate day and night, and can obtain land surface images under cloudy conditions.

Unsupervised classification is the easiest and most common way to classify satellite images that are entirely based on a mathematical process of grouping individual pixels with similar radiometric characteristics. After the classification, the classes must be identified and named, and this is where three main problems can occur, according to Van Niel and McVicar [7]: classification of meaningless groupings; idiosyncratic definition of an initial number of classes; and subjectivity involved in the combination of similar classes. Finally, classification accuracy depends on the level of spectral separation between crops, but the technique suffers from lower accuracy with mixed land cover signatures.

To overcome these problems and improve classification accuracy, supervised classification approaches, based on prior knowledge, have been developed over the last two decades. Most of these classification approaches [8,9] integrate the remote sensing, Geographic Information System (GIS), and field/historical databases. Hurbert-Moy et al. [10] mention that the method of classification should be based on existing landscape features, but usually it follows the same algorithm without considering this aspect. The scenario and outcomes differ between uniform large-scale landscapes compared to tiny land parcels with numerous different crops, with resulting variation in overall accuracy. Therefore, the options are either to divide the landscape into several more homogenous units or select a classification algorithm to give the best "averaged" results over the entire study area.

Gandia et al. [11] used TM images from Landsat 5 to identify four crop types, rice, vegetable, citrus orchards, and urban areas, in the agricultural area of the Spanish Mediterranean coast. Further, they showed that band 4 (NIR) and band 5 (MIR) offer greater suitability for the classification of rice due to the water absorption characteristic of MIR. Oetter et al. [2] used a number of Landsat TM images within a calendar year to characterize agricultural and related land cover in the Willamette River Basin. They used air-photographs to test and train sample areas with ancillary data in GIS. Then they derived a map of 20 land cover classes, combining a geoclimatic rule set and a regression analysis with unsupervised classification. Many researchers use vegetation indices, such as the Normalized Difference Vegetation Index (NDVI), as an additional or a single layer, to improve the classification accuracy, especially with irrigated lands. Among them, Lloyd [12] used a temporal NDVI profile to distinguish irrigated crops from nonirrigated crops in the Iberian Peninsula. Muchoney and Strahler [13] used multitemporal NDVI, temperature, and additional temporal metric and physical data to map vegetation cover in Central America.

Most of the approaches, such as those presented above, were developed for areas with large fields of single homogenous crops, but often fail to produce acceptable results for small field sizes and mixed cropping systems, which are common in South Asia. Ahmad et al. [14] used Système Pour l'Observation de la Terre (SPOT) XS imagery and field data to classify crops in southeastern Punjab, Pakistan. Gamage et al. [15] further refined this approach with the help of the Principle Component Analysis (PCA), and developed a semisupervised crop classification method for mixed cropping systems in South Asia. This approach was tested using Landsat 7 ETM+ imagery in Rechna Doab and the Pehur High Level Canal (PHLC) system in Pakistan.

18.2 METHODOLOGY

18.2.1 Description of the Study Area

Two irrigation command areas in Pakistan, Rechna Doab and PHLC, located in the middle and upper parts of the Indus, were selected for the study.

Rechna Doab is a land unit lying between the Ravi and Chenab rivers of Pakistan (Figure 18.1). The gross area of the Doab is about 2.97 Mha and currently around 80% of the area is cultivated. It has a maximum length of 403 km and a maximum width of 113 km, and lies between longitude 71°48′–75°20′ E and latitude 30°31′–32°51′ N. Climatologically, the area is subtropical, designated as semiarid, and is characterized by large seasonal fluctuations in temperature and rainfall. Summers are long and hot, lasting from April through September with temperatures ranging from 21 to 49°C. The winter season lasts from December through February with temperatures ranging from 25 to 27°C. The mean annual precipitation is about 650 mm in the upper Doab, falling to 375 mm in the central and lower areas. Almost 75% of the annual rainfall occurs during the monsoonal season from mid-June to mid-

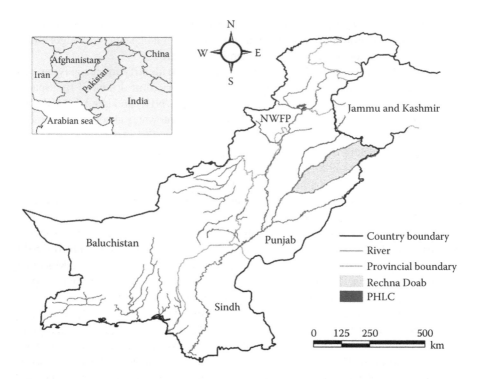

FIGURE 18.1   Location of PHLC and Rechna Doab in the Indus Basin irrigation system of Pakistan.

September. Due to scanty and erratic rainfall, successful agriculture is possible only with irrigation. Rechna Doab is one of the oldest and most intensively developed irrigated areas of the Punjab Province of Pakistan. Rice, cotton, and forage crops are the main kharif crops, whereas wheat and forage predominate in rabi. In central Rechna Doab, sugarcane is dominant and grows over 11 months of the year. Minor crops include oil seeds, vegetables, and orchards. Land distribution is highly fragmented and holdings are variable in size. About 56% of the Rechna Doab land is owned by small farmers, with farm sizes of 5 ha or less each, and 25% is held by large farmers owning 20 ha or more each [16]. Due to low canal discharges, farms consist of small fields of about 0.3 ha. To maximize gross margins, farmers adopt mixed cropping patterns.

The PHLC lies between the Malakand and Swabi hills at longitude 72°05′–72°49′ E and latitude 33°99′–34°10′ N. Climatologically, the area is similar to Rechna Doab, but much cooler in winter due to higher latitude and altitude. The mean annual precipitation is almost double that of Rechna Doab. The PHLC feeds into the Upper Swat Canal system, and takes its water from the Tarbela Dam in the upper Indus, then joins the Machai branch of the Upper Swat Canal, which is diverted from the Swat River. Due to additional flow from PHLC to Machai, it delivers more water to the tail-end farmers of Machai and increases the cropping intensity. It is unusual to have greater wave availability at the tail, compared to the head of large irrigation networks. In PHLC, maize, tobacco, and sugarcane are the major crops in summer and wheat in winter, with irrigation in both seasons. Additionally, vegetable and fodder are also cultivated, but these are highly mixed with perennial orchards. Groundnut and some other legumes could be cultivated in rain-fed areas adjusted to the command area. Land parcels in PHLC are smaller than those in Rechna Doab due to undulating topography.

18.2.2 Data Collection and Management

18.2.2.1 Satellite Imagery

For crop classification, one high spatial resolution (30 m) Landsat 7 ETM+ image per season was used for both areas. For Rechna Doab, two images (*Path 149* and *Row 038*) of 30 September 2001 and 25 March 2002 were selected to represent kharif (2001) and rabi (2001–02) seasons, respectively. For PHLC, two Landsat 7 ETM+ images (*Path 151* and *Row 036*) of 8 April 2002 and 15 September 2002 were selected to represent rabi (2001–02) and kharif (2002), respectively. These images were acquired when the crops were fully matured, after consulting typical local crop calendars.

18.2.2.2 Collection and Organization of Field Data

Information on crops was collected through crop surveys in both PHLC and Rechna Doab through a GPS survey, which also enabled good georeferencing of the satellite images.

In Rechna Doab, 11 sample watercourses were selected in the Punjab rice-wheat and sugarcane-wheat areas (Figure 18.2). Crop type, sowing date, and harvesting date were collected at the Killa (67×60 m cell size) level. On average, a Killa represents four Landsat 7 ETM+ pixels. This information was collected in previous studies conducted by the International Water Management Institute (IWMI). Although the selected watercourses are not well distributed across Rechna Doab (as required under an ideal scenario), they adequately represent the major crops in Rechna Doab. For image classification, Killa grids are generated for each sample watercourse and georeferenced in ARCGIS. This process allows the display of information on crops directly on the satellite imagery, in order to obtain training samples for classification. The net sample area totals about 0.14% of the whole study area, whereas in the PHLC, the field survey was conducted in the sample distributaries of Malakand and Yaqubi. Since, in that area, there is no grid (Killa) system, large homogenous groups of fields with one crop were identified, and GPS readings were taken at the center of these blocks to represent specific crop types.

18.2.3 Classification Approach

A semisupervised classification methodology was developed for crop classification (shown schematically in Figure 18.3). The main steps of the methodology are:

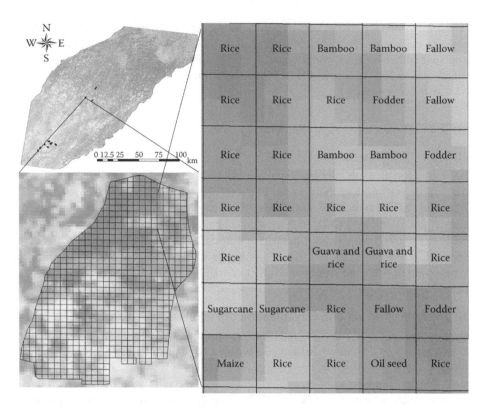

FIGURE 18.2 Location of a sample watercourse in the Rechna Doab. Each grid represents a Killa and information on crops was collected within that unit.

1. Conversion of digital numbers (DN) of a georeferenced image into reflectance. Since Landsat 7 ETM+ images were used, this conversion was done according to the methods outlined in the *Landsat 7 Science Data Users Handbook* [17]. Since the gain on each of the Landsat 7 channels can vary

   from high to low between successive overpasses, it is essential to convert apparent DNs to reflectance for consistency between different images.
2. Segregation of vegetated from nonvegetated areas. A commonly used vegetation index, NDVI, was calculated using the reflectance of NIR, λ_{NIR}, and Red band, λ_{red} (Equation 18.1).

$$NDVI = (\lambda_{NIR} - \lambda_{red}) / (\lambda_{NIR} + \lambda_{red}).$$ (18.1)

The field data attributes of the GIS coverage were overlaid on calculated NDVI and a threshold value for the presence of vegetation was estimated for each image. Next, using NDVI thresholds for each season, the nonvegetated areas were masked out. Nonvegetated areas mainly consist of barren

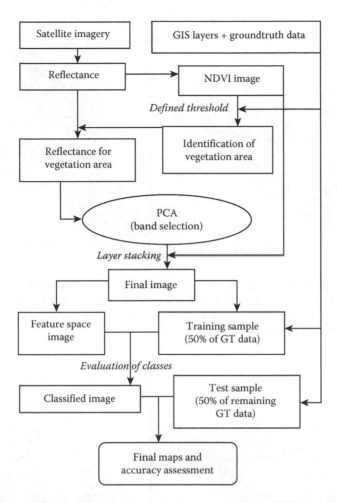

FIGURE 18.3 Schematic diagram of the semisupervised crop classification methodology.

and bare lands within urban areas. Since the images were selected in the middle of the season, it was easy to separate the two categories.

3. PCA, a mathematical transformation, which is based on the linear combination of the band measurements that reduce the number of bands used without losing spectrally useful variation, was performed to select the most useful bands for image classification. The bands selected through PCA are not correlated, and are often more interpretable than the source data [18,19]. The algorithm provided in ERDAS IMAGINE was used for this exercise.

4. The bands selected through PCA were stacked with NDVI. Thereafter, Feature Space Images (FSIs) were created for all band combinations. Of the georeferenced field crop data, 50% was used to obtain training samples for all land use classes. These training samples were evaluated based on their

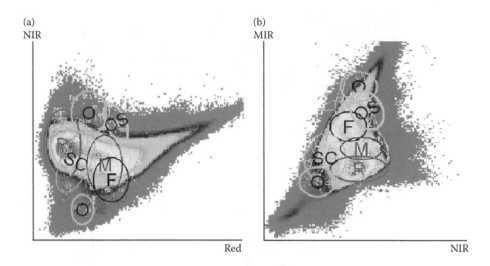

FIGURE 18.4 Feature space images between Red, (a) near-infrared (NIR), and (b) midinfrared (MID) bands of Landsat image of 30 September 2001. The first letters of the polygon names are used for the following: "R" for rice, "O" for orchard, "M" for maize, "F" for fodder, "OS" for oil seed/other vegetation, and "SC" for sugarcane.

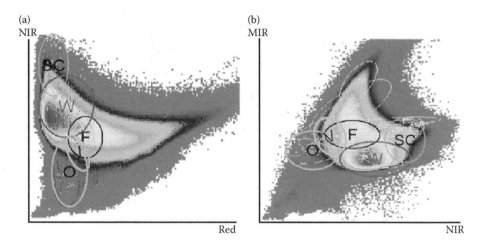

FIGURE 18.5 Feature space image between Red, (a) near-infrared (NIR), and (b) midinfrared (MID) bands of Landsat image of 25 March 2002. The first letters of the polygon names are used for the following: "W" for wheat, "O" for orchard, "F" for fodder, "V" for vegetables/other vegetation, and "SC" for sugarcane.

location on the various scatter plots (Figures 18.4a and b and 18.5a and b), and finally merged into broad crop categories. These final classes were used for supervised classification using the parametric rule of maximum likelihood.

5. Finally, using the unused field and classified images, contingency matrices were computed to access the classification accuracy.

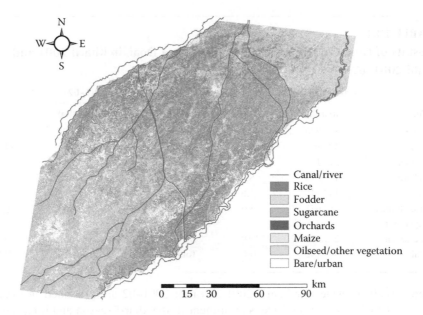

FIGURE 18.6 Land use/land cover (LULC) classification based on Landsat ETM+ image of 30 September 2001 (kharif-2001).

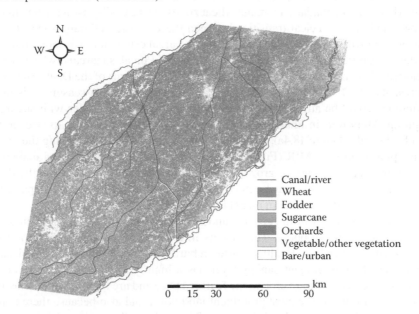

FIGURE 18.7 The LULC classification based on Landsat ETM+ image of 25 March 2002 (rabi 2001–02).

18.3 RESULTS AND DISCUSSIONS

18.3.1 RECHNA DOAB

TABLE 18.1
Results of Land Cover Classification of Rechna Doab in Kharif 2001 and Rabi 2001–02

	Kharif 2001			Rabi 2001–02	
Crop	Area (ha)	%	Crop	Area (ha)	%
Rice	825,930	42.4	Wheat	1,061,676	54.5
Fodder	324,191	16.6	Fodder	272,099	14.0
Sugarcane	210,199	10.8	Sugarcane	206,069	10.6
Orchards	168,175	8.6	Orchards	123,998	6.4
Maize	148,046	7.6			
Oilseed/other vegetation	57,677	3.0	Vegetable/other vegetation	102,803	5.3
Bare/urban	212,975	10.9	Bare/urban	180,547	9.3
Total	1,947,192	100	Total	1,947,192	100

The classified images for kharif 2001 and rabi 2001–02 are presented in Figures 18.6 and 18.7, respectively. Rice is prominent in the kharif season and it represents more than 42% of the total land area of the classified image (Table 18.1). In contrast, wheat is the dominant crop in rabi and covers 54.5% of the total area. The Punjab rice-wheat and the Punjab sugarcane-wheat zones can be easily distinguished in both images and there is a clear transition between these two agroclimatic zones. In both seasons and zones, all crops are planted, but the areal extent of rice is more than 60% in the Punjab rice-wheat zone and only 20% in the Punjab sugarcane-wheat zone. It is noticeable that sugarcane is prominent in the southern part of the Doab. Sugarcane represents 10.8% of the kharif crop area and 10.6% in the rabi season. This minor difference could be attributed to intercropping of wheat or fodder with sugarcane planting. Moreover, in kharif, sugarcane has a similar reflectance to rice in NIR and Red bands (Figure 18.4a), but separation was achieved by analyzing the feature space plot of NIR vs. MIR (Figure 18.4b). Kharif fodder (sorghum and maize) has a similar signature to maize grown for grain. This difference could also be differentiated in all FSIs. This may be the reason that area under fodder is comparatively larger in kharif (16.6%) than in rabi (14.0%).

Orchard areas (6.4%) in rabi are underestimated compared to those in kharif (8.6%), due to intercropping. This results in a mixed signature for pure orchards and orchards with wheat. Healthy, dense orchards are always clearly distinguishable, since their thick prominent canopy layer has a high NDVI. Sparse orchards could not be segmented as a pure crop in the feature space and are thus classified mostly as wheat because it was the most prominent land cover and also because there is little leaf cover in deciduous orchards in winter. This reduces the estimated percentage of orchards in the rabi season by 2%.

Due to a lack of field information, oil seeds, cotton, and vegetables were the most difficult to identify in both images and were often mixed with other vegetation. Cotton is not a major crop in the classified image and hence no field information could be obtained from the selected watercourses. Oil seeds and vegetables are typically grown on very small plots, therefore the sample size of oil seed and vegetable

TABLE 18.2
Contingency Table for the Classified Image of 30 September 2001 at Rechna Doab

Crops	Rice	Fodder	Sugarcane	Orchards	Maize	Oilseed/Other Vegetation	Total	User's Accuracy (%)
Rice	**580**	6	16	20	0	5	627	92.50
Fodder	3	**123**	14	2	2	6	150	82.00
Sugarcane	14	10	**151**	3	2	0	180	83.89
Orchards	3	8	2	**58**	2	0	73	79.45
Maize	0	6	1	0	**7**	0	14	50.00
Oilseed/other vegetation	6	7	3	3	0	**60**	79	75.95
Total	606	160	187	86	13	71	1,123	–
Producer's accuracy (%)	95.71	76.88	80.75	67.44	53.85	84.51	–	87.18

TABLE 18.3

Contingency Table for Classified Image of 25 March 2002 at Rechna Doab

Crops	Wheat	Fodder	Sugarcane	Orchards	Vegetable/ Other Vegetation	Total	User's Accuracy (%)
Wheat	**978**	28	116	16	24	1,162	84.17
Fodder	23	**52**	0	8	0	83	62.65
Sugarcane	4	0	**62**	5	1	72	86.11
Orchards	12	0	1	**54**	0	67	80.60
Vegetable/other vegetation	1	0	1	1	**16**	19	84.21
Total	1,018	80	180	84	41	1,403	–
Producer's accuracy (%)	96.07	65	34.44	64.29	39.02	–	82.82

areas was very small. Especially in kharif, there were no pure Killas of vegetables, thus we were unable to identify vegetables in the summer season. The masked bare/urban area of kharif (10.9%) is higher than that of rabi (9.3%) because the area of bare/fallow land in the kharif season increases slightly due to insufficiency in the availability of water for farming.

For accuracy assessment, contingency matrices were computed. The tables were prepared using the 50% of the field data not used for training (Tables 18.2 and 18.3). The overall classification accuracy was calculated by dividing the number of correctly classified samples by the total number of pixels checked. Two ways of measuring the accuracy of an individual class were used: the *producer's accuracy* and the *user's accuracy*. The producer's accuracy measures the probability of a reference site being correctly classified, whereas the user's accuracy is an indication of the probability that a site classified in the image actually represents that class on the ground. The user's accuracy is calculated by taking the number of correctly classified samples and dividing that number by the classified total. The producer's accuracy is found by dividing the number of correctly classified samples by the reference total for that class.

The overall accuracy of the kharif season supervised classification yielded 87% pixels correctly identified (Table 18.2). Except for orchards and maize, all other crops were classified with high accuracies. The accuracy of maize classification was 54% of producer's accuracy and 50% of user's accuracy. The apparent reason for this low accuracy is intercropping with orchards and the similarity in signatures of kharif fodder (sorghum and maize) and maize grown for grain. In the rabi season, the overall accuracy was 83% (Table 18.3). Wheat is the dominant crop in rabi and classifies with a higher accuracy: 96% of producer's accuracy and 84% of user's accuracy. Sugarcane and fodder have a lower producer's accuracy, which could be attributed to the intercropping of wheat with sugarcane. Vegetables were also classified with low accuracy, and as discussed in the previous section, this could be mainly due to

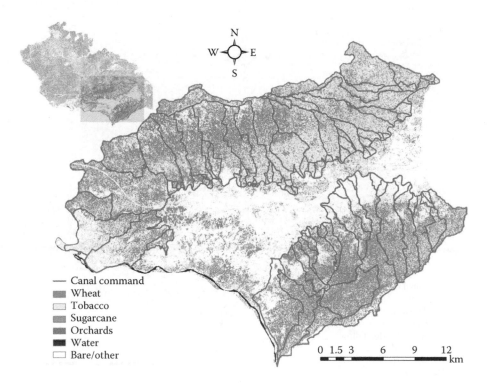

N
W—◇—E
S

— Canal command
▪ Wheat
☐ Tobacco
▪ Sugarcane
▪ Orchards
▪ Water
☐ Bare/other

0 1.5 3   6   9   12
                                 km

FIGURE 18.8 (See color insert following page 224.) The LULC classification of PHLC and USCC, based on Landsat ETM + image of 8 April 2002 (rabi 2001–02).

the very small plot size, small sample size, and more complex mixing, especially, of vegetable types.

18.3.2 PHLC Area

Classified images of the PHLC area are shown in Figures 18.8 and 18.9. Due to the similar conditions of the PHLC and neighboring Upper Swat Canal Command (USCC) area, the classification was extended to cover the USCC as well. Most of the rabi season area is wheat (Figure 18.8) in the PHLC and the southern part of the USCC, while the upper part has a mix of sugarcane and wheat. The extent of the wheat in the area is 38% (Table 18.4) of the total land cover during the rabi season. This is substituted by maize (32%) in the kharif and partially with sugarcane and then tobacco. Although the PHLC is located in the maize-wheat cropping system, tobacco is a major cash crop, which uses land that would otherwise be sown to wheat in rabi. It extends into the kharif season, but may be followed by a short-duration maize crop. Apart from these four major crops, vegetables and orchards are observed but, due to lack of groundtruth information, such areas were lumped in the "Other" class.

Contingency tables were prepared for the PHLC, as in the Rechna Doab crop classification, to assess the classification accuracy and are presented in Tables 18.5

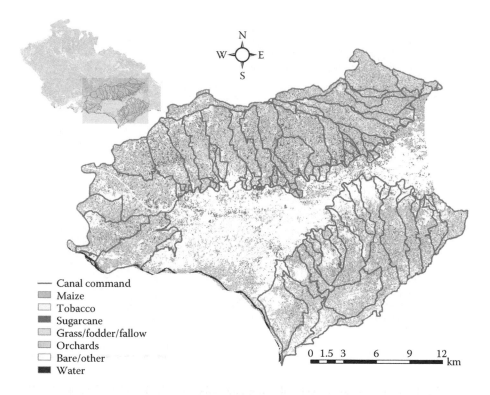

Legend:
— Canal command
▨ Maize
☐ Tobacco
■ Sugarcane
▨ Grass/fodder/fallow
▨ Orchards
☐ Bare/other
■ Water

0 1.5 3   6   9   12 km

FIGURE 18.9 **(See color insert following page 224.)** The LULC classification of PHLC and USCC based on Landsat ETM+ image of 15 September 2002 (kharif 2002).

TABLE 18.4

Results of Land Cover Classification of PHLC and USCC in Rabi 2001–02 and Kharif 2002

Rabi 2001–02			Kharif 2002		
Class	Area (ha)	%	Class	Area (ha)	%
Wheat	112,983	38.1	Maize	96,449	32.5
Tobacco	85,748	28.9	Tobacco	72,496	24.4
Sugarcane	13,738	4.6	Sugarcane	33,602	11.3
Other	84,101	28.4	Fodder/fallow	94,023	31.7
Total	296,570	100.0	–	296,570	100.0

and 18.6, for rabi 2001–02 and kharif 2002, respectively. Similarly, of the unused groundtruth data, 50% was used for this purpose. The overall accuracies of the rabi 2001–02 and kharif 2002 are 85% and 78%, respectively. It was noted that even

TABLE 18.5
Contingency Table of the Classified Image of 8 April 2002

Crops	Wheat	Tobacco	Sugarcane	Bare/Other	Totals	User's Accuracy (%)
Wheat	74	5	4	1	84	88.1
Tobacco	7	76	1	2	86	88.4
Sugarcane	3	0	13	0	16	81.3
Bare/other	5	4	0	18	27	66.7
Total	89	85	18	21	213	–
Producer's accuracy	83.1	89.4	72.2	85.7	–	84.97

TABLE 18.6
Contingency Table of the Classified Image of 15 September 2002

Crops	Maize	Tobacco	Sugarcane	Fodder/Fallow/Other	Total	User's Accuracy (%)
Maize	20	–	3	3	26	76.9
Tobacco	6	22	–	3	31	71.0
Sugarcane	1	–	12	2	15	80.0
Fodder/fallow/other	–	–	–	11	11	100.0
Total	27.0	22.0	15.0	19.0	83.0	
Producer's accuracy	74.1	100.0	80.0	57.9	–	78.31

though an accuracy assessment was carried out, the sample size of the data set was vey small in kharif 2002.

18.4 CONCLUSIONS AND FUTURE DIRECTIONS

This chapter demonstrates the development and application of a semisupervised classification approach for crop classification, which is suitable for small-scale and mixed cropping systems of South Asia. Results reveal that the methodology is sound enough to calculate land use/land cover (LULC) statistics with an improved accuracy level (87% in kharif and 83% in rabi) compared to other techniques: 74% [2] and 82% [14]. NDVI worked well in demarcating vegetated and nonvegetated areas and reducing spectral mixing, and facilitated higher overall accuracy. An attempt was made to clearly identify intercropping, but this needs at least two images per season, since intercropping consists of perennials intersown with seasonal crops. Moreover, the groundtruth data used in this study were not collected for the purpose of classification, and therefore better accuracy can be achieved using better geographic sampling. Lower accuracy results from relatively small numbers of reference samples, as with vegetables and maize. Very fine demarcation of crop types requires knowledge of the detailed cropping pattern, in order to select suitable image dates and interpret them accordingly.

ACKNOWLEDGMENT

The authors acknowledge the Remote Sensing and GIS team at IWMI Pakistan for their contribution in collecting ground data and preliminary image processing work. They also appreciate the field assistance of all support staff at IWMI Pakistan.

REFERENCES

1. Thenkabail, P.S. et al., *An Irrigated Area Map of the World (1999) Derived from the Remote Sensing*, Research Report 105, International Water Management Institute, Colombo, 2006.
2. Oetter, D.R. et al., Land cover mapping in an agricultural setting using Multiseasonal Thematic Mapper data, *Remote Sensing of Environment*, 76, 139–55, 2000.
3. Wegmuller, U., Signature research for crop classification by active and passive microwaves, *International Journal of Remote Sensing*, 14, 5, 871–83, 1993.
4. Anys, H. and Dong-Chen, H., Evaluation of textural and multipolarization radar features for crop classification, *IEEE Transactions on Geoscience and Remote Sensing*, 33, 5, 1170–81, 1995.
5. Bastiaanssen, W.G.M., *Remote Sensing in Water Resources Management: The State of the Art*, International Water Management Institute, Colombo, Sri Lanka, 1998.
6. Brisco, B. and Brown, R.J., Multidate SAR/TM synergism for crop classification in western Canada, *Photogrammetric Engineering and Remote Sensing*, 61, 8, 1009–14, 1995.
7. Van Niel, T.G. and McVicar, T.R., Current and potential uses of optical remote sensing in rice-based irrigation systems: A review, *Australian Journal of Agricultural Research*, 55, 155–85, CSIRO, 2004.
8. Janssen, L.L.F. and Middelkoop, H., Knowledge-based crop classification of a Landsat Thematic Mapper image, *International Journal of Remote Sensing*, 13, 15, 2827–37, 1992.
9. Ortiz, M.J., Formaggio, A.R., and Epiphanio, J.C.N., Classification of croplands through integration of remote sensing, GIS, and historical database, *International Journal of Remote Sensing*, 18, 1, 95–105, 1997.
10. Hurbert-Moy, L. et al., A comparison of parametric classification procedures of remotely sensed data applied on different landscape units, *Remote Sensing of Environment*, 75, 174–87, 2001.
11. Gandia, S., Caselles, V., and Gilaert, A., Spectral signature of rice fields using LANDSAT 5 TM in the Mediterranean coast of Spain, in *Proceedings of the Seventh International Symposium on Remote Sensing for Resource Development and Environmental Management, ISPRS Commission vii, Enschede*, A.A. Balkema, Rotterdam, 1986, 257–60.
12. Lloyd, D., A phenological description of Iberian vegetation using short wave vegetation index image, *International Journal of Remote Sensing*, 10, 827–33, 1989.

13. Muchoney, D.M. and Strahler, A.H., Pixel- and site-based calibration and validation methods for evaluating supervised classification of remotely sensed data, *Remote Sensing of Environment*, 81, 290–99, 2002.

14. Ahmad, M.D. et al., Use of satellite imagery for land use mapping and crop identification in irrigated systems in Pakistan, *17th Asian Conference of Remote Sensing (ACRS)*, Colombo, 1996.

15. Gamage, M.S.D.N., Ahmad, Mobin-ud-Din, and Turral, H., Semi-supervised technique to retrieve irrigated crops from Landsat ETM+ imagery for small fields and mixed cropping systems of South Asia, *International Journal of Geo informatics*, 3, 45–53, 2007.

16. Government of Pakistan, *Agricultural Census 2000, Province Report Vol. II Part-2*, Statistics Division, Agriculture Census Organization, Government of Pakistan, Lahore, Pakistan Punjab, 2001.

17. *Landsat 7 Science Data Users Handbook, Chapter 11-Data Products*, 2004, http://ltpwww.gsfc.nasa.gov/IAS/handbook/handbook_htmls/chapter11/chapter11.html. (accessed 15 June 2007).

18. Jensen, J.R., *Introductory Digital Image Processing: A Remote Sensing Perspective*, 2nd ed., Prentice-Hall, Englewood Cliffs, NJ, 1996.

19. Faust, N.L., Image enhancement, in Volume 20, Supplement 5 of *Encyclopedia of Computer Science and Technology*, A. Kent and J.G. Williams, Eds., Marcel Dekker, New York, 1989.

Section VII

Accuracies and Errors

19 Accuracy and Error Analysis of Global and Local Maps: Lessons Learned and Future Considerations

Russell G. Congalton
University of New Hampshire

CONTENTS

19.1 INTRODUCTION

Critical steps in analyzing any map generated from remotely sensed data are an error analysis and an accuracy assessment. The purpose of quantitative accuracy assessment is to identify and measure map errors. There are four major reasons for performing an accuracy assessment:

- To understand the errors in the map (so they can be corrected)
- To provide an overall assessment of the reliability or usefulness of the map [1]
- To use the map in a decision-making process
- Because the contract demands it

Assessing the accuracy of maps generated from digital remotely sensed data has a short history, beginning in the mid-1970s. Before this time, maps derived from analog (film) remotely sensed data (i.e., photo interpretation) were rarely evaluated using any kind of quantitative accuracy assessment. Field checking was performed as an important component of the interpretation process, but no overall map accuracy or other quantitative measure of quality was generally undertaken. It was only after photo interpretation began to be used as the reference data to compare digital remotely sensed data, did anyone begin to seriously consider how accurate the photo interpretation really was. In fact, the photo interpretation was often assumed to be 100% correct, resulting in some very poor digital image accuracies—in reality, some of these poor accuracies were due to errors in the photo interpretation and not due to the digital image analysis.

It is useful for discussion purposes to divide the history of accuracy assessment into four developmental stages or time periods. In the earliest stage, no real accuracy assessment was performed, but rather an "it looks good" philosophy prevailed. This approach is common with any new, emerging technology in which everything is changing very fast and little time is available to stop and evaluate the accuracy of the results. Despite the maturing of the technology over the last 30 years, some remote sensing and Geographic Information System (GIS) analysts still adhere to the "it looks good" philosophy. This philosophy is a necessary condition of any accuracy assessment. Why bother to further analyze a map that does not "look good?" However, it is not sufficient. The map should "look good" and then further error analyses should be performed. It is simply not enough for a map to "look good." The second stage in map accuracy history is called the age of non-site-specific assessment. During this period, total areas were compared between ground estimates and the map without regard for location [2]. This second age was short-lived and quickly led to the age of site-specific assessments. However, it should be noted that a non-site-specific assessment may be all that can be performed for some very large area/global mapping projects. In a site-specific assessment, specific locations on the ground are compared with the same place on the map and a measure of overall accuracy (i.e., percent correct) is computed. The fourth and current age of accuracy assessment is called the age of the error matrix. This time period includes a significant number of analytical techniques, the most common of which is the kappa analysis. A review of the techniques and considerations of the error matrix age can be found in Congalton [3] and in detail in Congalton and Green [4].

Since the early 1990s, the error matrix has been almost universally adopted as the means of reporting map accuracy. During this time, the use of remotely sensed data has steadily increased, and the size, complexity, and level of detail of these projects have grown tremendously. Recently, more interest has developed in regional and global map accuracy assessment. While most of the established accuracy assessment methodologies apply in these situations, additional factors must also be considered for these large projects. This chapter will present a review of the current methods for accuracy assessment and error analysis and will also discuss some of the specific issues for large-area assessments.

19.2   SOURCES OF ERROR

It is very important to be aware that error exists in every component of a mapping project. Failure to consider these errors does not minimize their impact. Congalton and Green [5] looked specifically at the error or sources of confusion between the remotely sensed classification and the reference data. They include:

1. Registration differences between the reference data and the remotely sensed map classification.
2. Delineation error encountered when the sites chosen for accuracy assessment are digitized.
3. Data entry error when the reference data are entered into the accuracy assessment database.
4. Error in interpretation and delineation of the reference data (e.g., photo-interpretation error, field observations, observer bias).
5. Changes in land cover between the date of the remotely sensed data and the date of the reference data (temporal error). For example, changes due to fires or urban development or harvesting.
6. Variation in classification and delineation of the reference data due to inconsistencies in human interpretation of heterogeneous vegetation.
7. Errors in the remotely sensed map classification.
8. Errors in the remotely sensed map delineation.

A paper by Lunetta et al. [6] took a broader approach and looked at a more inclusive list of errors and how they accumulate throughout a remotely sensed mapping project. This chapter demonstrates that each step in the project, image acquisition, data processing, data analysis, data conversion, error assessment, final product presentation, and implementation, adds errors. These errors accumulate and must be considered in any successful project. Given the many potential sources of error and the relative magnitude of each, it is critical that an accuracy assessment be an integral component of any mapping project. Failure to incorporate a valid assessment process into a mapping project dooms the map to misuse and misinterpretation, and the map user to potential trouble.

One method of analyzing the potential sources of error in a mapping project is the use of an error budget [7]. Our ability to quantify the total error in a spatial data set has developed substantially. However, little has been done to partition this error into its component parts and construct an error budget. Without this division into parts, it is not possible to evaluate the impact of a certain error on the whole. Therefore, it is not possible to determine which components contribute the most errors and which are most easily corrected. For many projects, there is a definite need to identify and understand: (1) error sources; and (2) the appropriate mechanisms for controlling, reducing, and/or reporting to the end users the magnitude of such errors.

Table 19.1 presents the results of performing an error budget analysis for a remote-sensing project, which includes monitoring land cover change over time. Table 19.1 is generated column-by-column, beginning with a listing of the possible sources of error for the project. Once the various components that comprise the total errors are listed, each component can then be assessed to determine its contribution to the overall errors. Next, our ability to deal with these errors is evaluated. It should be noted that some errors may be very large, but are easy to correct, while others may be rather small, but very difficult to fix. Of course, some very large errors may also

TABLE 19.1

Error Budget Analysis for a Change-detection Remote-sensing Project

Error Source	Error Contribution Potential	Implementation Difficulty	Implementation Priority	Error Assessment Technique
Systematic				
Sensor	Low	5	21	Instrumentation and analysis
Natural				
Atmosphere	Medium	4	20	Instrumentation and analysis
Input data				
Spectral radiometric imagery	Medium	3	19	Instrumentation and analysis, single date error matrix
Preprocessing				
Registration	Low	2	18	Position analysis (rms)
Cloud mask computation	Medium	3	7	Single date error matrix
Vegetation mask computation	Medium	3	8	Single date error matrix
Water mask computation	Medium	3	9	Single date error matrix
Forest mask computation	Medium	3	10	Single date error matrix
NDVI	Low	2	14	Single date error matrix
MSVI	Low	2	15	Single date error matrix
Brightness	Low	2	11	Single date error matrix
Greenness	Low	2	12	Single date error matrix
Wetness	Low	2	13	Single date error matrix
TC4	Low	2	16	Single date error matrix
Turbidity	Medium	4	17	Single date error matrix
Change-detection technique	High	1	3	Change-detection error matrix
Change-detection labeling scheme	High	2	2	Change-detection error matrix
Sampling scheme	High	2	1	Spatial variation analysis, change-detection error matrix
Interpretation				
Change-detection results	High	1	4	Change-detection error matrix
Training data	High	1	5	Training
Field data	High	1	6	Training

Note: Implementation difficulty—ranked from 1: not very difficult to 5: extremely difficult.

be very difficult to correct. It is precisely this combination of factors that this method is designed to reveal.

Next, an implementation priority or order to try and deal with these errors can be derived by combining the magnitude of the errors with our ability to fix them. Obviously, the largest errors that are easiest to correct should be dealt with first. Lastly, Table 19.1 lists the error assessment technique used in evaluating the various errors.

19.3 ACCURACY

There are two types of accuracy that are very important when discussing the accuracy of a map generated from remotely sensed data. These are (1) positional accuracy, and (2) thematic accuracy. In addition, there are two other types of accuracy that must be especially considered when dealing with large area/global mapping projects. These are (1) radiometric accuracy, and (2) temporal accuracy. We will deal with these two first.

19.3.1 Radiometric Accuracy

Radiometric accuracy is a measure of the effect of various factors on the digital number (DN) recorded by a digital sensor. Many digital images contain one byte or eight bits of information; therefore, the DN ranges from 0 to 255, depending on the amount of light reflected in that particular band or wavelength by the object of interest. The newer digital sensors collect 11 or 12 bits and the imagery is recorded as two bytes. In addition, factors such as scene illumination, sensor characteristics, the atmosphere, and the viewing angle can alter the DN values. When dealing with a single scene or when only relative differences are important, radiometric accuracy is less critical. However, when using multiple scenes, as in an image mosaic for a large region, or when comparing multiple dates for change detection, radiometric accuracy becomes very important.

19.3.2 Temporal Accuracy

Temporal accuracy is a measure of time. Time is an important factor in large area/ global mapping projects as there can be (1) gaps between the times when images are collected, and (2) gaps between the times when the imagery is acquired and when the reference data used in the accuracy assessment are collected. Both of these time gaps can cause major problems in the project and must be considered and accounted for throughout the mapping process.

19.3.3 Positional or Geometric Accuracy

Positional or geometric accuracy is a measure of how close the imagery matches the ground. It is critical in any accuracy assessment that the same exact location can be determined both on the image and on the ground. If not, then the poor positional accuracy may result in a false thematic error. The major factor influencing geometric accuracy is topography. Sensor characteristics and viewing angles can also have some effect. It is commonly accepted that a positional accuracy of one half a pixel is sufficient for sensors such as Landsat Thematic Mapper and Système Pour l'Observation de la Terre (SPOT) multispectral imagery. As sensors increase in spatial resolution, such as the 4-m multispectral IKONOS data, positional accuracy increases in importance, and new standards need to be established.

Like radiometric accuracy, geometric accuracy is an important component of thematic accuracy. If an image of 30-m pixel resolution is registered to the ground to within half a pixel and a Global Positioning System (GPS) unit is used to locate the place on the ground to within about 15 m, then it is impossible to use a single pixel as the sampling unit for assessing the thematic accuracy of the map. If positional accuracy is not up to the standard or GPS is not used to precisely locate the point on the ground, then these factors increase in importance and can significantly affect the thematic accuracy assessment.

19.3.4 THEMATIC ACCURACY

Thematic accuracy refers to the accuracy of a mapped land cover category at a particular time in reference to what was actually on the ground at that time. Clearly, in order to perform a meaningful assessment of accuracy, land cover classifications must be assessed with reference to data that are believed to be correct. Therefore, it is vital to have at least some knowledge of the accuracy of the reference data before using them for comparison against the remotely sensed map. Congalton [3] summarized as follows: "Although no reference data set may be completely accurate, it is important that the reference data have high accuracy or else it is not a fair assessment. Therefore, it is critical that the collection of ground or reference data be carefully considered in any accuracy assessment."

Accuracy assessment begins with the generation of an error matrix (Table 19.2), a square array of numbers, or cells, set out in rows and columns, which express the number of sample units assigned to each land cover type as compared to what is on the ground. The columns in the matrix usually represent the reference data (actual land cover) and the rows usually represent assigned (mapped) land cover types. The major diagonal of the matrix indicates agreement between the reference data and the interpreted land cover types. The process of generating an error matrix is summarized in Figure 19.1.

The error matrix is useful for both visualizing image classification results and, perhaps more importantly, for statistically measuring the results. An error matrix is the most effective way to compare two maps *quantitatively*. A measure of overall accuracy can be calculated by dividing the sum of all the entries in the major diagonal of the matrix by the total number of sample units in the matrix [8]. In the ideal situation, all the nonmajor diagonal elements of the error matrix would be zero, indicating that no sample units had been misclassified [9]. The error matrix also provides accuracies for each land cover category as well as both errors of exclusion (errors of omission) and errors of inclusion (errors of commission) present in the classification [3,4,10].

TABLE 19.2
Example of an Error Matrix

		Reference Data				row total
		D	C	AG	SB	
	D	63	4	20	16	103
	C	6	79	8	3	96
Classified Data	AG	0	11	85	5	101
	SB	4	7	1	90	102
column total		73	101	114	114	402

Land Cover Categories

D = deciduous

C = conifer

AG = agriculture

SB = shrub

OVERALL ACCURACY = (63+79+85+90)/402 = 317/402 = 79%

PRODUCER'S ACCURACY

D = 63/73 = 86%
C = 79/101 = 78%
AG = 85/114 = 75%
SB = 90/114 = 79%

USER'S ACCURACY

D = 63/103 = 61%
C = 79/96 = 82%
AG = 85/101 = 84%
SB = 90/102 = 88%

Errors of omission can be calculated by dividing the total number of correctly classified samples, or cells, in a category by the total number of samples in that category from the reference data (the column total) [3,8]. This measure of the error of omission may be referred to as "producer's accuracy," because from this measurement, the producer of the classification will learn how well a certain area was classified [3]. For example, the producer may be interested in learning how many times "deciduous" was in fact classified as deciduous (and not, say, conifer). To determine this, the 63 correctly classified deciduous samples (see Table 19.2) would be divided by the total 73 units of deciduous from the reference data, for a producer's accuracy of 86%. In other words, deciduous forests were correctly identified as deciduous 86% of the time.

Errors of commission, on the other hand, are calculated by dividing the number of correctly classified samples for a category by the total number of samples that were classified in that category [3,4,8]. This is also called "user's accuracy," indicating for the user of the map the probability that a sample unit classified on the map actually represents that category on the ground [4,8]. So, while the producer's accuracy for the deciduous category is 86%, the user's accuracy is only 61%. That is, only 61% of the areas mapped as deciduous are actually deciduous on the ground. However, because each omission from the correct category is a commission to the wrong category, it is critical that both producer's and user's accuracies are considered, since reporting only one value can be misleading [4,8].

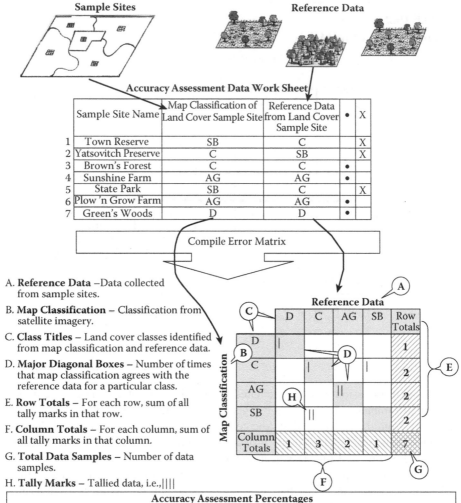

FIGURE 19.1 A summary of the process for generating an error matrix.

19.3.4.1 Factors to be Considered in Assessing Thematic Accuracy

An accuracy assessment based on an error matrix that is not properly generated can be meaningless, since the improperly generated error matrix may not be truly representative of the map. To properly generate an error matrix, the following factors must be considered [3]:

1. Collection of reference data
2. Classification scheme
3. Sampling scheme [11–14]
4. Spatial autocorrelation [15,16]
5. Sample size and sample unit [4,11–13]

Failure to consider even one of these factors could lead to significant shortcomings in the process of accuracy assessment.

19.3.4.1.1   Collection of Reference Data

Collection of reference data is the first step in any assessment procedure, and may be the single most important factor in accuracy assessment, since an assessment will be meaningless if the reference data cannot be trusted.

Reference data have been collected in many ways, including photo interpretation, aerial reconnaissance from a helicopter or airplane, video, drive-by surveys, and visiting the area of interest on the ground [17]. Ground visits, themselves, have ranged from visual observation by walking through the area, to detailed measurements of parameters, such as forest species, size class, and crown closure [17]. Despite these varied approaches, reference data have been historically assumed to be 100% accurate, or at least substantially more accurate than the map [10]. Congalton and Green [5], however, have shown that differences between photo-interpreted reference data and mapped data are often caused by factors other than the map/classification error, including photo-interpretation error. While photo interpretation remains a widely used means of collecting reference data, Congalton and Biging [17] demonstrated that at least some reference data should be collected using field measurements. Indeed, they found that field measurements coupled with other visual estimates provided the most efficient reference data.

In addition to verifying a remotely sensed classification, discovering any confusion between cover types, and perhaps improving the classification, reference data are also used initially to "train" the classification algorithm. In the not so distant past, many assessments of remotely sensed maps were conducted using the same data set used to train the classifier [3]. For example, in a supervised classification, spectral characteristics of known areas were used to train the algorithm that then classified the rest of the image, and these known areas were also used to test the accuracy of the map. This training and testing on the same data set resulted in an improperly generated error matrix that clearly overestimated classification accuracy [3]. In order for accuracy assessment procedures to be valid and truly representative of the classified map, data used to train the classifier should not be used for accuracy assessment. These data sets must be independent.

The information used to assess the accuracy of remotely sensed maps should be of the same general vintage as those originally used in map classification (consider temporal accuracy). The greater the time period between the date of the image acquisition and the collection of the reference data used in assessing map accuracy, the greater the likelihood that differences are due to change in vegetation (from harvesting, land use changes, etc.) rather than due to misclassification. Therefore, collection of ground data should occur as close as possible to the date of the remotely sensed data.

As can be deduced from the discussion above, the collection of good reference data can be challenging for many reasons. This challenge is especially difficult for large area/global mapping projects because of the issues with covering the entire project area. In most instances, collection of reference data for projects with a large area is limited to interpretation of photography or digital imagery because of the prohibitive costs of visiting the ground over such a large area. Therefore, the analyst may have less confidence in the correctness of the reference data than if more intensive collection of reference data were possible. A good solution to this challenge is to employ fuzzy accuracy assessment techniques [4,18].

The use of a fuzzy error matrix allows the analyst to compensate for situations where classification scheme breaks represent artificial distinctions along a continuum of land cover and/or where observer variability is often difficult to control [4]. While it is difficult to control observer variation, it is possible to use a fuzzy assessment approach to compensate for differences between reference data and the map, which are not caused by map error, but by variation in interpretation [1]. While one of the assumptions of the traditional or deterministic error matrix is that a reference data sample site can have only one label, this is not the case with a fuzzy error matrix. Instead, to account for variation in interpretation, the accuracy assessment analyst completes a special form (Figure 19.2) for every reference data sample site. Each site is evaluated for the likelihood of being identified as each of the possible map classes. First, the analyst determines the most appropriate ("good") label for the site and enters this label in the appropriate box under the "classification" column on the form. This label determines which row of the matrix the site will be tallied, and is used for calculation of the deterministic error matrix. After assigning the label for the site, the remaining possible map labels are evaluated as either "acceptable" or "poor" candidates for the site's label. For example, a site might fall near the classification scheme margin between deciduous forest and evergreen forest because of the exact mix of species and/or the difficulty in interpreting the exact mixture on the reference data imagery. In this instance, the analyst might rate the deciduous forest as the most appropriate, but the evergreen forest as "acceptable" (Figure 19.2). In this case, no other map classes would be acceptable; all the others would be rated as "poor."

This fuzzy error matrix analysis allows for the analyst to compensate for interpreter variability and difficulty in determining only a single label for each reference data sample site. While this method can be used for any assessment, it works best when there are issues in collecting good reference data because of limitations in the collection of reference data methods. Therefore, this approach is especially useful for situations where extensive collection of reference data, such as for large area/global projects, as opposed to intensive collection for smaller area projects is conducted.

Computation of the overall, producer's, and user's accuracy statistics for the fuzzy error matrix follows the same methodology as the traditional deterministic error matrix with the following additions. Nondiagonal cells in the matrix contain two tallies, which can be used to distinguish class labels that are uncertain or that fall on class margins, from class labels that are most probably erroneous. The first number represents those sites in which the map label matched an "acceptable" reference

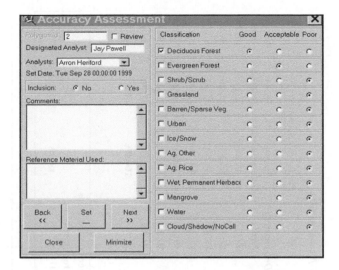

FIGURE 19.2   Form for labeling accuracy assessment reference sites.

label in the fuzzy assessment (Table 19.3). Therefore, even though the label was not considered the most appropriate, it was considered acceptable given the fuzziness of the classification system and/or the minimal quality of some of the reference data. These sites are considered a "match" for estimating fuzzy accuracy. The second number in the cell represents those sites where the map label was considered poor (i.e., an error). The fuzzy assessment overall accuracy is estimated as the percentage of sites where the "good" and "acceptable" reference labels matched the map label. Producer's and user's accuracies are computed in the traditional way, but again instead of just using the value on the major diagonal ("good"), the value in the first off-diagonal position ("acceptables") is also included (Table 19.3).

19.3.4.1.2 Classification Scheme

Classification schemes or systems categorize remotely sensed map information into a meaningful and useful format. The rules used to label the map must therefore be rigorous and well defined. One way to ensure this precision is to define a classification system that is totally exhaustive, mutually exclusive, and hierarchical [4].

A totally exhaustive classification scheme ensures that every place in the image fits into a category; i.e., nothing is left unclassified. A mutually exclusive classification scheme further ensures that everything in the image fits into one and only one category, i.e., one object in an image can be labeled only once. These criteria rely on two critical components: (1) a set of labels (e.g., white pine forest, oak forest, nonforest, etc.); and (2) a set of rules (e.g., white pine forest must comprise at least 70% of the stand). Without these components, the image classification would be arbitrary and inconsistent. It is critical that the rules for the classification scheme are well thought out and complete.

Finally, hierarchical classification schemes—those that can be collapsed from specific categories into more general categories—can be advantageous [4]. For

TABLE 19.3

Example of a Fuzzy Error Matrix Showing Both Deterministic and Fuzzy Accuracy Assessments

REFERENCE	MAP Decid. Forest	EG Forest	Scrub/ Shrub	Grass	Barren/ Sparse	Urban	Agric.	Water	Deterministic Totals	Producer's Accuracies Percent Deterministic	Fuzzy Totals	Percent Fuzzy
Deciduous Forest	48	24,7	0,1	0,3	0,0	0,1	0,11	0,18	48/113	42.5%	72/113	63.7%
Evergreen Forest	4,0	17	0,1	0,0	0,0	0,0	0,1	0,3	17/26	65.4%	21/26	80.8%
Shrub/Scrub	2,0	0,1	15	8,1	0,0	0,0	2,2	0,0	15/31	48.4%	27/31	87.1%
Grassland	0,1	0,0	5,1	14	0,0	0,0	3,0	0,0	14/24	58.3%	22/24	91.7%
Barren/Sparse Veg	0,0	0,0	0,2	0,0	0	0,0	0,1	0,0	0/3	0.0%	0/3	0.0%
Urban	0,0	0,0	0,0	0,0	0,0	20	2,0	0,0	20/22	90.9%	22/22	100.0%
Agriculture	0,1	0,1	7,15	18,6	0,0	2,0	29	1,2	29/82	35.4%	57/82	69.5%
Water	0,0	0,0	0,0	0,0	0,0	0,0	0,0	8	8/8	100.0%	8/8	100.0%
User's Accuracies												
Deterministic Totals	48/56	17/50	15/47	14/50	NA	20/24	29/51	8/33				
Percent Deterministic	85.7%	34.0%	31.9%	28.0%	NA	83.3%	56.9%	24.2%				
Fuzzy Totals	54/56	41/50	27/47	40/50	NA	22/24	36/51	10/33				
Percent Fuzzy	96.4%	82.0%	57.4%	80.0%	NA	91.7%	70.6%	30.3%				

Overall Accuracies

Deteministic	Fuzzy
151/311	230/311
48.6%	74.0%

example, if it is discovered that white pine, red pine, and hemlock forest cannot be reliably mapped, these three categories could be collapsed into one general category called coniferous forest.

19.3.4.1.3 Sampling Scheme

An accuracy assessment almost never involves a complete census of the classified image (i.e., every pixel in the image), since this is too large a data set to be practical [11,12,19]. Creating an error matrix to evaluate the accuracy of a remotely sensed map therefore requires sampling to determine if the mapped categories agree with the reference data categories [20].

In order to select an appropriate sampling scheme for accuracy assessment, some knowledge of the distribution of the vegetation/land cover classes should be known. For example, if there are cover types that comprise only a small area of the map, simple random sampling and systematic sampling may undersample this land cover class, or completely omit it. In this instance, simple random sampling or systematic sampling may create an error matrix that is not truly representative of the map. Stratified random sampling, by which a minimum number of samples are selected from each land cover category, may be more appropriate.

Ginevan [21] suggested general guidelines for selecting a sampling scheme for accuracy assessment, where the sampling scheme should (1) have a low probability of accepting a map of low accuracy, (2) have a high probability of accepting a map of high accuracy, and (3) require a minimum number of reference data samples.

Stratified random sampling has historically prevailed for assessing the accuracy of remotely sensed maps [11,22]. Stratified sampling has been shown to be useful for adequately sampling important minor categories, whereas simple random sampling or systematic sampling has tended to oversample categories of high frequency and undersample categories of low frequency [10,23]. Systematic sampling, however, has been less widely agreed upon. Cochran [4] found that the properties of the mapped data greatly affected the performance of systematic sampling in relation to that of stratified or simple random sampling. He found that systematic sampling could be extremely precise for some data and much less precise than, say, simple random sampling for other data.

No single method can be recommended universally for accuracy assessment. Rather, the best choice of sampling a scheme requires a knowledge of the structure and distribution of characteristics in the map, so that the map classes will be sampled proportionately and a meaningful error matrix can be generated. This issue is of special importance for any large area/global mapping project. It is almost certain that some stratification be employed to identify representative areas to be selected in which additional sampling can be performed. Green and Congalton [18] suggest a multistage approach in which first stage sample units (15-minute quadrangles or similar mapping scales) were selected to ensure that the diversity and complexity of the areas were covered and then second stage reference data samples were selected within the first stage sample.

19.3.4.1.4   Spatial Autocorrelation

Because of sensor resolution, landscape variability, and other factors, remotely sensed data are often spatially autocorrelated [16]. Spatial autocorrelation is defined as a dependency between neighboring pixels, such that a certain quality or characteristic at one location has an effect on that same quality or characteristic at neighboring locations [16,25]. Spatial autocorrelation can affect the result of an accuracy assessment if an error in a certain location can be found to positively or negatively influence errors in surrounding locations.

Spatial autocorrelation is closely linked with a sampling scheme. For example, if each sample carries some information about its neighborhood, that information may be duplicated in random sampling where some samples are inevitably close, subsequently violating the assumption of sample independence [26]. However, spatial autocorrelation could also be responsible for periodicity in the data, which could dramatically affect the results of a systematic sample [4]. Indeed, a spatial autocorrelation analysis performed by Pugh and Congalton [27] indicated that systematic sampling should not be used when assessing error in Landsat TM data for New England. Therefore, while spatial autocorrelation may not be avoidable, its effects should be considered when selecting an appropriate sample size and sampling scheme. Because of the large area covered in global mapping projects, spatial autocorrelation should be less of an issue than for smaller area projects, assuming that proper consideration is given to the sampling design.

19.3.4.1.5   Sample Size

A sufficient number of samples are essential in order for the error matrix to be meaningful. Too few samples can result in an error matrix that is not statistically valid. That is, small samples that imply there are no errors can be deceptive, since the error-free result may have occurred by chance when, in fact, a large portion of the classification was in error [11]. For this reason, sample sizes of at least 30 per category have been recommended [23], while others have concluded that fewer than 50 samples per category are not appropriate [12]. Sample sizes can be calculated from equations from the multinomial distribution, ensuring that a sample of appropriate size is obtained [28]. In general, however, sample sizes of 50–100 for each cover type are recommended, so that each category can be assessed individually [4]. It is especially important to note that, for large area/global mapping projects, if an error matrix is created based on some sample size calculation for the entire project, the assessment holds for the entire project. However, if the accuracy of a subproject area is warranted, a separate accuracy assessment with the appropriate sample size must be conducted for the subarea.

In addition to determining the appropriate sample size, an appropriate sample unit must be chosen. Selection of the proper sample unit is the best way to deal with the issue of positional or geometric accuracy. If the sample unit cannot be precisely co-located on the image and the ground, then using that sample unit will result in a thematic error that is actually the fault of the positional accuracy of the sample. Historically, the sample unit may be a single pixel, a cluster of pixels, a polygon, or a cluster of polygons. For raster data, a single pixel has always been a poor, but often selected, choice of sample unit [4], since it is an arbitrary delineation of the land cover and may have little relation to the actual land cover delineation. Further, it is nearly impossible to align one pixel in an image to the exact area in the reference data. A cluster of pixels (e.g., a 3×3 pixel square) is thus often a better choice for the sample unit, since it minimizes registration problems. A good rule of thumb is to choose a sample unit whose area most closely matches the minimum mapping unit of the reference data. For example, if the reference data have been collected in 2-hectare (ha) minimum mapping units, then an appropriate sample unit may be a 2-ha polygon. Failure to consider the proper sample unit can add significant problems/errors to the accuracy assessment process.

19.3.4.2 Error Matrix Analysis Techniques

Once the error matrix has been properly generated considering all of the factors discussed in the previous section, then there are additional analysis techniques that may prove useful as part of the accuracy assessment.

One of these analysis techniques is to "normalize" or standardize the error matrix using a technique known as "MARGFIT." This technique uses an iterative proportional fitting procedure that forces each row and column in the matrix to sum to one. The rows and column totals are called marginals, hence the technique name, MARGFIT. In this way, differences in sample sizes used to generate the matrices are eliminated and therefore, individual cell values within the matrix are directly comparable. Also, because as part of the iterative process the rows and columns are totaled, the resulting normalized matrix is more indicative of the off-diagonal cell values (i.e., the errors of omission and commission) than the original matrix. The major diagonal of the normalized matrix can be summed and divided by the total of the entire matrix to compute a normalized overall accuracy.

Another discrete multivariate technique of use in accuracy assessment is called KAPPA [29]. The result of performing a KAPPA analysis is a KHAT statistic (an estimate of KAPPA), which is another measure of agreement or accuracy. The KHAT statistic is computed as:

$$\hat{K} = \frac{N \sum_{i=1}^{r} x_{ii} - \sum_{i=1}^{r} (x_{i+} * x_{+i})}{N^2 - \sum_{i=1}^{r} (x_{i+} * x_{+i})},$$

where r is the number of rows in the matrix, x_{ii} is the number of observations in row i and column i; x_{i+} and x_{+i} are the marginal totals of row i and column i, respectively; and N is the total number of observations [30]. The equations for computing the variance of the KHAT statistic and the standard normal deviate can be found in Congalton et al. [9], Rosenfield and Fitzpatrick-Lins [31], Hudson and Ramm [32], Congalton [3], and Congalton and Green [4], to list just a few. It should be noted that the KHAT equation assumes a multinomial sampling model and that the variance is derived using the Delta method.

The power of the KAPPA analysis is that it provides two statistical tests of significance. Using this technique, it is possible to test to see if an individual land cover map generated from remotely sensed data is significantly better than if the map had been generated by randomly assigning labels to areas. The second test allows for the comparison of any two matrices to see if they are statistically, significantly different. In this way, it is possible to determine whether one algorithm is different from another, and based on a chosen accuracy measure (e.g., overall accuracy), to conclude which is better.

19.4 DISCUSSION AND FUTURE DIRECTIONS

Error analysis and accuracy assessment are critical components of any map generated from remotely sensed data. The key is to balance statistically valid methods with what is practically achievable. Many projects have flawed accuracy assessments because the assessment was not considered throughout the project, but rather not until the end. Assessments are expensive and it is essential that the methodology and considerations presented in this chapter be well understood from the beginning of the project.

Large area/global mapping projects have additional issues that must be considered to perform a valid error analysis and accuracy assessment. Timely collection of reference data is a big issue. The use of the fuzzy error matrix can be of tremendous benefit here. Proper selection of the sampling scheme and sample unit is also vital for generating a valid error matrix and minimizing registration error.

There is no single solution or simple recipe for conducting a valid assessment. It is important that the assessment be designed from the beginning of the project, and that each factor be carefully considered in order to balance the need for a statistically valid assessment with what can actually be achieved given all the constraints of the project. Given the rapid increase in the use of remotely sensed imagery and the complexity of the mapping projects using this imagery, there is a great need for further research and development of methodologies to effectively analyze error and assess the accuracy of these maps.

REFERENCES

1. Gopal, S. and Woodcock, C., Theory and methods for accuracy assessment of thematic maps using fuzzy sets, *Photogrammetric Engineering and Remote Sensing*, 60, 181–88, 1994.
2. Meyer, M. et al., ERTS data applications to surface resource surveys of potential coal production lands in southeast Montana, *IARSL Research Report 75-1 Final Report*, University of Minnesota, Minneapolis, MN, 1975.
3. Congalton, R.G. and Green, K., *Assessing the Accuracy of Remotely Sensed Data: Principles and Practices*, 2nd ed., CRC/Taylor & Francis, Boca Raton, FL, 2009.
4. Congalton, R.G. and Green, K., *Assessing the Accuracy of Remotely Sensed Data: Principles and Practices*, Lewis Publishers, Boca Raton, FL, 1999.

5. Congalton, R. and Green, K., A practical look at the sources of confusion in error matrix generation, *Photogrammetric Engineering and Remote Sensing*, 59, 641–44, 1993.
6. Lunetta, R. et al., Remote sensing and geographic information system data integration: Error sources and research issues, *Photogrammetric Engineering and Remote Sensing*, 57, 677–87, 1991.
7. Congalton, R. and Brennan, M., Error in remotely sensed data analysis: Evaluation and reduction, *Proceedings of the Sixty Fifth Annual Meeting of the American Society of Photogrammetry and Remote Sensing*, Portland, OR, 1999, 729–32 (CD-ROM).
8. Story, M. and Congalton, R.G., Accuracy assessment: A user's perspective, *Photogrammetric Engineering and Remote Sensing*, 52, 397–99, 1986.
9. Congalton, R.G., Oderwald, R.G., and Mead, R.A., Assessing Landsat classification accuracy using discrete multivariate statistical techniques, *Photogrammetric Engineering and Remote Sensing*, 49, 1671–78, 1983.
10. Card, D., Using known map category marginal frequencies to improve estimates of thematic map accuracy, *Photogrammetric Engineering and Remote Sensing*, 48, 3, 431–39, 1982.
11. van Genderen, J.L. and Lock, B.F., Testing land use map accuracy, *Photogrammetric Engineering and Remote Sensing*, 43, 1135–37, 1977.
12. Hay, A.M., Sampling designs to test land-use map accuracy, *Photogrammetric Engineering and Remote Sensing*, 45, 529–33, 1979.
13. Congalton, R.G., A comparison of sampling schemes used in generating error matrices for assessing the accuracy of maps generated from remotely sensed data, *Photogrammetric Engineering and Remote Sensing*, 54, 593–600, 1988.
14. Stehman, S., Comparison of systematic and random sampling for estimating the accuracy of maps generated from remotely sensed data, *Photogrammetric Engineering and Remote Sensing*, 58, 1343–50, 1992.
15. Campbell, J., Spatial autocorrelation effects upon the accuracy of supervised classification of land cover, *Photogrammetric Engineering and Remote Sensing*, 47, 355–63, 1981.
16. Congalton, R.G., Using spatial autocorrelation analysis to explore errors in maps generated from remotely sensed data, *Photogrammetric Engineering and Remote Sensing*, 54, 587–92, 1988.
17. Congalton, R.G. and Biging, G.S., A pilot study evaluating ground reference data collection efforts for use in forest inventory, *Photogrammetric Engineering and Remote Sensing*, 58, 12, 1669–71, 1992.
18. Green, K. and Congalton, R., An error matrix approach to fuzzy accuracy assessment: The NIMA geocover project, in *Remote Sensing and GIS Accuracy Assessment*, R.S. Lunetta and J.G. Lyon (Eds.), CRC Press, Boca Raton, FL, 2004, 163–72.
19. Stehman, S., Estimating the Kappa coefficient and its variance under stratified random sampling, *Photogrammetric Engineering and Remote Sensing*, 62, 4, 401–402, 1996.
20. Rosenfield, G.H., Fitzpatrick-Lins, K., and Ling, H., Sampling for thematic map accuracy testing, *Photogrammetric Engineering and Remote Sensing*, 48, 131–37, 1982.
21. Ginevan, M.M.E., Testing land-use map accuracy: Another look, *Photogrammetric Engineering and Remote Sensing*, 45, 1371–77, 1979.
22. Jensen, J.R., *Introductory Digital Image Processing: A Remote Sensing Perspective*, Prentice Hall, Upper Saddle River, NJ, 1996.
23. van Genderen, J.L., Lock, B.F., and Vass, P.A., Remote sensing: Statistical testing of thematic map accuracy, *Remote Sensing of Environment*, 7, 3–14, 1978.
24. Cochran, W.G., *Sampling Techniques*, 3rd ed., John Wiley, New York, 1977.
25. Cliff, A.D. and Ord, J.K., *Spatial Autocorrelation*, Pion, London, 1973.

26. Curran, P.J. and Williamson, H.D., Sample size for ground and remotely sensed data, *Remote Sensing of Environment*, 20, 31–41, 1986.

27. Pugh, S. and Congalton, R.G., Applying spatial autocorrelation analysis to evaluate error in New England forest cover type maps derived from Thematic Mapper Data, Proceedings, Sixty-Third ASPRS Annual Meeting, Seattle, Washington, *American Society of Photogrammetry and Remote Sensing*, 3, 648–57, 1997.

28. Tortora, R., A note on sample size estimation for multinomial populations, *The American Statistician*, 32, 100–102, 1978.

29. Cohen, J., A coefficient of agreement for nominal scale, *Educational and Psychological Measurement*, 20, 37–46, 1960.

30. Bishop, Y., Fienberg, S., and Holland, P., *Discrete Multivariate Analysis – Theory and Practice,* MIT Press, Cambridge, MA, 1975.

31. Rosenfield, G. and Fitzpatrick-Lins, K., A coefficient of agreement as a measure of thematic classification accuracy, *Photogrammetric Engineering and Remote Sensing*, 52, 223–27, 1986.

32. Hudson, W. and Ramm, C., Correct formulation of the Kappa Coefficient of agreement, *Photogrammetric Engineering and Remote Sensing*, 53, 4, 421–22, 1987.

Section VIII

Way Forward in Mapping Global Irrigated and Rain-Fed Croplands

20 Remote Sensing of Global Croplands for Food Security: Way Forward

Prasad S. Thenkabail
U.S. Geological Survey

John G. Lyon
Environmental Protection Agency

CONTENTS

This book opens a new pathway for global mapping that is focused on a specific land use theme, such as irrigated or rain-fed croplands and classes within these themes. Since croplands use most of the water consumed by humans, specific knowledge of irrigated and rain-fed croplands will be critical for precise estimates of water use. At present and in the coming decades, irrigated and rain-fed cropland area mapping is crucial for food security studies. Throughout this book, various subjects pertaining to global croplands are discussed comprehensively.

The book begins by looking at the history of irrigated areas of the world. Chapter 2 takes us back to the beginning of settled agriculture some 23,000 years ago to present times. All early great civilizations thrived on agriculture and irrigation for their advancement. Yet, global irrigated area was a meager eight million hectares (Mha) in the beginning of the year 1800, reached around 90 Mha at the end of the Second World War, and then swiftly rose to 278 (Siebert et al., 2006) to 399 Mha (Thenkabail et al., 2008) by the end of the last millennium. The reasons for uncertainties in areas are discussed later in this chapter and throughout this book.

Chapter 3 produces the first ever satellite-sensor-based Global Irrigated Area Map (GIAM) and discusses many methods and protocols for doing so. It provides irrigated area maps and statistics for the end of the last millennium. Chapter 15 does the same for the rain-fed croplands of the world. Together, Chapters 3 and 15 provide a complex set of methods and approaches for mapping irrigated and rain-fed croplands using time-series, remote-sensing data. They demonstrate that the only possible

461

way for a consistent mapping of global croplands is through remote sensing. The current state of global irrigated and rain-fed croplands, using remote-sensing data and methods, has been reported in the form of maps and statistics. The results show spatial distribution of global croplands and specific locations of irrigated and rain-fed croplands. Overall, the book shows how to make use of time-series, well-calibrated satellite sensor data from various sensors to map agricultural crops—irrigated and rain-fed. Chapters 3 and 15 show several innovative methods, such as spectral matching techniques (SMTs) developed to analyze time-series multispectral megafile data cubes akin to hyperspecral data, Ideal Spectral Data Bank (ISDB), and Space Time Spiral Curves (STSCs), as well as the use of well-known techniques such as decision trees and clustering algorithms. These chapters also provide guidelines for practical approaches to computing subpixel areas (SPAs), crucial for obtaining actual areas from coarse-resolution imagery. Throughout the book, various methods and approaches for cropland mapping have been extensively discussed and illustrated. For example, Chapter 16 discusses multiangle measurements using Multiangle Spectroradiometer (MISR) sensor data, Chapter 11 demonstrates spectral mixture analysis, Chapter 6 uses multiple data sets in ArcAvenue scripts, Chapters 3 and 15 apply decision trees, Chapter 17 explores stepwise linear discriminant analysis, and Chapter 18 applies the semisupervised classification approach to detect irrigated lands. These studies use a wide array of satellite sensor data, such as Advanced Very High Resolution Radiometer (AVHRR) time series, Système Pour l'Observation de la Terre Vegetation (SPOT VGT) time series, Moderate Resolution Imaging Spectroradiometer (MODIS), Indian Remote Sensing Satellite (IRS), and Landsat. Overall, the book provides a one-stop overview of global irrigated and rain-fed cropland areas, advanced data sets required to produce them, methods and approaches involved in producing cropland areas and their statistics, and models for determining cropland water use.

As per GIAM (Chapter 3), the total area available for irrigation (or net irrigated area that does not account for intensity of irrigated areas) at the end of the last millennium was 399 Mha. The Annualized Irrigated Area (AIA) (or gross area that accounts for intensity of irrigated areas) during the same period was 467 Mha. The Asian continent has 79% of all global AIA. In comparison, the irrigated areas in other continents are low: Europe (7%), North America (7%), South America (4%), Africa (2%), and Australia (1%). China (32%), India (28%), and the United States (5%) together have about 65% of all global AIAs. Elsewhere, in Africa, irrigated area is just about 2% of the global AIAs, but has 14% of the global population (Figure 20.1). There is a real opportunity for irrigated area expansion in the African continent that can contribute to Africa's green revolution. The total rain-fed croplands, as per the global map of rain-fed cropland areas (GMRCA) (Chapter 15), constitute 1.13 billion hectares (Bha). The rain-fed croplands are dominant in the United States (11.8%), followed by Russia (10.1%), China (8.1%), Brazil (7.72%), India (4.31%), Australia (3.25%), Canada (3.09%), and Argentina (3.03%). China and India, with a population of 2.6 billion, depend, primarily, on irrigated double crops to feed their populations. The United States and Europe depend primarily on rain-fed single crops to feed a combined population of about 1.3 billion. They also export their grains to many other nations. The African continent has about 17% of the total global rain-fed croplands (Figure 20.2), but has low productivity. There is

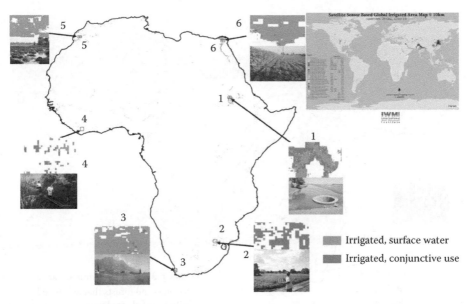

FIGURE 20.1   (See color insert following page 224.) Irrigated cropland areas of Africa. The global Annualized Irrigated Area (AIA) in the African continent is only about 2% compared to 14% of the global population. There is a real opportunity to expand irrigated areas in Africa to facilitate green and blue revolutions.

real opportunity to irrigate a significant proportion of these lands (Figure 20.2) to increase productivity, which can further contribute to Africa's green revolution. The total global cropland areas (irrigated and rain-fed), as estimated by remote-sensing methods (Chapters 3 and 15), are 1.53 Bha for the end of the last millennium, which is about the same as that estimated by the Aquastat of the Food and Agriculture Organization (FAO) of the United Nations for the same period. The FAO Aquastat is based on national statistics. However, the FAO and University of Frankfurt (FAO and UF) have estimated the areas equipped for irrigation (but not necessarily irrigated) as 278 Mha (Siebert et al., 2006), a value much lower than the GIAM estimated 399 Mha (Thenkabail et al., 2008).

The uncertainties in irrigated and rain-fed croplands are due to a number of causes that were identified in various studies and reported in this book. These include, but not limited to, factors such as: (a) definition issues; (b) inadequate accounting of minor irrigation (groundwater, small reservoirs, tanks); (c) resolution (or scale); (d) methods and approaches; (e) subjectivity involved in census data collection; (f) difficulties in precise estimate of area fractions; and (g) minimum mapping unit. There are maps that depict only major surface water irrigated areas from large and medium dams as irrigated (e.g., CBIP, 1994). As we see in Chapter 9, in temperate climates such as in the UK, irrigation is supplemental to rainfall. Yet, it is water use through artificial means that needs to be accounted for. Yet again, we see many maps and statistics that consider even substantial supplemental irrigated areas as rain-fed (Agrawal et al., 2004). The remote-sensing approach provides the best step forward for a globally consistent digital database on irrigated and rain-fed areas. Therefore, we recommend

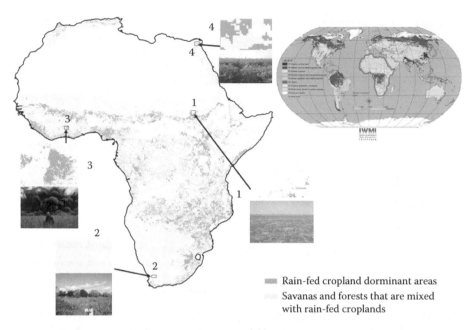

Rain-fed cropland dorminant areas
Savanas and forests that are mixed
with rain-fed croplands

FIGURE 20.2 **(See color insert following page 224.)** Distribution of rain-fed croplands in the African continent. The total global rain-fed cropland area (Global: 1.13 Bha) in Africa is 17% and its productivity can be improved through irrigation and better land management practices to help move towards a green revolution.

continued work using higher-resolution remotely sensed data, in the immediate future using Landsat TM data in conjunction with time-series MODIS data. The Landsat data provides the spatial details needed for precise area estimations. However, when intensity is considered, the need for temporal resolution will be crucial, therefore, we recommend using MODIS time series in conjunction with Landsat. Chapters 4 through 6 highlight uncertainties in reported irrigated and rain-fed cropland areas from various sources in China, India, and the United States, respectively. Chapters 3 through 5 show the importance of informal (or minor) irrigated areas from groundwater, small reservoirs, and tanks. They provide evidence to show that inadequate accounting of minor irrigation is one of the biggest probable causes of uncertainty in areas. Other important causes, listed in the beginning of this paragraph, lead to uncertainty in areas. In India, for example, the number of tube wells have increased from a meager 100,000 in the early 1960s to about 25 million at the end of the last millennium, an overwhelming proportion of which are used for irrigation. This has resulted in massive groundwater pumping. In addition, small reservoirs built across lower order streams have proliferated to such an extent that they are resulting in virtually near-zero outflow from basins (Biggs et al., 2006). Irrigated areas from minor sources (groundwater, small reservoirs, and tanks) are exceeding those from very large reservoirs. Yet, irrigated areas from these sources are inadequately accounted for and, in many cases, not accounted for at all. The main reason for this is that minor sources are owned and operated privately by farmers, and governments do not keep adequate records of these. In both China and India, the remote-sensing-based GIAM accounts for these minor sources, even though uncertainties still exist due to the resolution (or scale of mapping). The quantum of water used by minor sources is estimated to even exceed that from major sources in India. The authors of Chapter 7 used field-plot data from the central-pivot irrigation systems from 108 locations in Texas and Kansas and found the errors of omission in the GIAM product were 20.4% and those of commission were 18.5%. Given that these errors can nearly cancel each other, the areas could remain quite accurate, even though uncertainties exist due to errors of omission and commission. Chapter 6 identifies resolution and irrigated area fractions for class (IAFs) as possible causes of uncertainties in areas in GIAM. However, it does not provide an answer to whether the fragmented (yet significant) irrigated areas are accounted for in the national statistics reported by USDA and in MODIS-based remote-sensing estimates by the authors. In Chapter 8, we learn that turf grass with 4.5–9.6 Mha is the biggest irrigated area in conterminous United States, more than any single agricultural crop. Again, the uncertainty in estimating areas is high. So, it is not only important to estimate irrigated and rain-fed croplands accurately, but also other land use, such as turf grass, that uses a significant proportion of the water. As long as we have uncertainties in area estimations, uncertainties will remain in water use estimations. The book has made great advances in providing an understanding of mapping irrigated and rain-fed areas and reducing uncertainties in doing so. Nevertheless, the need for high-resolution global-irrigated and rain-fed cropland area maps using, in the least, Landsat 30-m data, is crucial in developing precise knowledge of irrigated areas and their water use. The high-resolution data need, however, to be fused with MODIS 250/500-m data to obtain time-series knowledge of irrigated area classes. The combined high-resolution one-time and coarse-resolution continuous time series provide cropland types (e.g., rice crop in Chapter 11), and/or cropland type dominance, and their precise geographic location. This will facilitate accurate estimates of water use by crops.

One of the strengths of this book is in the various regional studies of mapping and studying the irrigated and the rain-fed cropland areas. This has been done using a wide array of methods, approaches, and data sets in different environments, such as urban (Chapter 8), temperate (Chapter 9), semiarid (Chapter 10), arid (Chapter 10), and tropical climates (Chapter 11).

The book highlights one of the critical issues facing humanity at present: the issue of food security and how we are going to make more land and water available for a food-secure world. What is clear from the various chapters in the book is that the choice is not in increasing land and water resources to produce more food. In Chapter 10, we see how fully irrigated lands are turned into supplementing irrigated lands to conserve water, but at the same time grow more food. In Chapter 9, we see GIS modeling approaches required to better allocate water and strategize for any climatic variability. The choice will be in innovation, adaptability, and efficiency. A central question here is: how can we produce more food and continue to feed populations from existing land and water? This will require us to increase land productivity (to grow more food per unit of land) and water productivity (WP) (to grow more food per unit of water). The green revolution era focused on land productivity. The world is waiting for a blue revolution to evolve and focus on higher WP of irrigated and rain-fed croplands, so that we can continue to feed the ballooning populations, in the coming decades, without having to allocate higher amounts of land and water. The need for increasing WP is also becoming crucial in a changing climate and there is a need for adaptation to these climatic changes, as demonstrated in a simple water balance model and a General Circulation Model (GCM) in Chapter 12. This chapter shows that, for the Northeastern Fertile Crescent (southern Turkey, Syria), with progress of time to years 2020, 2050, and 2080, there is a progressive decrease in water availability in rivers, but an increase in irrigation water needs. One can expect populations to grow during this period and a subsequent increased food demand. All these issues point to a need for increasing land and WP as the best option to feed the populations. Water use assessment requires surface energy balance models. These have been provided using a Simplified Surface Energy Balance (SSEB) model in Chapter 13 and using the Surface Energy Balance Algorithm for Land (SEBAL) model in Chapter 14. The SSEB model is very useful in a developing country setup, where data are scarce. The SEBAL requires frequent (daily data preferred) images and good ground-based measurement sites. Remote sensing provides the best opportunity to produce WP maps (obtained by dividing the land productivity with water use). This will enable us to pinpoint areas of high and low WP of agricultural croplands. The outcome will be an estimate of the proportion of areas under low WP, and planning a strategy to improve productivity of these low WP areas. Such an outcome will help us to produce more food from existing land and water allocations for croplands, leading to food security in coming decades. However, for this to happen, we need to accurately estimate cropland areas and determine their precise location. This book has provided an exhaustive road map for achieving this goal.

REFERENCES

Agrawal, S., Joshi, P.K., Shukla, Y., and Roy, P.S. (2004). Spot-vegetation multi temporal data for classifying vegetation in South Central Asia. *Current Science*, 84 (11), 1440–48.

Biggs, T., Thenkabail, P.S., Krishna, M., GangadharaRao, P., and Turral, H. (2006). Vegetation phenology and irrigated area mapping using combined MODIS time-series, ground surveys, and agricultural census data in Krishna River Basin, India. *International Journal of Remote Sensing*, 27 (19), 4245–66.

CBIP (Central Board of Irrigation and Power). (1994). *Central Board of Irrigation and Power Irrigated Area Map of India for 1994*. New Delhi, India: CBIP.

Loveland, T.R., Reed, B.C., Brown, J.F., Ohlen, D.O., Zhu, J, Yang, L., and Merchant, J.W. (2000). Development of a Global Land Cover Characteristics Database and IGBP DISCover from 1-km AVHRR data. *International Journal of Remote Sensing*, 21 (6/7), 1303–30.

Siebert, S., Hoogeveen, J., and Frenken, K. (2006). *Irrigation in Africa, Europe, and Latin America. Update of the Digital Global Map of Irrigation Areas to Version 4.* Frankfurt Hydrology Paper 05. Germany: University of Frankfurt (main) and Rome, Italy: FAO.

Thenkabail, P.S., Biradar, C.M., Noojipady, P., Dheeravath, V., Li, Y.J., Velpuri, M., Gumma, M. et al. (2009). Global irrigated area map (GIAM) for the end of the last millennium derived from remote sensing. *International Journal of Remote Sensing* (accepted, in press).

Index

T - #0353 - 101024 - C47 - 234/156/30 - PB - 9781138116559 - Gloss Lamination